The Last Frontier

The Last Frontier

Imagining Other Worlds,
from the Copernican Revolution
to Modern Science Fiction

KARL S. GUTHKE

TRANSLATED BY
HELEN ATKINS

Cornell University Press

ITHACA AND LONDON

English translation first published 1990 by Cornell University Press.

The publisher gratefully acknowledges the assistance of Inter Nationes in defraying part of the cost of translation.

Library of Congress Cataloging-in-Publication Data

Guthke, Karl Siegfried, 1933–
  [Mythos der Neuzeit. English]
  The last frontier : imagining other worlds, from the Copernican revolution to modern science fiction / Karl S. Guthke.
    p. cm.
  Translation of: Der Mythos der Neuzeit.
  ISBN 0–8014–1680–9 (alk. paper).—ISBN 0–8014–9727–2 (pbk.: alk. paper)
  1. Science—History. 2. Science—Philosophy. I. Title.
Q125.G96513 1990
509—dc20   89–46165

We of the higher primates have delved into the half-known cosmic facts deeply enough to recognize also the need of cosmic fancies when facts are delayed.

—Harlow Shapley

[The scientist] must depend on the poetic myth.

—Charles I. Glicksberg

In the second half of the twentieth century one can hardly be a complete human being without at least occasionally calling to mind that community of rational beings, as yet unknown, to which we presumably belong.

—Stanislaw Lem

"What are these Martians?"
"What are we?"

—H. G. Wells

The irruption of the planetary consciousness into our tellurian world must be one of the greatest subjects that there have ever been or that can possibly be conceived of.

—Fritz Usinger

The effort to understand the universe is one of the very few things that lifts human life a little above the level of farce, and gives it some of the grace of tragedy.

—Steven Weinberg

Contents

Preface

The belief that "we are not alone"—that there must be planets, not necessarily in our own solar system, inhabited by forms of life resembling our own or at least possessing intelligence—has in our time become a religion or quasi religion with, undeniably, a certain popular following. Yet in the sixteenth and seventeenth centuries it was an esoteric heresy, one of the beliefs for which, presumably, Giordano Bruno was burned at the stake by the Inquisition. This book seeks to trace the stages by which the dangerous heresy has become a new gospel.

The impulse for the idea that the universe might have other inhabitants came from the "New Science" of the sixteenth and seventeenth centuries. Copernicus and Galileo demonstrated that the Earth is not unique, but merely one planet among several. Since then the question about the identity of mankind has increasingly been defined in relation to those higher forms of life on other planets whose existence and nature have been and are being extrapolated by analogy from scientific facts. Not only theology, philosophy, and literature have responded to the challenge posed by this "Copernican consciousness," as Fritz Usinger has called it; but in the sciences themselves the existence of extraterrestrial life is still the subject of serious investigation and is, indeed, receiving more attention now than ever before. "Experience shows," wrote Johann Jakob Breitinger in his *Critische Dichtkunst* (Critical treatise on poetics, Zurich 1740), "that man has a far stronger desire to investigate things that are possible and that lie in the future than to acquaint himself with what is real and exists in the present" (1:61). On the borderline between reality and possibility there is a domain of imaginative writing that is in the broadest sense philosophical and that is no mere jeu d'esprit but a serious contribution to the study of humanity in the light of what is known about the universe at a given time.

The subject of such imaginative investigation is—to apply the technical term used for centuries—"the plurality of worlds," that is, the presumed existence of more than one humankind in the universe. The imaginative treatment of this idea may give rise to a comforting or to a nightmarish fictive world; but by retaining, for all that, a rational basis, it bridges the gulf between the "two cultures" in a way that answers the aspirations of the Romantics (Friedrich Schlegel, Wordsworth, Shelley, and others): the standpoints of science and the humanities are resolved in a single coherent vision. Schlegel, in his *Gespräch über die Poesie* (Conversation on poetry, 1800), termed this kind of fusion or symbiosis a "new mythology," and in our own day too it is often claimed that science fiction and other works based on imaginative extrapolation from scientific facts now fulfill the same function as myth in preindustrial societies.[1] For, like the old myths, they provide symbols that help to clarify our understanding of ourselves and of the world, particularly when they touch on areas at the outer limits of our knowledge, above all, of our cosmologic knowledge. Science has given us many such symbols or modern "myths," and if the advent of the Copernican theory is seen as the dawn of the modern age, then surely the post-Copernican concept of the plurality of worlds may count as a "myth," or even *the* myth, of this age. Once the medieval period, with its cosmology based on a single inhabited world, had been left behind, the conviction that we are to be defined in imaginative "encounters of the third kind" could be seen as a "realistic" mode of thought, and it exerted its fascination on ever-wider circles as the Copernican outlook gradually gained ground. The history of the "plurality" idea shows clearly how modern science has fundamentally changed Western man's understanding of himself.

To call the concept of a plurality of inhabited worlds a "myth" (Jung similarly referred to flying saucers as "a modern myth") may arouse misgivings, since the term is both ambiguous and controversial. On the one hand there is the suspicion that thinking in mythological terms represents an infantile regression into irrationality inappropriate to the level of consciousness our civilization has reached. The

[1]See chap. 1, sec. 3; also Walter D. Wetzels, ed., *Myth and Reason* (Austin, Tex., 1973), pp. 162, 164; Michael P. Carroll, "La pensée scientifique: Myth and the Popularity of Scientific Theories," *Structuralist Review* 2 (1980), 49–58, esp. p. 50.

contrary view, expressed by Schlegel and reaffirmed in recent times by Bruno Snell and Claude Lévi-Strauss, is that creating myths and orienting oneself by them by no means implies a rejection of reason but is a fundamental characteristic of the human condition that can never be outgrown.[2] Those who are suspicious of myth can certainly point to writers from Richard Wagner to Alfred Rosenberg who subjected both term and concept to obscurantist distortion. In their hands, myth became a seductive magical voice promising a people redemption through the folk memory of its archaic past. But such deplorable misuse of the term should not overshadow the legitimate way in which people in modern industrial society can still use mythological images to help define themselves. Myths are not only something to which we have looked back, relics of primeval antiquity. Myths have also come into being within the historical period and are still springing up spontaneously in widely differing societies today. Such modern myths (in the sense of images that help to clarify our understanding of ourselves and of the world) need not imply a leap into the unconscious, possibly as a result of manipulation by special-interest groups. These myths may equally well be elements of what Schelling called a "mythology based on reason" (Mythologie der Vernunft), and of this there is perhaps no better illustration than what I call "the myth of the modern age," the myth about our neighbors in the universe. For, imaginative though it is, it is closely bound up with scientific findings and can properly claim to represent scientifically respectable extrapolation from established facts and valid theories. Far from being purely speculative fantasy, a dream of utopia, gothic horror, or visionary euphoria, it offers, in Kant's phrase, "well-founded probability." (At present the "myth of the modern age" is being trivialized and drawn into the sphere of occultism by such mass phenomena as the belief in flying saucers and the cult of Erich von Däniken's gods. This is a cause for concern on various counts, but the blame for it should not automatically be attributed to the sound belief in the plurality of worlds, supported nowadays by acknowledged scientists. In any case, these mass phenomena are outside the scope of this book.)

To impose some sort of order on the wealth of material dealt with, ranging from theological treatises to light fiction, I have grouped the material under the chronological headings—Renaissance, baroque, and so on—that are commonly used in the history of literature and

[2]On what follows see Wetzels, *Myth and Reason*, esp. the discussion printed in the app.; also Theodore Ziolkowski, "Der Hunger nach dem Mythos," in *Die sogenannten Zwanziger Jahre*, ed. Reinhold Grimm and Jost Hermand (Bad Homburg, 1970), pp. 169–201.

thought. The value of these labels is disputed, and I use them faute de mieux. And yet we find that in each of these major periods the theme of man in relation to a plurality of inhabited worlds does, in essence, appear in a guise peculiar to that period (in spite of all "Gleichzeitigkeit des Ungleichzeitigen"), so that perhaps this study of a little-known subject may also add to our knowledge of the characteristic features of each period.

Unless otherwise indicated in the notes, quotations are taken from the first editions referred to in the text.

I thank Walter Schatzberg for his help in obtaining the latest studies in the border area between science, philosophy, and literature; Wallace Tucker for conversations that gave me a clearer understanding of relevant areas of contemporary science; David Silas, of Widener Library, for obtaining photocopies of rare books and articles; and the National Endowment for the Humanities for financing a semester's leave.

K.S.G.

Harvard University
September 1981

Postscript, August 1989

Rather than simply translate the title of the German edition, *Der Mythos der Neuzeit*, I have chosen as an English title a term that has become widely current in connection with the "adventure of reason" that is described in this book.

I am most grateful to Michael J. Crowe, author of *The Extraterrestrial Life Debate, 1750–1900* (Cambridge University Press 1986), for reading the typescript of the translation and making helpful suggestions. His comments allowed me to modify a number of statements. He has been the most generous of colleagues. We are agreed that as a result of their difference in scope and perspective, his book and mine complement rather than compete with each other.

Last, but by no means least, I thank Helen Atkins for the resourcefulness and thoughtfulness with which she went about her unenviable task. I have reviewed the translation and made some minor changes and additions.

K.S.G.

Translator's Note

There are numerous quotations from works written in French, German, and other languages. Where I have used published translations, I have so indicated in a note; otherwise the translations are mine.

H.A.

CHAPTER 1

Introduction:
"Are We Alone?"

1. *The Plurality of Worlds in Modern Science*

On 24 December 1968 the first images showing the Earth as a whole appeared on countless television screens, beamed down by the crew of the Apollo 8 spacecraft. For millions of people this event marked the opening of a new chapter in the history of human consciousness. The pictures showed the Earth, with its familiar continents, oceans, and cloud formations, rising over the moon's horizon and then floating in space. For the first time we saw our planet as a whole, as a celestial body, from a vantage point outside it, and this meant that we saw ourselves, too, from outside. This new view of the Earth gave fresh impetus to the conservation movement, but that was not all. Through telescopic observations and photography, we had long been familiar with the appearance of other planets; now the Earth, seen from this new viewpoint, looked essentially similar. Of all the achievements of that momentous mission, it was this image, and its implications, that came as the most startling revelation. Faced suddenly with visible proof of the similarity between the Earth and other planets, many of us were inescapably struck by the thought that the analogy might go further—that just as we could observe other planets from outside, so our own planet might be visible not only to the astronauts in the American spacecraft but also to observers on other planets—intelligent extraterrestrial beings akin to ourselves. One of the Apollo 8 astronauts told television viewers of his thoughts as he looked back at the Earth floating in space: inevitably, he had found himself wondering, "What would a traveller from another planet think of Earth?"[1]

And what would he think of us, and we of him? And how would this affect the way we see ourselves? For our reflections on the possibility of intelligent life in outer space—a prospect that can inspire either hope or fear—are ultimately a part of our attempt to understand ourselves. It is this attempt that enters a new phase once we accept the possibility of other "worlds." Research programs using radio telescopes to try to detect signals from intelligent extraterrestrial beings—programs on which governments of whatever political color have for years been prepared to lavish considerable sums of money—are often justified by scientists in "humanistic" terms as being part of the search for ourselves. Are we humans the only intelligent form of life in the universe? If so, who are we? And if not, who are we compared with "them"? The two possible conclusions, that we are alone or that we are not, are both equally astounding in the light of present-day science. Either way, it is a "miracle," to quote the highly unscientific expression used by both the leading U.S. radio astronomer, Carl Sagan, and his Soviet counterpart, I. S. Shklovskii.[2] The universe has a radius of between 15 and 18 billion light-years and contains about 100 billion galaxies like our Milky Way; and if one thinks of the period of its existence as a year, then *Homo sapiens* appeared on our planet only within the last few minutes of the year. It is amazing to think that in all this space and time, man may be a unique phenomenon. Yet it is equally amazing to think that his may be just one of millions of planetary civilizations, some of which may be the manifestation of beings as far superior to us in intelligence as we are to the amoeba.

The English science fiction writer, philosopher, and physicist Arthur C. Clarke, who at one time headed the British Interplanetary Society, is not exaggerating when he remarks that the question of whether there is one world or more than one, whether we are alone in the universe or not, is "one of the supreme questions of philosophy."[3] The point science has now reached makes it one of compelling importance. Clarke and many others believe that it places us on the threshold of a new era in the history of human consciousness. He

[1]James Lovell, quoted in Ian Ridpath, *Worlds Beyond* (London, 1975), p. 15.

[2]Sagan in a conversation with other scientists, *Newsweek Focus: Mysteries of the Cosmos*, June/July 1980, p. 76; Shklovskii quoted by Stanislaw Lem, *Summa Technologiae* (Frankfurt, 1976), p. 123.

[3]*Report on Planet Three and Other Speculations* (1972; New York: Signet Books, 1973), p. 79.

imagines how the present time might be seen by a historian writing in the year 3000:

> To us a thousand years later, the whole story of Mankind before the twentieth century seems like the prelude to some great drama, played on the narrow strip of stage before the curtain has risen and revealed the scenery. For countless generations of men, that tiny, crowded stage—the planet Earth—was the whole of creation, and they the only actors. Yet towards the close of that fabulous century, the curtain began slowly, inexorably to rise, and Man realised at last that the Earth was only one of many worlds; the Sun only one among many stars. The coming of the rocket brought to an end a million years of isolation. . . . the childhood of our race was over and history as we know it began.[4]

But as the Earth emerges from its isolation into a new stage of consciousness, what will this consciousness actually be like? The National Aeronautics and Space Administration (NASA) commissioned a study from the Brookings Institution on the human implications of space exploration. The section of the study dealing with the possible discovery of higher extraterrestrial life forms sounds a clear note of warning: "Anthropological files contain many examples of societies, sure of their place in the universe, which have disintegrated when they had to associate with previously unfamiliar societies espousing different ideas and different life ways; others that survived such an experience usually did so by paying the price of changes in values and attitudes and behavior."[5] Clarke is more optimistic. He believes the search for contact with extraterrestrial intelligence, which is being so strenuously pursued worldwide, could lead to one all-important achievement, the replacement of man's traditionally earthbound outlook by a new cosmic perspective: "For on that quest, whatever else [men and nations] lose or gain, they will surely find their souls."[6]

Of course the question of whether ETI (extraterrestrial intelligence) exists has probably been asked—and answered in the affirmative—for as long as mankind has existed. Buddhism, for example, accepts the idea of a plurality of worlds; the Pythagoreans believed that the moon and stars were peopled by humans or similar beings; in the

---

[4]*The Exploration of Space* (New York, 1951), p. 195.
[5]Donald N. Michael, ed., *Proposed Studies on the Implications of Peaceful Space Activities for Human Affairs* (Washington, 1961), p. 215.
[6]*Voices from the Sky* (New York, 1965), p. 184.

Middle Ages, thinkers as different as Nicholas of Cusa and Teng Mu reflected on the individual characteristics of the inhabited planet-worlds. But in all these cases the idea was a matter of faith, based, as it was, wholly on metaphysical speculation and not on the empirical natural sciences. For centuries Platonists, Aristotelians, and Christians in *their* metaphysical speculation refused to accept that there were other "Earths" analogous to our own. But only when the existence of other planets was *scientifically* established did the dream of other inhabited worlds cease to be merely a product of the imagination; for then it could be shown to be logically extrapolated from known data. It was the Scientific Revolution of the sixteenth and seventeenth centuries that for the first time made the existence of ETI more than a matter of pure speculation. Copernicus found, by analysis of observational data, that the Earth was only one of the planets revolving around the sun, and Galileo's telescopic observations showed that the planet Jupiter was another "Earth." To argue on the basis of these facts that there was life on other planets, either in our own solar system or in another (since Copernicus's discovery made it possible to assume other, analogous, systems) was to speak more or less as a scientist, not merely playing with ideas but aspiring to establish a truth. Anyone doing this, of course, risked a confrontation with the truth as propounded by Christian doctrine, which admitted of only one world and one humankind, regarded as the pinnacle of creation. Such was the fate of Giordano Bruno: on 17 February 1600, on the Campo de' Fiori in Rome, he was burned to death as a heretic—by the same church that in the mid-fifteenth century had made Nicholas of Cusa, who held very similar views, a cardinal. And yet a century and a half after Bruno's death, the heresy had become a new gospel: Christian laymen of varying degrees of enlightenment found it possible to hear the universe resounding with thousands upon thousands of planet-worlds singing in praise of their Creator. Finally, in 1977, the Roman Catholic priest and doctor of theology Kenneth J. Delano published, with the imprimatur of his bishop, a book entitled *Many Worlds, One God* (Hicksville, N.Y.). In it he attempts to break down the theological parochialism of one world and one humankind on the basis of the modern scientific world view while nevertheless leaving the church's central teachings intact.

Clearly, then, the Apollo 8 pictures of the Earth broke new ground in the history of human consciousness only in one specific sense: through the mass medium of television they opened the eyes of a vast number of people to the Copernican and post-Copernican view of the cosmos. For the first time ever, millions saw with their own eyes images that vividly confirmed what scientists since Copernicus had known. And just as scientifically minded thinkers ever since the sixteenth and seventeenth centuries had taken the step from recognizing that the Earth was merely one of the planets to entertaining the idea of a plurality of worlds, so these millions were now able to take that same step for themselves. By including ETI in our conception of the world and of ourselves, we are, as the American astrophysicist A. G. W. Cameron has said, "completing the Copernican intellectual revolution."[7] The Apollo 8 mission brought this process to a point where the thinking of ordinary people (or "popular culture") could be said to have caught up with that of educated thinkers.

It was a propitious moment. Since about 1950 the subject of ETI had once again become respectable, particularly among scientists, as it had been up to the turn of the century. The nebular hypothesis of Kant and Pierre Simon de Laplace, which makes the formation of planetary systems plausible, had become discredited in the first decades of the twentieth century but was rehabilitated in 1944 by Carl Friedrich von Weizsäcker. The theory that microorganisms—and thus, life—developed inevitably and spontaneously from matter (an idea that makes the emergence of life on other planets more probable) had been thoroughly refuted by Louis Pasteur in 1860; the experimental work of Stanley Miller and Harold C. Urey from 1953 onward, while not proving it to be correct, did show it to be at least theoretically possible, though with a different understanding of the process involved. Using inorganic substances, Miller and Urey simulated the physical and chemical conditions of the Earth's surface at a period before life existed, and they succeeded, by means of electric discharges, in synthesizing organic molecules, including various amino acids and later also nucleotides, which are constituent parts of proteins and nucleic acids respectively and are thus among the building blocks of cellular life.[8] By 1970 at the latest, radiation research had shown that various organic molecules known on Earth were also present in clouds of gas and dust in interstellar space, and the same molecules were discovered in meteorites. Lastly, by the late fifties, scientists in the field of radio astronomy were in a position to pick up electromagnetic signals transmitted by any advanced extraterrestrial civilizations there might be, recognize the nature of the signals, and respond to them; they could assert: "communication is now possible."[9] In 1958,

[7]A. G. W. Cameron, ed., *Interstellar Communication* (New York, 1963), p. 1.

[8]Bernard Lovell, *In the Center of Immensities* (New York, 1978), p. 59.

[9]Carl Sagan, ed., *Communication with Extraterrestrial Intelligence (CETI)* (Cambridge, Mass., 1973), p. 304.

Harlow Shapley, in his popular book *Of Stars and Men*, presented the first calculation in modern times of the probable number of inhabited planets; and the next year the journal *Nature* carried the now-famous appeal by Giuseppe Cocconi and Philip Morrison for a search program to be initiated (pp. 844–46). "Few will deny," the two physicists concluded, "the profound importance, practical and philosophical, which the detection of interstellar communications would have." The effect was sensational. Many scientists today insist that if we could succeed in making radio contact with extraterrestrials, the significance of this event would be without equal in the history of mankind. "Standing on the threshold of such profound knowledge makes our times a unique and most exciting period in the history of man," writes the astrophysicist Michael D. Papagiannis.[10]

From the 1960s onward the question of the existence of ETI and of ways and means of discovering it has been one of the most intensively debated issues of modern science, with astrophysicists and biochemists working equally hard toward a solution. The technology of space travel is less closely involved, since the speeds that can be achieved within the foreseeable future will not be sufficient to cover the necessary distances within a reasonable portion of a human life-span. Photon-drive rockets might make it possible to achieve relativistic speeds, that is, speeds approaching the speed of light. According to Einstein's theory, confirmed by experiment, such speeds would appreciably shorten the length of time that passed on board the spacecraft, but even this would not get us much further: if we suppose that a spacecraft capable of traveling at 99 percent of the speed of light makes the journey to a planetary civilization at a distance of one hundred light-years (and there is not likely to be one much closer), then for the crew the round trip would take only twenty-eight years, but by the time they returned, two hundred years would have passed on Earth.[11] It is widely thought, however, that such a speed cannot be achieved in any case, because the necessary launch mass would be too great. Only radio astronomy offers any hope of contact. The electromagnetic signals that could be received and responded to travel through space at the speed of light and could theoretically arrive at any time.

Always central to any discussion of how to make contact with ETI is Frank Drake's formula for calculating the likely number of

---

[10]"Search for Other Life Goes On," *Christian Science Monitor*, 6 February 1980, p. 16.
[11]Ronald N. Bracewell, *The Galactic Club: Intelligent Life in Outer Space* (San Francisco, 1975), p. 108. See also Günter Paul, *Unsere Nachbarn im Weltall* (Düsseldorf, 1976), chap. 5.

technologically advanced civilizations in the universe. It takes the form of an equation, on one side of which is the unknown, N, the number of extraterrestrial civilizations, intelligent, able, and willing to communicate, that exist in our galaxy at a given time; on the other side are seven factors that, multiplied together, yield a value for N. All the factors involve calculations of probability, and the mathematical values for them are obtained by answering the questions How many stars are there in the Milky Way? How many of them possess a planetary system? How many planets in such a system are ecologically suitable for the emergence of life? On how many of these does life in fact occur? How often does the life that does occur evolve to produce creatures of human or comparable intelligence? How many such intelligent "races" develop the ability and the desire to take up radio contact with other civilizations in the galaxy? and What is the typical life-span of such a technological civilization?

It is clear that the Drake formula is an equation made up entirely of unknowns. The calculation of each probability is bound to involve subjective estimates.[12] Let us consider, for instance, the last factor,[13] a crucial one, given the enormous distances involved: there is little point in replying to a radio signal unless it seems probable that when the answering signal finally arrives, the communicating civilization will still be in existence and interested in interstellar contact. The numerical value of this factor L (life-span of a civilization) is decisively important in determining the value of N: N is often set at one-tenth of L. But we have no evidence on which to base our estimate of L except for the duration of technological civilization on our own planet, and as we have entered the era of radio astronomy only within the last few decades, this evidence is clearly of no practical value. Some scientists

[12]Sagan, Communication, pp. 357–61.
[13]On the "life-span" factor, see in particular Gerrit L. Verschuur, "We Are Alone!" Astronomy 3:12 (1975), 47–49. The factors used in the Drake equation were discussed at the international ETI conference in Byurakan, Armenia, in 1971; these discussions form the content of the work cited in n. 9. The authors of several academic and popular scientific books have used these factors, directly or indirectly, as a basis for organizing and subdividing their material; see, in addition to the books cited in nn. 1, 7, 11, and 15, Carl Sagan, The Cosmic Connection (1973; New York: Dell Books, 1975); Walter Sullivan, We Are Not Alone, 2d ed. (New York, 1966); I. S. Shklovskii and Carl Sagan, Intelligent Life in the Universe (San Francisco, 1966); Roger A. MacGowan and Frederick I. Ordway, Intelligence in the Universe (Englewood Cliffs, N.J., 1966); John W. Macvey, Whispers from Space (New York, 1973); Cyril Ponnamperuma and A. G. W. Cameron, eds., Interstellar Communication: Scientific Perspectives (Boston, 1974); Donald Goldsmith and Tobias Owen, The Search for Life in the Universe (Menlo Park, Calif., 1980); Michael D. Papagiannis, ed., The Search for Extraterrestrial Life: Recent Developments (Dordrecht, 1985); Edward Regis, Jr., ed., Extraterrestrials (Cambridge, 1985). See also the more specialized studies referred to in the following paragraphs.

set the life-span of a technological civilization at dozens of years; others, at hundreds of millions. As a result, there are widely varying estimates of the number of technological civilizations in existence at a given time, and our experience on our own planet gives us no guide as to which estimate is the most accurate.

The other factors in the Drake equation are equally difficult to determine. Estimates of the number of stars in the Milky Way vary between 100 and 400 billion. Whether other stars besides our sun have their own planets has not yet been established with certainty, and even if in the near future infrared detectors were to provide definitive proof, our estimates of how many such planetary systems exist would still rest on vague calculations of probability. One of the most uncertain factors in the Drake equation is the likelihood of life as we know it (carbon-based life) occurring on other planets. Would such an event be inevitable, given certain conditions, or a mere freak of chance? For life to come into being, certain organic molecules, the proteins and nucleic acids (which are extremely complex formations in themselves), have to combine in one specific way (which again is highly complex and statistically most unlikely even over astronomical periods of time) in order to form the most basic unit of life, the cell capable of reproducing itself and evolving. On our planet this process took place just once, three and a half billion years ago. Some biochemists consider this a more or less automatic process. Others rate the likelihood of its occurrence virtually at zero; in their view, exobiology is the study of something that does not exist. The origin of life, even of life on Earth, remains an unsolved problem for biochemists.[14] Until the process can be reproduced in the laboratory we cannot assess the probability of its being repeated elsewhere in the universe.

But, supposing that at some time, somewhere, single-celled organisms have come into existence: we next face the question of whether they are bound to undergo a process of evolution with the same outcome as that on Earth, the emergence of an intelligent life-form analogous to man. Different biologists and paleontologists, with the same facts at their disposal, have reached opposite conclusions. Thus, in *This View of Life* (New York 1964), George Gaylord Simpson points

[14]See Gerald Feinberg and Robert Shapiro, *Life beyond Earth* (New York, 1980), p. 113. See also Lovell, *In the Center*, chap. 4; Ridpath, *Worlds Beyond*, pp. 65–81; Sagan, *Communication*, pp. 42–67; Ponnamperuma and Cameron, *Interstellar Communication*, chap. 3; Cyril Ponnamperuma, *The Origins of Life* (London, 1972); Michael H. Hart, "Atmospheric Evolution, and an Analysis of the Drake Equation," in *Extraterrestrials— Where Are They?* ed. Michael H. Hart and Ben Zuckerman (privately distributed, 1981; New York, 1982).

to the countless occurrences of mutation and selection that have led, by a highly devious route, to the emergence of *Homo sapiens*. It is highly improbable, indeed virtually impossible, he argues, that this tortuous path and its outcome could be duplicated (pp. 253–71). On the other hand, Carl Sagan, in *The Dragons of Eden* (New York 1977), speaks for those who would confidently expect any evolutionary process taking place over billions of years to produce intelligence and who take it equally for granted that intelligent beings would achieve technological prowess and would wish to communicate with other civilizations (pp. 229–38). The philosopher Roland Puccetti and the scientist N. J. Spall have gone still further, arguing that the physical appearance of the intelligent beings produced by evolution is bound to be "more or less" similar to that of *Homo sapiens*.[15]

Faced with these problems, the layman—and perhaps not only the layman—is tempted to give up in despair, particularly when he sees respected scientists proceeding, undeterred, to name an approximate numerical value for the unknown, N, in the Drake equation. Their estimates, based on very complicated mathematical calculations, are astonishingly far apart because of diverging views on how far the development of life on Earth may be seen as typical and capable of being reproduced analogously (the "principle of mediocrity"). The number of advanced civilizations in the Milky Way was set at 500 million by James R. Wertz in 1976, at 3 million by J. Freeman in 1975, at one million by Carl Sagan in 1973, at a single one—our own—by Frank J. Tipler in 1980, and at theoretically less than one by Michael H. Hart in 1981.[16] Of course these last two calculations do not exclude the possibility of a plurality of worlds; they suggest only that we are alone *in the Milky Way*. Hart stresses that as the number of galaxies is "infinite" (current research suggests a figure of about 100 billion galaxies), we may assume that an "infinite" number of advanced civilizations exist, albeit at a distance of millions of light-years, which rules out any possibility of making contact and engaging in communication.

---

[15]Puccetti, *Persons: A Study of Possible Moral Agents in the Universe* (London, 1968), pp. 96–97; Spall, "The Physical Appearance of Intelligent Aliens," *Journal of the British Interplanetary Society* 32 (1979), 99–102.

[16]Wertz, "The Human Analogy and the Evolution of Extraterrestrial Civilizations," *Journal of the British Interplanetary Society* 29 (1976), 445–64; J. Freeman and M. Lampton, "Interstellar Archeology and the Prevalence of Intelligence," *Icarus* 25 (1975), 368–69; Sagan, *Cosmic Connection*, p. 202; Tipler, "Extraterrestrial Intelligent Beings Do Not Exist," *Quarterly Journal of the Royal Astronomical Society* 21 (1980), 267–81; Hart, "Atmospheric Evolution." See also Donald Goldsmith, ed., *The Quest for Extraterrestrial Life* (Mill Valley, Calif., 1980).

In the face of such disagreement among experts in the relevant disciplines, the agnostic position adopted by the American physicist Freeman Dyson appears still—indeed more than ever—to be the only appropriate one. Dyson argued in 1972 that research in astrophysics, biochemistry, and evolutionary biology had simply not yet reached the point where one could judge whether the existence of ETI was probable or not.[17] The fact that scientists nevertheless continue to form judgments, and arrive at widely differing estimates, is bound to raise the suspicion (which has indeed often been voiced) that their estimates owe much to *philosophical* preconceptions—to the hope or the fear that we are alone, or that we are not . . .

"Unprejudiced observation," advocated by Dyson himself in 1972, offers a way out of the morass of uncertainties. Whether this observation, this search for electromagnetic signals from space, is not just as much inspired by hopes and fears is a question we need not pursue. (In 1978, Nigel Calder said in all seriousness that current search programs were being carried out in the hope of proving that "they are not there.")[18] The challenge thrown out by Cocconi and Morrison at the close of their sensational article of 1959 still stands: "If we never search, the chance of success is zero."

But how is one to search? As work on interstellar communication has intensified, so the theoretical and practical problems have multiplied. Thus for decades those working in the field of radio astronomy took it for granted that the means used by either side to establish contact would be electromagnetic radiation in the radio spectrum, but this is now questionable. Should they not rather be considering laser or maser beams? Even if one does keep to radio waves, it is no good searching at random. One has to decide which frequency range and which wavelength to monitor or to use for transmission. The hydrogen frequency of 1,420 megahertz (1,420 million vibrations per second, wavelength 21 cm) is now no longer universally favored. Does this mean that our only hope of successful contact lies in monitoring *all* frequencies? This is the view of Drake, who in 1960 directed the first search program, Project Ozma; but such an undertaking would present enormous practical problems.[19] And at which out of all the stars—at least 100 billion in the Milky Way alone—should the radio telescope antennae be pointed?

[17]See Ridpath, *Worlds Beyond*, p. 30.
[18]*Spaceships of the Mind* (New York, 1978), p. 137. On the philosophical preconceptions see sec. 2 of this Introduction.
[19]"On Hands and Knees in Search of Elysium," *Technology Review* 78:7 (1976), 25–

The chief obstacle to interstellar communication, however, is the time factor, the length of time it takes for a radio signal racing across space at the speed of light to travel from one civilization to another. On 16 November 1974, Drake transmitted a signal using the radio telescope at Arecibo, Puerto Rico, to the globular star cluster M13: it will arrive there in twenty-four thousand years' time. Supposing that Sagan is roughly correct in his assumption that there are one million technological civilizations in the Milky Way, then these would be on average three hundred light-years apart.[20] If a signal from us to one of these were instantly understood and answered, we should still have to wait six hundred years for that response, always assuming that our civilization was still in existence and had not by then lost interest in interstellar communication. Or suppose that there is only one civilization per galaxy: communicating with the Andromeda Nebula, one of the nearest other galaxies to our own, would take millions of years. But, again, these calculations presuppose that the civilizations have already discovered each other. Given these immense distances in time and space, how can they possibly discover one another, even supposing that they used the same frequency for transmitting and receiving signals? Even if they were to transmit and receive continuously for hundreds of years, there is little likelihood that the sender and the receiver of signals would ever "meet," so to speak. It could only happen if a very large number of civilizations were continuously transmitting and listening for vast periods of time on all wavelengths, covering every possible distance in every direction, an assumption so unrealistic it has never been seriously put forward.

The astrophysicist Gerrit L. Verschuur has worked through a number of possible scenarios (though he excludes coincidence, which should perhaps not be discounted!), and the title of his article sums up his conclusion: "We are alone."[21] By this he means "effectively alone" in the Milky Way. If, for instance, the life-span of a technological civilization is 100–200 years, then we have at a given time ten to twenty such civilizations in the Milky Way at an average distance from Earth of 2,000 light-years. So if we were able to locate a civilization

---

26, 29. See also Robert C. Cowen, "The Cosmic Haystack," *Christian Science Monitor*, 28 December 1977, p. 21; Michael D. Papagiannis, "Strategies for the Search for Life in the Universe," *Cosmic Search* 2 (1980), 24–27; idem, ed., *Strategies for the Search for Life in the Universe* (Dordrecht, 1980).

[20]Sagan, "The Quest for Intelligent Life in Space Is Just Beginning," *Smithsonian* 9:2 (1978), 43.

[21]Verschuur, "We Are Alone." See also Cameron, *Interstellar Communication*.

among the dozens of millions of stars, the time it would take for a signal to reach it would be many times the life-span of that civilization. Alternatively, if we follow Sagan in setting that life-span at 10 million years, then we have a million civilizations in the galaxy and an estimated 600 years to wait for a reply to a signal, so that interest in interstellar communication would need to be maintained for at least that length of time. But in the meantime, technological and evolutionary development would continue, both here and on the other planet. Surely in many imaginable cases the *qualitative* distance between the two would remain unchanged. Would the two civilizations be able to "understand" each other, to "speak" to one another, to recognize each other's signals as signals—or would contact between them be like an encounter between orangutans and humans, between Neanderthal men and New Yorkers, between smoke signals and radio pulses? If we achieve "contact" of this kind, we should still be effectively alone in the Milky Way. Calculations made at the ETI conference in Byurakan, Armenia, led to the conclusion that two-way communication with a technological civilization more than a thousand years in advance of our own would not be possible.[22] Sagan formulates as follows the dilemma posed by the Drake equation: it is inconceivable that an extraterrestrial civilization is close to us in both space and time (i.e., relative stage of development). For if there were a technological civilization near to us, then there would be very many such civilizations in the Milky Way, and these must have a very long life (since their number, N, is a function of their life-span, L); this in turn means that that civilization would necessarily be far older than ours, far more advanced, and therefore unlikely to be interested in contact with us, except possibly in the spirit of archaeological curiosity! Conversely, a technological civilization at a stage of development comparable to our own would be one of very few and therefore very far away, practically impossible to locate, and because of its presumed short life-span, impossible to reach.[23]

The uncertainties and conflicting opinions of the "galactic demographers," and the logical and practical difficulties in communicating with ETI, do not mean that the subject of ETI is itself unscientific, frivolous, *purely* speculative, or concerned with something that does not exist. They simply underline the paradoxical nature of the problem. In the felicitous formula coined by Martin Rees, "absence of

[22]Sagan, *Communication*, p. 211.
[23]Ibid., p. 254; see pp. 212–14. On the second possibility see also N. H. Langton, "The Probability of Contact with Extra-Terrestrial Life," *Journal of the British Interplanetary Society* 29 (1976), 465–68.

evidence [of ETI] is not evidence of absence [of ETI]."[24] The hypothesis of the existence of ETI is based on extrapolation from facts and theories in astrophysics and biochemistry and is thus, in principle, capable of proof and therefore scientifically legitimate. The one criterion it cannot fulfill is Karl Popper's principle that a statement is scientific only if it is potentially capable of being proved false. Empirical proof of its falsehood would obviously be impossible to obtain. As the philosopher Puccetti has pointed out, this could be accomplished only by a personal journey through all the galaxies, since an extraterrestrial civilization is not necessarily transmitting signals. Puccetti concludes that ETI is one of those topics for which the fact that they are based on extrapolation from facts and theories belonging to the basic sciences is sufficient to make them scientifically legitimate. Although, therefore, the subject is inherently speculative, it is of a different order from a religious or metaphysical belief.[25]

It is possible that increasing confidence in the scientific validity of the subject of plurality of worlds has been implicit in some recent speculations in astrophysics that may strike the layman as outrageously unscientific. Their starting point, like Puccetti's, is the perplexing fact that, though some calculations suggest the Milky Way is densely populated, no contact has yet taken place between "them" and us. Recently the idea has been much bandied about that any extraterrestrial civilization would, in accordance with what is alleged to be a natural tendency in all living creatures, *necessarily* have spread out gradually over the whole galaxy, like algae in a neglected swimming pool or voyagers in the South Seas. This, we are told, would have happened within less than 10 million years,[26] a minute fraction of the age of the Milky Way (10–15 billion years), and, moreover, millions of years ago. By now, therefore, the Milky Way should be teeming with life. How, then, are we to explain that the numerous search programs undertaken from 1960 onward, using a variety of strategies, have so far failed to detect these extraterrestrials and that they for their part have not made themselves known to us, either by admitting us to the "galactic club" or by colonizing the Earth? (Flying saucers play no part in the discussion: astrophysicists do not as a rule

[24]Quoted in Richard Berendzen, ed., *Life beyond Earth and the Mind of Man* (Washington, 1973), p. 1.
[25]Puccetti, *Persons*, pp. 82–85.
[26]Michael D. Papagiannis, "Are We All Alone, or Could They Be in the Asteroid Belt?" *Quarterly Journal of the Royal Astronomical Society* 19 (1978), 278. Tipler, "Extraterrestrial Intelligent Beings," p. 268, estimates 300 million years. See also Papagiannis, "Could We Be the Only Advanced Technological Civilization in Our Galaxy?" in *Origin of Life*, ed. Haruhiko Noda (Tokyo, 1978), pp. 583–95.

take them seriously, and if they pay them any attention at all, it is only to conclude that on the basis of current knowledge they cannot exist.)[27] "Where is everybody?" Enrico Fermi is reported to have asked. Theorists such as Hart and Tipler reply that the fact there has been no encounter as yet proves that the premise is wrong:[28] there *are* no extraterrestrials in the Milky Way, and we are not just effectively, but actually alone in the galaxy. (In the universe as a whole, we are effectively alone, since the distances measured in millions of light-years would make any attempt at contact pointless.) Other theorists, such as David Viewing and D. G. Stephenson, consider it possible that our solar system may be a sort of carefully protected nature reserve, in which perhaps the intelligent cosmic beings observe us just as we observe the African wild animals in Serengeti National Park. Perhaps, suggests Michael Papagiannis, their vantage point is the asteroid belt of the solar system.[29] Other related theories are the so-called zoo hypothesis advanced by the astrophysicist John A. Ball and the idea put forward by the anthropologist Ashley Montagu that man's nuclear experiments make him so dangerous that his species is being kept in isolation by the "others," like a source of infection. Papagiannis speculates that the extraterrestrials are still undecided as to how to deal with the Earth.[30]

Of course these ideas are all too human, based as they are on the unconscious assumption that "they" think like us. Faced with such thought experiments seriously advanced by responsible scientists, many skeptics may be drawn to the alternative theory that we are indeed "alone." Perhaps ours really is the only advanced civilization in the Milky Way. This suspicion, rarely voiced by scientists in the heady years of the sixties and early seventies, now seems to be gaining ground among specialists.[31] But how can we possibly account for the fact, if it is true, that we are alone in possessing intelligence,

[27]Philip J. Klass, *UFOs Explained* (New York, 1974); Donald H. Menzel and Ernest H. Taves, *The UFO Enigma* (Garden City, N.Y., 1977); see also n. 46 below.

[28]Tipler, "Extraterrestrial Intelligent Beings"; Hart, "Atmospheric Evolution"; idem, "An Explanation for the Absence of Extraterrestrials on Earth," *Quarterly Journal of the Royal Astronomical Society* 16 (1975), 128–35.

[29]Viewing, "Directly Interacting Extra-Terrestrial Technological Communities," *Journal of the British Interplanetary Society* 28 (1975), 735–44; Stephenson, "Factors Limiting the Interaction between Twentieth Century Man and Interstellar Cultures," ibid. 30 (1977), 105–8; Papagiannis, "Are We All Alone."

[30]Ball, "The Zoo Hypothesis," *Icarus* 19 (1973), 347–49; Montagu, in Berendzen, *Life beyond Earth and the Mind of Man*, p. 21; Papagiannis, "Are We All Alone"; idem, "Could We Be."

[31]See Tipler, "Extraterrestrial Intelligent Beings," p. 267; Papagiannis, "Are We All Alone," p. 277.

when we are merely the inhabitants of a planet belonging to a totally unremarkable star thirty thousand light-years away from the center of an equally unremarkable galaxy containing hundreds of billions of other suns? Scientists attempting an explanation can do little more than paraphrase the alleged "fact" and offer an interpretation that seems, almost inevitably, teleological: we are told that the galaxy had to develop precisely in the way it did in order that, just once, intelligent life might occur. This is, once again, the "anthropic principle" A. R. Wallace outlined around 1900.[32] What we are offered now is in effect a new, scientific form of anthropocentrism, be it absolute or, if we assume analogous developments in other galaxies, "effective."

Clearly, these scientific questions lead us into the realm of philosophy.

2. Philosophical Perspectives of Modern Science

We have thus arrived at a paradox: it is virtually inconceivable that in a universe with about 100 billion galaxies like the Milky Way, there should be *no* intelligent beings besides ourselves; but it is almost equally unthinkable, or at any rate improbable, that we can ever establish contact with them. Despite all the disagreement among scientists on the various aspects of the ETI problem, their imagination is often haunted by this paradox, which is essentially a philosophical one and one that invests man's reflections on himself with amazement and wonder. A leading astronomer, Harlow Shapley, has even suggested that at such a stage in science it is actually necessary for the imagination to intervene, for "cosmic fancies" to supply what we still lack in factual knowledge.[33] It is, however, not entirely unlikely that in principle there may be no getting beyond this stage, that this is the definitive limitation of our "human condition" in the cosmic context. Stanislaw Lem has remarked that "in the second half of the twentieth century one can hardly be a complete human being without at least occasionally calling to mind that community of rational beings, as yet unknown, to which we presumably belong."[34]

[32]Tipler, "Extraterrestrial Intelligent Beings," pp. 279–80, and the literature cited there; Reinhard Breuer, *Das anthropische Prinzip* (Vienna, 1981); John D. Barrow and Frank J. Tipler, *The Anthropic Cosmological Principle* (Oxford, 1986).

[33]*Of Stars and Men* (Boston, 1958), p. 156. The passage referred to is the one used as a motto for this book.

[34]*Summa Technologiae*, p. 130.

But it is not only those in the fields of philosophy and literature who call upon their imagination to help them deal with the paradoxical outcome of scientific research. As we saw at the end of the preceding section, the scientist, searching as he must for the *meaning* of his empirically obtained facts, is equally obliged to resort to his imagination. As Charles I. Glicksberg says, "He must depend on the poetic myth as the means of unifying his vision and affirming his faith, whatever it may be."[35] He may write science fiction, like Fred Hoyle, or set out his galactic philosophy in books aimed at a wide readership, as Carl Sagan has done. Or he may never actually formulate the beliefs underlying his scientific work; yet a stray adjective or use of the subjunctive will afford a glimpse of them. Thus, each in his own way, scientists too are now trying the unfamiliar but necessary path of letting philosophical imagination play a role in their quest. As soon as NASA had been created by an act of Congress as an agency of the U.S. government, it commissioned the already-mentioned report from the Brookings Institution. This report recommended, among other things, further studies into the changes in our view of ourselves that followed from the recognition that intelligent extraterrestrial life was a scientifically plausible possibility.[36] The international summit conference held at Byurakan in 1971, under the auspices of the U.S. and Soviet Academies of Sciences, discussed the issues relating to communication with ETI and passed a resolution that referred to the "enormous philosophical significance" of interstellar contact.[37]

This philosophical significance remains essentially the same whether or not contact is in fact achieved. The mere fact that belief in the existence of "aliens" can be scientifically justified and cannot be proved wrong suffices, particularly in this age of mass communications, to achieve a momentous broadening of human horizons to a "cosmic" perspective. Indeed, it is precisely because empirical proof of ETI is still lacking that this new perspective invites us to attempt that wholly worthwhile scientific-cum-philosophical intellectual experiment which for many promises to be the greatest possible adventure of the human spirit. For in thinking about ourselves vis-à-vis other intelligent life-forms the universe may contain, we have reached our last frontier, our "Childhood's End," to quote the title of an Arthur C. Clarke novel, one of the most widely read of its kind. Philosophy,

---

[35]"Science and the Literary Mind," *Scientific Monthly* 70 (1950), 357.
[36]Michael, *Proposed Studies*, pp. 215–16.
[37]Sagan, *Communication*, p. 353.

science, and literature are all joining forces in the attempt to delineate the unique change in our consciousness expressed in that phrase.

This new way of thinking about ourselves—this "Milky Way speculation," as Thomas Mann called it[38]—causes fewest problems to the theologians. Jainists, Buddhists,[39] Mormons, and some others do in any case assume the existence of a plurality of worlds. One might expect that Christian theologians, on the other hand, would find it hard to relinquish the idea of man on this planet as unique and therefore uniquely important—made in God's image and redeemed by Christ. Today, however, the opposite is true. Exochristology seems to be less problematic than exobiology. Theologians are able to reconcile the new data with the familiar concepts by interpreting these flexibly and less "parochially." Protestants Paul Tillich and Krister Stendahl, the Anglican C. S. Lewis, and Roman Catholics Kenneth J. Delano and T. J. Zubek, for example, all agree in this matter.[40] The kind of reasoning they employ has not changed much since the fifteenth century (though at that time it was esoteric and soon branded as heresy). The assumption that God created many worlds and humankinds only serves to enhance his glory; those "races" (like the angels) may either be "fallen" or not; if they are, either they have been redeemed by Christ or the incarnation of another savior, or they have not been redeemed; in the latter case, they are either damned, like the fallen angels, or capable of being saved at the end of time, like heathens in distant lands not reached by the gospel.

Unlike the theologians, scientists have no framework of prescribed, unquestionable beliefs within which to work, and the diversity of their reactions when faced with the idea that "we are not alone" reveals profound uncertainty. Supposing that an electromagnetic sig-

---

[38]*Gesammelte Werke* (Frankfurt, 1974), 9:447.

[39]The Buddhist doctrine of the plurality of worlds developed along a number of different lines, particularly in Chinese thought. This aspect of the topic cannot be examined in the present study, which is confined to Western tradition. See Wolfgang Bauer, *China und die Hoffnung auf Glück* (Munich, 1971), esp. pp. 148–49, 262–64, 326–28 (Teng Mu), 444–52; Joseph Needham, *Science and Civilization in China*, vol. 2 (Cambridge, 1956), pp. 219–21, 419, 438–42; Edward H. Schafer, *Pacing the Void: T'ang Approaches to the Stars* (Berkeley and Los Angeles, 1977).

[40]Tillich, *Systematic Theology*, vol. 2 (Chicago, 1957), pp. 95–96; Stendahl, in Berendzen, *Life beyond Earth and the Mind of Man*, p. 29; Lewis, "Onward, Christian Spacemen," *Show* 3:2 (1963), 57, 117; idem, *Shall We Lose God in Outer Space?* (London, 1959); Delano, *Many Worlds, One God* (Hicksville, N.Y., 1977); Zubek, "Theological Questions on Space Creatures," *American Ecclesiastical Review* 145 (1961), 393–99. See also Sullivan, *We Are Not Alone*, pp. 283–89; William Hamilton, "The Discovery of Extraterrestrial Intelligence: A Religious Response," in *Extraterrestrial Intelligence: The First Encounter*, ed. James L. Christian (Buffalo, N.Y., 1976), pp. 99–127.

nal from outer space were to be received and deciphered: this fact in itself would, in the view of many scientists, be more significant than the content of the message. For as well as indicating that we are not the only form of intelligent life in the universe, it would also mean that we are the most backward member of the galactic community— the "dumbest," to use Sagan's forthright expression,[41] or among the dumbest. After all, it is only in the last few decades that we have been in a position to attempt interstellar communication, and any signals would in all probability come from a planet at least a hundred light-years away. This means that in any such encounter with aliens, we would be technologically less mature, and thus inferior to them. Scientists are agreed on this but differ on whether it makes contact with the aliens something to be desired or feared. Their views on the probable outcome of such contact are often expressed in very vivid predictions, some of them influenced by the obvious parallel with the meeting of Europeans and Indians. The Nobel prizewinner Sir Martin Ryle, one of the founders of radio astronomy, advised against responding to a signal, or indeed initiating electromagnetic communication ourselves;[42] there are in fact some who fear that we might become the exploited colonial subjects of an extraterrestrial master race that does not share our conception of morality. Others, however, confidently anticipate that contact with a more advanced extraterrestrial civilization would bring with it an increase in scientific and technical knowledge that would turn our Earth, endangered by its own technology, into a paradise in which all problems, from cancer to the threat of atomic self-annihilation, would be solved.

Of greater interest are the less concrete, more philosophical questions raised by the idea of such an encounter between man and a counterpart who is superior. Here again the answers given by scientists, whether explicitly or only between the lines, reveal an unbridgeable gulf between fear and euphoria. Both reactions stem from considerations we may term *religious*, in the sense of Tillich's "ultimate concern"; and both are based on the premise that man, however secularized his outlook, will be conscious that he can no longer regard himself as the pinnacle of creation and is bound to feel this as a blow to his self-esteem.[43] In George Wald, Nobel prizewinner in biochemistry, this prompted the following fear: If we were to gain access all at once

---

[41]See for instance Sagan, *Cosmic Connection*, chap. 31; Berendzen, *Life beyond Earth and the Mind of Man*, passim.

[42]According to Calder, *Spaceships of the Mind*, p. 131. See also p. 3 above.

[43]See Christian, *Extraterrestrial Intelligence*, esp. pp. 30–31, 84–85, 298–99; Bob Parkinson, "The Starship as a Philosophical Vehicle," *Journal of the British Interplanetary*

to an incomparably more advanced state of knowledge, if the answers to all questions were suddenly simply handed to us, then all human endeavor would be paralyzed and all sense of our dignity and the meaning of our existence would be hopelessly lost. We would be faced with the alarming prospect of the end of human culture; the reins, as it were, would have been taken from our hands.[44] But Harlow Shapley, an equally eminent authority, saw the effect of our demotion as healthy and even liberating, because our success in discovering ETI would be not least a triumph of our intelligence and because we could be confident in future that we would no longer need to have recourse to the supernatural when confronted with scientific problems such as, for instance, the origin and development of life (*Of Stars and Men*, pp. 16, 157). The further point is made, again by scientists, that what would be truly frightening and unendurable would be not the encounter with ETI but its opposite, the knowledge that we were either effectively or in fact alone in a cosmos of unimaginable vastness. The presence of neighbors in the universe with the ability to communicate would be reassuring; it would, according to Sullivan, afford us undreamed-of spiritual and philosophical enrichment. We would also have the consolation that, were we to die out or, as is quite thinkable, destroy ourselves, life might have a future elsewhere; our intellectual achievements might thus be preserved from oblivion, so that in that sense we would be immortal.[45]

It is obvious that such feelings of fear or hope attached to cosmic powers superior to ourselves can easily slide over into religious mysticism. As we have seen, the astrophysicist Papagiannis puts forward the idea that extraterrestrials perhaps already present in the asteroid belt may be hesitating to reveal themselves to us because, in view of our discovery and use of nuclear technology, they are still uncertain whether they should extinguish our species or offer it development aid. From this kind of speculation it is but a small step to a mystical belief in some immeasurably superior forces on whom we are totally dependent.

On the fringes of science, as we know, millions have already taken

Society 28 (1975), 745–50; also Ridpath, *Messages from the Stars* (New York, 1978), p. 137; Shapley, *Of Stars and Men*, pp. 1, 143, and passim; Donald K. Stern, "First Contact with Non-human Cultures: Anthropology in the Space Age," in *Cultures beyond the Earth*, ed. Magoroh Maruyama and Arthur Harkins (New York, 1975), p. 45.

[44]In Berendzen, *Life beyond Earth and the Mind of Man*, p. 17.

[45]Sullivan, *We Are Not Alone*, p. 289; Bracewell, *Galactic Club*, p. 30; Sagan, *Communication*, pp. 339, 344; Papagiannis, "Are We All Alone," p. 277; Christian, *Extraterrestrial Intelligence*, p. 76; Lem, *Summa Technologiae*, p. 129; Puccetti, *Persons*, p. 118; Arthur Koestler, *Janus* (New York, 1978), pp. 283–84; Clarke, *Report on Planet Three*, p. 95.

this step. Since 1947 vast numbers of people have believed in flying saucers; and since the 1960s Erich von Däniken has been building up a large following with his speculations about "gods," extraterrestrial astronauts supposed to have visited the Earth within historical time and even to have made a genetic contribution to the evolution of *Homo sapiens*. Such mass cults can only be explained as attempts to find a substitute for religion, as instances of the search for a new form of dependence on something quite different from ourselves, now that the traditional religious ties have been weakened—not least as a result of modern science's influence. In Evanston, Illinois, a distinguished astrophysicist, J. Allen Hynek, headed an institute for UFO studies; Däniken's books claim to report historical facts and present scientific theories; and it is precisely because both these "cults" believe they have a more or less scientific basis that millions of convinced laypeople accept them as fitting doctrines for our time. They are variations of that archetypal religious phenomenon, the desire for redemption through the intervention of a supernatural, superhuman intercessor, a Savior who by his power, wisdom, and goodness gives meaning to human life and elevates it to a higher level of consciousness.[46] Also present as an undercurrent in these cults is the apocalyptic motif of the end of the world, its destruction by monsters from outer space. The popularity of science fiction horror movies shows that this motif, too, has a strong hold on the popular imagination. The clearest demonstration occurred on 30 October 1938, when Orson Welles's production of *The War of the Worlds* was broadcast and panic broke out on the eastern seaboard of the United States: listeners believed they were hearing a genuine news report of an invasion from Mars.[47]

We cannot leave the topic of these scientifically inspired mass phenomena and the needs they fulfill without also mentioning the intense interest the press, radio, films, and television show in the subject of life on other planets. Also, the number of "popularizing" books on the subject is legion, and they are to be found in some of the poorest countries in the world. *Star Wars*, a kind of galactic cowboys-and-Indians, and its many sequels and imitations, both in book form and on the screen, convey to children ideas about the universe

[46]See for example Carl Sagan and Thornton Page, eds., *UFO's: A Scientific Debate* (Ithaca, N.Y., 1972), p. 272; C. G. Jung, *Ein moderner Mythus: Von Dingen, die am Himmel gesehen werden* (Zurich, 1958). On Däniken's "religious" significance, see for instance Bracewell, *Galactic Club*, p. 103; also Ronald Story, *The Space Gods Revealed: A Close Look at the Theories of Erich von Däniken* (New York, 1976).

[47]See Hadley Cantril, *The Invasion from Mars: A Study in the Psychology of Panic* (Princeton, N.J., 1947).

with quasi-religious overtones; to today's children, extraterrestrials are more familiar than Robinson Crusoe. Almost four centuries after Giordano Bruno's death at the stake, a whole generation has grown up with the television series *Star Trek;* the novels on which it is based, by James Blish and others, can be bought at airport bookstalls in packs of six. The science fiction film *Close Encounters of the Third Kind,* a sort of gospel celebrating the landing of benevolent gods in the United States, is said to have been one of the greatest box-office successes in the history of film-making. Much the same may be true of *E.T.*

These popular interpretations of scientific findings, like the philosophical views either expressed or implied by the scientists themselves, have one vital element in common: they assume, without exception, that the intelligence of the extraterrestrials is of the same kind as ours, that we and they can understand and relate to one another, even share the same feelings, motives, and attitudes. This tends to support the Freudian explanation of the current, many-faceted preoccupation with ETI. The psychiatrist Robert Plank has seen in it the search for a guardian, for a father figure. While this is of course also an attempt to find a substitute for religion, it is more specifically a manifestation of the Oedipus complex, that is, primarily the product of an unresolved problematic relationship between child and father. The extraterrestrials with their superior power would be the father, who appears either as an ideal or as a bogey, either loving and protective or so deeply feared that he takes on the guise of a monster.[48] In either case, however, he would be essentially like "us," with familiar characteristics and understandable on a human level. Whether benevolent or threatening, our image of the extraterrestrial bears the imprint of the human psyche.

At the conference in Byurakan the suspicion was voiced more than once that the presumed extraterrestrial civilizations and the signals they might transmit could be so far outside our experience as to be quite incomprehensible to us. Can we be so sure there is common ground even in the scientific domain, let alone a shared psychological and philosophical "wavelength"?[49] It was thought quite possible that there is no such meeting ground. Stanislaw Lem in particular has focused on this point. He thinks it probable that the universe contains beings endowed with various kinds of intelligence, some of which

[48]*The Emotional Significance of Imaginary Beings* (Springfield, Ill., 1968). From a Marxist point of view, too, the utopian and science fiction encounter with higher beings from other planets tends to be seen as a regression into infantility and regarded as decidedly undesirable; see Werner Krauss, *Perspektiven und Probleme* (Neuwied, 1965), pp. 364–66.

[49]Sagan, *Communications,* pp. 341–44.

we would not even recognize as such.[50] If this were so, then some interstellar signals might reach the Earth but simply go over our heads, as it were, leaving us just as effectively alone and isolated as before the contact took place—if anything, more so. Might we not merely be part of a "Lonely Crowd" on a universal scale?

3. Science Fiction and the "Myth of the Modern Age"

Reflections like these take us deep into that no-man's-land beyond the bounds of our scientific knowledge. It is no coincidence that Stanislaw Lem, who points us toward these uncharted regions, is himself not only an eminent theoretical astrophysicist but also a highly regarded writer of science fiction novels, and in this respect he is by no means unique.

Science fiction is in fact a special form of philosophical literature that allows a writer grappling with the philosophical questions thrown up by scientific advances to extrapolate more boldly and give freer rein to his imagination than those who write only as physicists are able to do. To appreciate this we must rid ourselves of the view, still widely held, that all science fiction is simply lightweight popular fiction and merits attention, if at all, only as the reading matter of the masses in industrial society. Of course nine-tenths or more of what is sold as science fiction fully deserves this judgment. There is a historical basis for it too: science fiction as "consumer literature" (the term *science fiction* was coined in 1929 by Hugo Gernsback, an American immigrant from Luxembourg) was created when Gernsback founded his pulp magazine *Amazing Stories* in 1926, and for decades after that the genre was associated with similar popular magazines, characterized by lurid presentation, high circulation figures, and mediocre literary quality. These were joined by small-format magazines containing such serials as the Perry Rhodan stories. The production of science fiction thrillers for young people and adults developed into a vast and frenetic industry. Today all over the world hundreds of new titles appear each year, as well as dozens of anthologies of stories; radio, films, and television keep millions of people supplied with these consumer products. There is an obvious parallel here with the detective thriller. But just as that genre has in recent times attracted writers of the caliber of Friedrich Dürrenmatt and Jorge Luis Borges, so authors such as Stanislaw Lem and Arthur C. Clarke, Ray Bradbury and Ernst Jünger, C. S. Lewis

---

[50]*Summa Technologiae*, pp. 117–21.

and Ursula Le Guin—and, incidentally, Dürrenmatt and Borges too—have raised science fiction to a level of maturity that has earned it, certainly since the 1950s, undisputed recognition as "literature."

Most so-called science fiction falls into set stereotypes: the escapism of futuristic fantasies about space rockets and wonder weapons; intergalactic action stories; critiques or utopian visions of earthly society in the transparent disguise of a cosmic setting; or fantasy per se, whether spine chilling or exotic. The works of the writers mentioned, and a few—a very few—others, offer more than this. The reason they are read in schools and discussed in university courses and at international conferences is that here we have science fiction living up to its real claim to significance, which is its role of interpreting for the nonscientist the implications the present state of scientific knowledge has for our view of man—of showing what the new discoveries might mean for *us*. That is how, for instance, Aldous Huxley, a novelist with a scientific background, saw the role of science fiction.[51] It may be seen as the branch of serious literature best able to deal intellectually and artistically with the philosophical issues and implications inherent in scientific advance. It is a product of the kind of responsibly channeled imagination that, as no less a thinker than Jacob Bronowski has repeatedly reminded us,[52] is to be found at work in both science and literature: imagination underpinned by a scientific training, experimenting with ideas in a way which is of indisputable intellectual significance. The best products of the genre genuinely bridge the chasm that divides the "two cultures" both in the English-speaking world and in those countries whose education has been shaped by the humanistic German tradition.[53]

This assessment of the kind of literary creativity that underlies serious science fiction—that portion of the genre Raimund Borgmeier has termed *Weltbildliteratur*, meaning that it is concerned with finding

---

[51]*Literature and Science* (New York, 1963).

[52]*A Sense of the Future* (Cambridge, Mass., 1977), esp. pp. 6–31.

[53]There is a vast literature on the history of science fiction and specific aspects of it. Its "subliterary" phase as simply light popular fiction is described for example by Anthony Boucher, "The Publishing of Science Fiction," in *Modern Science Fiction: Its Meaning and Its Future*, ed. Reginald Bretnor, 2d ed. (Chicago, 1979), pp. 23–42. A useful reference book that also contains a critical bibliography of secondary literature is Neil Barron's *Anatomy of Wonder: Science Fiction* (New York, 1976). See also James Gunn, *Alternate Worlds* (Englewood Cliffs, N.J., 1975); Brian Ash, *The Visual Encyclopedia of Science Fiction* (New York, 1977); Robert Scholes and Eric S. Rabkin, *Science Fiction: History, Science, Vision* (London, 1977); David Ketterer, *New Worlds for Old* (Garden City, N.Y., 1974); Eike Barmeyer, ed., *Science Fiction: Theorie und Geschichte* (Munich, 1972); Ulrich Suerbaum, Ulrich Broich, and Raimund Borgmeier, *Science Fiction: Theorie und Geschichte, Themen und Typen, Form und Weltbild* (Stuttgart, 1981).

or expressing a philosophical view of the universe—has in recent years undeniably won general acceptance among critics, even though the claims made for the philosophical value of science fiction are often wildly overstated. Darko Suvin's definition of science fiction (as opposed to mere fantasy) as "literature of cognitive estrangement" has become a byword of recent criticism.[54] One of the most eminent authorities to express such high regard for science fiction as philosophical literature is the theologian, literary historian, and science fiction writer C. S. Lewis. In a conversation with Kingsley Amis and Brian Aldiss, he expressed the opinion that science fiction deals with incomparably more serious issues and greater themes than so-called realistic fiction, which, he says, really represents only a very narrow world.[55]

In Lewis's own science fiction novels and in many others from around the middle of this century, these themes are even presented in explicitly religious terms. *Out of the Silent Planet* (1938), *Perelandra* (1943), and *That Hideous Strength* (1945), conceived as a sequence, differ from most of the others, however, in that Lewis still seeks to reconcile the assumption of inhabited planetary worlds (in our solar system) with Christian doctrine. On Mars, the Fall of Man has not taken place; on Venus it is thwarted; sinful terrestrial man is isolated by divine decree from the rest of Creation. In other science fiction novels and stories of our period, the themes are more loosely theological. In these works, which are regularly cited as evidence of the intellectual stature of modern "literary" science fiction,[56] the fact that the universe is perceived as containing many worlds brings one up against those

[54]*Metamorphoses of Science Fiction* (New Haven, Conn., 1979), p. 4. See also (to cite a few out of dozens of examples) Lois Rose and Stephen Rose, *The Shattered Ring: Science Fiction and the Quest for Meaning* (Richmond, Va., 1970); Mark Rose, ed., *Science Fiction* (Englewood Cliffs, N.J., 1976), p. 3; Karen Ready and Franz Rottensteiner, "Other Worlds, Otherworldliness: Science Fiction and Religion," *Christian Century* 90 (1973), 1192–95. Almost all the studies mentioned in this section see science fiction in this light; see esp. Suerbaum, Broich, and Borgmeier, *Science Fiction*, chap. 6 (Borgmeier).

[55]C. S. Lewis, *Of Other Worlds* (London, 1966), pp. 89–91.

[56]For instance Suerbaum, Broich, and Borgmeier, *Science Fiction*, pp. 151–62; Ash, *Visual Encyclopedia of Science Fiction*, pp. 222–36; Theodore Sturgeon, "Science Fiction, Morals and Religion," in *Science Fiction, Today and Tomorrow*, ed. Reginald Bretnor (Baltimore, 1975), pp. 98–113; Patricia Warrick and Martin H. Greenberg, eds., *A New Awareness: Religion through Science Fiction* (New York, 1975), pp. 1–21; Sam Moskowitz, *Strange Horizons: The Spectrum of Science Fiction* (New York, 1976), pp. 3–21; J. Norman King, "Theology, Science Fiction, and Man's Future Orientation," in *Many Futures, Many Worlds*, ed. Thomas D. Clareson (Kent, Ohio, 1977), pp. 237–59; John Rothfork, "Science Fiction as a Religious Guide to the New Age," *Kansas Quarterly* 10:4 (1978), 57–66; Tom Woodman, "Science Fiction, Religion and Transcendence," in *Science Fiction: A Critical Guide*, ed. Patrick Parrinder (London, 1979), pp. 110–30; Heinrich Krauss, "Religiöse Themen und Kategorien in der Science Fiction," *Neugier oder Flucht? Zu Poetik, Ideologie und Wirkung der Science Fiction*, ed. Karl Ermert (Stuttgart, 1980), pp. 95–105.

"ultimate questions" that have otherwise largely disappeared from literature. Franz Werfel's *Der Stern der Ungeborenen* (The star of the unborn, 1946) is an early example; others are Bradbury's *The Martian Chronicles* (1950), Eric Frank Russell's "Second Genesis" (1951), Clarke's "The Nine Billion Names of God" (1953) and "The Star" (1955), Lester del Rey's "For I Am a Jealous People" (1954), Blish's *A Case of Conscience* (1958), Walter M. Miller's *A Canticle for Leibowitz* (1960), Robert A. Heinlein's *Stranger in a Strange Land* (1961), Frank Herbert's *Dune* (1965), and Michael Moorcock's *In His Image* (1968). These works are theological in the sense that even if they no longer set out to affirm the truth of Christian doctrine, they are still explicitly concerned with familiar theological themes. Many other works are religious in a more general sense: like theology, they too focus on man's relationship with some all-embracing cosmic "other" quite different from himself. Under this heading would come the whole range of "apocalyptic" subjects examined by David Ketterer in *New Worlds for Old* (Garden City, N.Y., 1974) and also the many depictions of the attainment of what Pierre Teilhard de Chardin calls the omega point, the dawning of a new phase of human consciousness, which for the individual is an experience of mystical enlightenment and religious salvation, as for instance in Clarke's *Childhood's End* (1953). This metaphysical element helps to explain how it is that science fiction, though so often disparaged and dismissed as trivial, is at the same time recognized, for better or for worse, as a modern substitute for religion.

Still, the criticism is often made that because science fiction is largely "space literature" and its world view primarily a cosmology of the numerous presumed worlds, it is therefore not "humanistic" literature in the sense of being principally concerned with man. Even C. S. Lewis admitted that a reader looking for psychological studies of character would be disappointed, precisely because broader themes were at issue (*Of Other Worlds*, p. 65). But preoccupation with these "large themes" (p. 89) need not mean that man is lost sight of. After all, the themes are "large" precisely because of their relevance to man: they are man's way of expressing his reactions to the new cosmic situation. Alexander Pope saw the study of man as the proper concern of literature (see chap. 2, sec. 6 below), and we hear an echo of Pope in Reginald Bretnor's dictum "To science fiction, man is the proper study of the writer." Commenting on this passage, Judith Merril, the pacesetter of the "new wave" of more philosophically oriented science fiction of the sixties and seventies, arrived at this succinct yet telling formulation: "You cannot define or describe a man except in terms of the universe of which he is aware; you cannot define or describe the universe except in terms of man's orientation within it."[57] This conception of science fiction as a form of philosophical literature of both cosmologic and "humanistic" orientation has been perfectly expressed by Brian Aldiss in his "true history of science fiction," *Billion Year Spree* (New York 1974): "Science fiction is the search for a definition of man and his status in the universe which will stand in our advanced but confused state of knowledge" (p. 8).

Science fiction's philosophical and religious preoccupation with defining man's place in his world finds its most vivid expression in one theme, which is so frequent and so central that it represents the essence of the greater part of science fiction. This is the meeting between *Homo sapiens* and his extraterrestrial counterpart, the encounter with aliens—the very point the philosophical questions of modern scientists also lead up to. Puccetti depicts the aliens as intelligent beings similar to humans, while Clarke imagines creatures physically and psychologically as different from ourselves as termites.[58] But even the most original descriptions of the encounter, by writers aware of the seriousness of the philosophical issues involved, do not, whatever some authors may think, add to our knowledge of aliens (any more than theology adds to our knowledge of "God"). They do, however, add significantly to our knowledge of man—man, who creates an image of aliens, of one kind or another, as a result not of playful fantasizing but of serious exploratory thought, the impulse for which often comes from existential anxieties.

How, then, does science fiction depict the "first contact"? Unfortunately Clarke is right when he says that treatments of this theme generally break off just as they are getting interesting.[59] Still, we may begin by saying that there is always an element of shock at the first meeting. Two of the stories in Damon Knight's anthology *First Contact* (New York 1971) show this with particular forcefulness. The shock is expressed in a question rather than an answer, but the question has the effect of making man profoundly unsure of his status in the world.

---

[57]Bretnor, *Modern Science Fiction*, p. 279; Merril, "What Do You Mean—Science Fiction?" *Extrapolation* 8 (1966), 12. See also Mark Mumper, "SF: A Literature of Humanity," ibid. 14 (1972), 90–93.

[58]Puccetti, *Persons*, pp. 96–97; Clarke, *Report on Planet Three*, pp. 98–99.

[59]Clarke, *Report on Planet Three*, p. 89. Cf., however, the following more recent studies: Eberhard Bauer, "Aus einer anderen Welt: Die Begegnung mit ausserirdischen Lebewesen im Lichte der Science Fiction," *Neue Wissenschaft* 15 (1967), 32–53; Jörg Hienger, *Literarische Zukunftsphantastik* (Göttingen, 1972), pp. 46–65; Gary K. Wolfe, *The Known and the Unknown: The Iconography of Science Fiction* (Kent, Ohio, 1979), pp. 201–8; Patrick Parrinder, "The Alien Encounter: Or, Ms Brown and Mrs Le Guin," in Parrinder, *Science Fiction*, pp. 148–61.

In Robert A. Heinlein's story "Goldfish Bowl" (1942), the following conversation takes place about the nature of the superior beings on Planet X:

> "Maybe we are just—pets." . . .
>
> "It's . . . it's *humiliating* from an anthropocentric viewpoint. But I think it may be true. . . ."
>
> "But they hunt us!"
>
> "Maybe. Or maybe they just pick us up occasionally by accident. A lot of men have dreamed about an impingement of nonhuman intelligences on the human race. Almost without exception the dream has taken one of two forms, invasion and war, or exploration and mutual social intercourse. Both concepts postulate that nonhumans are enough like us to fight with us or talk to us—treat us as equals, one way or the other.
>
> "I don't believe that X is sufficiently interested in human beings to want to enslave them, or even exterminate them. They may not even study us, even when we come under their notice. They may lack the scientific spirit in the sense of having a monkey-like curiosity about everything that moves. For that matter, how thoroughly do *we* study other life forms? Did you ever ask your goldfish for their views on goldfish poetry or politics? . . ."
>
> "You are joking."
>
> "No, I'm not. Maybe the life forms I mentioned don't have such involved ideas. My point is: if they did, or do, we'd never guess it. I don't think X conceives of the human race as intelligent." (In Knight, *First Contact*, pp. 156–57)

In Theodore Sturgeon's short story "The Hurkle Is a Happy Beast" (1949), a similar insight into the nature of Jupiter's inhabitants begins to dawn on the two astronauts from Earth:

> "And so they decided we were aliens. What next?"
>
> "If we weren't Jovians, then, in their eyes, we weren't people. It turned out that a non-Jovian was 'vermin' by definition."
>
> Orloff's automatic protest was cut off sharply by Birnam. "In their eyes, I said, vermin we were; and vermin we are. Moreover, we were vermin with the peculiar audacity of having dared to attempt to treat with Jovians—with *human beings*. Their last message was this, word for word—'Jovians are the masters. There is no room for vermin. We will destroy you immediately.' I doubt if there was any animosity in that message—simply a cold statement of fact. But they meant it."
>
> "But why?" . . .

"Do you feel competent to pass on Jovian psychology? Do you know just *how* alien Jovians must be physically? . . . It's thoroughly incomprehensible. Do you expect their mentality, then, to be any more understandable? Never! Accept it as it is. They intend destroying us. That's all we know and all we need to know." (In Knight, *First Contact*, pp. 80–81)

In these cases we are not taken beyond the first bewildered questioning; other articulations of the theme reveal definite conclusions regarding the nature and intentions of aliens. These judgments typically fall into two distinct groups, and the division is reflected in the contrasting titles of two further anthologies of short stories on the "encounter" theme: on the one hand Groff Conklin's *Enemies in Space* (London 1962), and on the other Hans Stefan Santesson's *Gentle Invaders* (New York 1969). It is understandable that when our centuries-old belief in our special and supreme position in the universe is shattered, the first reaction should be one of fear: the "others" are more powerful and are hostile toward us; we are threatened, exploited, subjected to them. This is the situation we find in novels as varied as Jack Finney's *The Body Snatchers* (1954), Heinlein's *The Puppet Masters* (1951), Chad Oliver's *Shadows in the Sun* (1954), and, as far back as 1897, H. G. Wells's *The War of the Worlds*. What is almost more damaging to the self-esteem of *Homo sapiens* is for the extraterrestrials to show complete indifference to us "primitives," as in Clarke's *The City and the Stars* (1956) and *Rendezvous with Rama* (London 1973). In *Rendezvous with Rama* we are told: "They had used the Solar System as a refuelling stop—as a booster station—call it what you will; and had then spurned it completely, on their way to more important business. They would probably never even know that the human race existed; such monumental indifference was worse than any deliberate insult" (p. 256).

Such painful thoughts can be avoided, of course, by retreating to the comforting notion of man's uniqueness in the universe. But in the light of modern astrophysics this would leave man terrifyingly isolated in the immeasurable vastness of space, with the support of traditional beliefs slipping away. It is only natural therefore that the imagination should grasp at an alternative possibility, the hope that the "others" will adopt a role similar to that of the angels of Christian mythology. Accordingly, they are sometimes shown as powerful and wise protectors, like the Regent in Ernst Jünger's *Heliopolis* (1949); they act as heralds of salvation or "midwives" to man as he enters a new phase in his development, reaching an unimaginably high level of collective consciousness or even achieving "apotheosis," as in Clarke's *Child-*

hood's End (1953). Or, again, they are superior but benevolent and helpful neighbors, as in *2001: A Space Odyssey* (1968). We find such a relationship as early as 1897, in Kurd Lasswitz's *Auf zwei Planeten* (On two planets), where the inhabitants of Mars, intellectually, morally, and emotionally far superior to ourselves, call themselves "Nume."

Obviously the "answers" given in works of literature, as in the sciences, put a human interpretation on what is nonhuman, incommensurable. Man is the measure, and the "totally other" is imagined as a friendly or hostile human counterpart. Nevertheless, all the works mentioned convey an inkling of the miraculous nature of the encounter, a sense of wonder at events unheard-of and yet conceivable. They do not descend into banality as do, for instance, Murray Leinster's "First Contact" (1945) and Chad Oliver's "Scientific Method" (1953)— two stories often mentioned in studies of science fiction—where it turns out that the extraterrestrials are all too human, even down to characteristics like peasant cunning and a taste for dirty jokes. And of course the works in question are also very far removed indeed from the type of science fiction in which the inhabitants of Earth are cast as galactic imperialists, bringers of civilization and salvation to less highly developed "races." This kind of colonialist self-affirmation, found for instance in E. E. Smith's *Lensmen* series in the thirties, forties, and fifties, nowadays lives on only in cheap magazines.

Alongside the "encounter" novels by Clarke, Lasswitz, Wells, and others which make the assumption of "their" similarity to "us", there are others which accept that no norms of ours can be applied to those totally alien beings—which makes them all the more disturbing—and this approach is in some ways intellectually more attractive. In Fred Hoyle's *The Black Cloud* (1957) and Lem's *Solaris* (1961), Earth's inhabitants are unable to achieve any communication whatever with extraterrestrial beings, who are decidedly nonhuman, and we are shown the bewildering and devastating effect of this failure. Lee Correy's less well known story about a UFO, "Something in the Sky" (1957), ends with the words: "Something still watched from the sky. It did not understand those below any more than those below might have understood it—had they known. . . . Lacking a true understanding of the nature of mankind, it made its decision."[60] Encountering "something" that has overwhelming intellectual superiority—a cloud, an oceanic planet, a flying saucer—man finds that he is both the prisoner of his own mental limitations and the object of a mysterious cosmic power. This causes frisson but not in the way that the horror movies with

[60]In *Encounters with Aliens*, ed. George G. Earley (Nashville, Tenn., 1978), p. 118.

their bug-eyed monsters are frightening. What sends a shiver down the spine is the awesome experience of boldly facing up to the philosophical question of man's place in the universe. It is the imaginative experience of the "Last Frontier."

To repeat: a certain, very limited, number of twentieth-century works of space science fiction deserve to be classed as "serious literature" rather than merely as light fiction, and the main qualities that lift these above the common run are the philosophical approach and original ideas they bring to bear on the theme of the encounter with aliens, or the plurality of worlds. And yet, the very fact that this theme is so closely identified with science fiction distinguishes and separates the genre from the remainder of that "mainstream" literature with which it so earnestly aspires to be united. In *2001: A Space Odyssey*, Clarke gives us a forceful reminder of this distinctiveness of theme and hence of the claim of science fiction to independent status. He says of the cosmonaut traveling through endless space: "At first, needing the companionship of the human voice, he had listened to classical plays—especially the works of Shaw, Ibsen, and Shakespeare—or poetry readings from *Discovery's* enormous library of recorded sounds. The problems they dealt with, however, seemed so remote, or so easily resolved with a little common sense, that after a while he lost patience with them."[61] *Space Odyssey* itself, by contrast, takes as its theme the kinds of questions that are the domain of science fiction, and these questions are implicitly accorded higher value than the traditional ones, which relate to human beings in a narrower context.

It is in dealing with these space themes—with what Michel Butor has called the "fundamentally new themes" such as "unknown worlds" and "unexpected visitors"[62]—that science fiction keeps pace with and reflects current thinking about the universe, as its authors justifiably claim. Indeed, in Fritz Usinger's opinion, the thematic focus on "the changed frame of reference within which we live," "the planetary" rather than "the tellurian," is the hallmark of all "literature which is truly of our time."[63] By this definition, science fiction *is* modern literature. Goethe, Usinger says, lacked any sense of "man's new situation in the universe," of "planetary consciousness"—of what it means for the individual to be conscious of living on just one planet in infinite space. It may well be, however, says Usinger,

---

[61](New York: Signet Books, 1968), p. 175.
[62]In Barmeyer, *Science Fiction: Theorie und Geschichte*, pp. 82, 78, 80.
[63]"Tellurische und planetarische Dichtung," *Tellurium: Elf Essays* (Neuwied, 1966), pp. 137, 148.

that someone in ages to come will be quite unable to enter into the mentality of a man such as Goethe. Greek culture with its sculptures and temples will be a beautiful fairy tale whose truth no one any longer feels obliged to believe in. Perhaps writers and poets of our terrestrial past will no longer be read, because the Earth-bound narrowness of their ideas makes them no longer valid. The very essence of our existence will change, as the whole pattern of our ideas adjusts to encompass things far beyond traditional perspectives, and as cosmic infinity directly affects our everyday lives. (P. 149)

Through its "new themes" and the "truth" they embody, science fiction seeks to help us orient ourselves in our "new cosmic situation." Not surprisingly, therefore, it has become a commonplace to ascribe to works of "scientific fantasy" the role of a modern mythology or modern myth. Indeed, science fiction has been seen as the *only* mythology appropriate to our time, in the sense that science has been drawn into the poetic myth-making process to form a single, fully rounded vision of the world, thus realizing the ideal of Romantics such as William Wordsworth and Friedrich Schlegel.[64] Precisely what is meant by this is not always made clear. Still, two separate strands of thought may be distinguished. One is that science fiction unmistakably revives motifs present in the authentic myths studied by folklorists, religious historians, and in the case of some motifs, psychologists. Both Jung and the Freudian Robert Plank recognize, in the UFO phenomenon and in science fiction, such archetypal mythological motifs as the end of the world, the epiphany, rebirth, the superman, the savior, and the guardian angel.[65]

The second and more significant strand of thought, however, focuses on the essential similarity of function between authentic myth and the literary "mythology" of science fiction. Both attempt to provide man with symbols to guide him in his quest for self-knowledge. Such symbols are formed by the creative imagination as it grapples with what is beyond the frontiers of knowledge, especially knowledge

[64]A few examples: Butor, in Barmeyer, *Science Fiction: Theorie und Geschichte*, p. 82; Bretnor, *Science Fiction, Today and Tomorrow*, p. 11; Philip Wylie, in Bretnor, *Modern Science Fiction*, p. 239; Gail Landsman, "Science Fiction: The Rebirth of Mythology," *Journal of Popular Culture* 5 (1972), 989–96; Igor Bogdanoff and Grichka Bogdanoff, *La Science Fiction* (Paris, 1976), p. 312; Thomas C. Sutton and Marilyn Sutton, "Science Fiction as Mythology," *Western Folklore* 28 (1969), 230–37; John Radford, "Science Fiction as Myth," *Foundation* 10 (1976), 28–33. "Science Fiction: The New Mythology" was also the theme of a panel discussion held at the 1968 conference of the Modern Language Association of America, printed in *Extrapolation* 10 (1969), 69–115.

[65]See for instance Vera Graaf, *Homo futurus: Eine Analyse der modernen Science Fiction* (Hamburg, 1971), pp. 185–88; Rothfork, "Science Fiction as a Religious Guide." See also Jung, *Moderner Mythus*; Plank, *Emotional Significance*.

of the universe. According to one writer on the theory of science
fiction, the image captured at this frontier represents an interpretation
of the world, put forward as an answer to "the fundamental question,
what is man?"[66] Very similar terms are used by one of the most
influential specialists in the study of myth, Joseph Campbell. Writing
on *his* subject, the authentic myth, he calls it an "image of the universe"
that helps to orient man in his search for knowledge. It is by its very
nature an image of that which has no image; a form representing
what is formless; a pointer, similar to a religious symbol, toward that
"ultimate mystery transcending names and forms, 'from which . . .
words turn back.' "[67] In each case we are dealing with ciphers in
pictorial form, which, to paraphrase Hugo von Hofmannsthal, explain
the unexplainable ("die was nicht deutbar, dennoch deuten"). These
are symbols by which and toward which man directs his life, not only
in the "primitive" societies that are the province of the ethnologist but
also in modern industrial society: there is no significant difference in
function between "authentic" and "literary" myth. To orient oneself
by a mythological symbol is not necessarily an infantile regression
into irresponsible "irrationalism." As William Irwin Thompson says,
"Myth is not an early level of human development, but an imaginative
description of reality in which the known is related to the unknown
through a system of correspondences in which mind and matter, self,
society, and cosmos are integrally expressed in an esoteric language
of poetry and number which is itself a performance of the reality it
seeks to describe."[68]

The claim made for science fiction, then, is that the best examples
of the genre are mythical creations of that kind; their questions and
answers begin where scientific knowledge comes to a halt. And we
cannot lightly dismiss the view that the mythology of science fiction
has a grip on the minds of many people today comparable to that of
myth in other societies in the past or at the present time. To quote
Ben Bova, "Myths are a sort of codification on an emotional level of
man's attitudes towards life, death, and the whole vast and sometimes
frightening universe. . . . Science fiction, when it's at its very best,
serves the functions of a modern mythology."[69]

This modern mythology may be briefly summed up as the myth
of the plurality of worlds or the plurality of humankinds. It confronts

[66]Radford, "Science Fiction as Myth," p. 28.
[67]Joseph Campbell, *The Masks of God: Creative Mythology* (New York, 1968), pp. 611,
609. See also Rothfork, "Science Fiction as a Religious Guide," pp. 63–65.
[68]*At the Edge of History* (New York, 1971), p. 136.
[69]*Science Fiction, Today and Tomorrow*, pp. 9, 11.

man not with the emptiness of the heavens (which leads to the mythology of nihilism or of the "world without God" and its images of the evil or weak God[70]), but instead with a heaven abundantly filled—with "worlds" and their inhabitants, human or similar to humans. Whereas the mythology of the "world without God" is a product of theological questioning, this "modern mythology" is inspired by science. As we have seen, there are both direct and indirect links between science fiction and theology, not only in the twentieth century but occasionally in earlier times too; nonetheless this modern mythology has its own unique range of themes.

How modern, though, is the myth of the plurality of worlds? References to the "new themes" of science fiction usually relate to the flowering of the genre from the mid-twentieth century onward. The literary historian's memory goes back further. When he reads, for instance, Pierre Boulle's La planète des singes (1963), it will occur to him that perhaps the seed of the idea underlying the book may be found in Pope, where he voices the suspicion that "superior beings . . . shewed a NEWTON as we shew an Ape" (Essay on Man 2, lines 31–34). The historian of science will, as we have seen, trace the idea of the plurality of worlds back to the event that made it possible, the scientific revolution brought about by Copernicus's heliocentric conception of the universe, which showed the Earth to be just one planet among several. In the context of intellectual history, the "modern myth" is thus a myth, or the myth, of the modern era, that is, of postmedieval times. Usinger is right to regard this new view of man's place in the universe as being in effect the heritage of Copernicanism (he refers to it as "Copernican consciousness" or the "Copernican situation,"—"Tellurische und planetarische Dichtung," pp. 141, 143), even though he considers that the "planetary" perspective has achieved its real breakthrough only in the twentieth century. In our day, to be sure, there is massive popular interest in science fiction as a consumer product, the UFO craze, and Däniken's substitute religion (dating back to 1926, 1947, and 1968 respectively); but behind this interest is a whole history, less well known, that stretches back to the sixteenth century. In this period we find themes being treated in philosophical and literary writings that reflect the disquiet caused by the empirical "new science" inaugurated by Copernicus (and also by

[70]Karl S. Guthke, Die Mythologie der entgötterten Welt (Göttingen, 1971). There are few instances of overlap—"negative theodicy" in science fiction—and these occur only in twentieth-century literature, which is not covered by this book (Clarke, "The Star"; Olaf Stapledon, The Star Maker; Clifford Simak, A Choice of Gods; George Zebrowski, "Heathen God").

Galileo): What if other planets in our solar system, or in other solar systems, are inhabited? How do we, as planet dwellers, compare with the inhabitants, human or akin to humans, of other planets? What are we, the pinnacle of Creation or the product of a failed experiment that was more successful elsewhere? Possessors of the highest, "divine," or at least godlike intelligence—or the least intellectually developed beings in the universe? The final outcome of evolution—or a mere stage in the process, long since overtaken elsewhere?

The chapters that follow trace the history of this theme, the myth of the modern age. They start from Copernicus's *De Revolutionibus Orbium Coelestium* (1543), which made it possible for the idea to develop as it did (see sec. 1 above), and they end on the threshold of twentieth-century science fiction, with two contrasting treatments of the motif of an invasion by Martians. Kurd Lasswitz, in his novel of 1897, portrayed the invaders as enlightened guardian angels, while H. G. Wells in the same year depicted them as unfeeling technological monsters. Both their novels have had a strong influence on twentieth-century science fiction; indeed, they can be said to have served as archetypes or paradigms for the development of the genre. It was no accident that Gernsback's magazine *Amazing Stories*, founded in 1926, included reprints of stories by Wells in early issues, while Gernsback's own work bore the mark of Lasswitz's influence. It is equally significant that Lasswitz's *Auf zwei Planeten* and Wells's *The War of the Worlds* are still in print, in more than one edition. So historians of science fiction have good reason to fix the birth of science fiction at 1926, when *Amazing Stories* first appeared, but to acknowledge as its "fathers" or ancestors Lasswitz, Wells, and perhaps, with respect to the purely technological nature of much science fiction, Jules Verne.[71] The following chapters, tracing the development of the most prominent theme of science fiction back as far as the sixteenth century, occasionally touch on literary accounts of space journeys—starting with Johannes

[71]For example Franz Rottensteiner, "Science Fiction," in *Pfade ins Unendliche*, ed. Franz Rottensteiner (Frankfurt, 1972), pp. 7–9; Klaus Peter Klein, *Zukunft zwischen Trauma und Mythos: Science Fiction* (Stuttgart, 1976), pp. 28–29; Bauer, "Aus einer anderen Welt," p. 40; Hienger, *Literarische Zukunftsphantastik*, pp. 17–22; Suerbaum, Broich, and Borgmeier, *Science Fiction*, pp. 38, 44. See also the discussion of the "beginnings" in chap. 1 of Kingsley Amis's *New Maps of Hell* (New York, 1960). Robert M. Philmus places the beginnings of science fiction in the seventeenth century, but his concern is not with the history of the genre but with defining it; his characterization of the genre, incidentally, includes a "mythic" factor: *Into the Unknown* (Berkeley and Los Angeles, 1970). On the influence of Lasswitz upon Gernsback see Franz Rottensteiner, "Kurd Lasswitz," in *Science Fiction: The Other Side of Realism*, ed. Thomas D. Clareson (Bowling Green, Ohio, 1971), p. 297.

Kepler's *Somnium* (1609, 1634)—that could well be classed as science fiction novels. But this is not to suggest that the history of the genre begins at such an early date. It may have roots that reach so far back, but actual novels on this theme, works that may at most be regarded as anticipating the genre, make only a fairly sporadic appearance. The subject of the following chapters is in any case the history not of a genre but of a theme: it is a study in intellectual history. Occasional references to features in earlier works that anticipate the conventions of the science fiction novel (and such elements can already be found in Lucian's satiric account of a fantastic journey *Vera historia*, or *True History*) are only incidental to this primary aim. The book sets out to portray a thematic tradition from which science fiction, but not only science fiction, is derived.

When we view the philosophical and literary development of the "plurality" theme against the backdrop of intellectual history in general, what emerges at first sight is a succession of attacks that shatter or at least threaten man's self-esteem. Above all, of course, the Christian view of man as the pinnacle of Creation is severely shaken. But it is not only in the Christian context that the assumption of man's supremacy is challenged. A basic tenet of the Enlightenment, that man is the measure of all things, is threatened equally by the philosophical and anthropological ideas extrapolated from the "new science"; and in the nineteenth century, man's eventual secular pride in himself as the culmination of evolution is similarly undermined. When we view the development as a whole, however, it becomes unmistakably clear that the idea of plurality has not always been a heresy and a challenge to accepted beliefs. As early as the mid-eighteenth century, and again in our own day—after a period of doubt linked with a revival of anthropocentric thinking in the nineteenth and early twentieth centuries—the idea has quite clearly become a new gospel, though in this area as in others one finds different strands of thought existing simultaneously. Around 1900, Percival Lowell's conviction that there was intelligent life on Mars shocked many people and met with widespread, though not universal, resistance, apparently on secular as well as religious grounds; today, by contrast, the plurality of worlds is a dream held dear by the masses.[72] Just as other ideas that threaten man's self-esteem, notably those of Darwin and Freud, have been assimilated one way or another, so the more fundamental challenge unwittingly initiated by Copernicus has been painlessly integrated into our late-twentieth-century view of ourselves.

[72]See Plank, *Emotional Significance*, pp. 44–46.

4. Antecedents

It was the "new science" with its heliocentric view of the universe that opened the way for the development of thought about the plurality of worlds. The new cosmologic model, in which a plurality of planets all revolved round the sun, deprived the Earth of any special status. It superseded the Aristotelian-Ptolemaic-Christian cosmology, dominant throughout antiquity and the Middle Ages, which had placed the Earth both physically and symbolically at the center of the universe. Man, though sinful, was the preeminent object of God's concern and love, and the physically central position of the Earth was the outward sign of this love—even if that symbolism became fully apparent only in retrospect, in the new situation created by Copernicus. Of course neither the theme of the plurality of worlds nor heliocentric cosmology made its first appearance in the sixteenth century. If, nonetheless, it is legitimate to see Copernicanism as having provided the stimulus for the philosophy and literature of the "pluralité des mondes" (to quote the title of a best seller of the seventeenth and eighteenth centuries), the justification lies in the fact, also touched upon already, that with the advent of Copernicanism, the notion of plurality acquired a different philosophical status. Those thinkers, especially in the ancient world, who appear to anticipate the modern conception of plurality are, however, sometimes spoken of in post-Copernican times in a way that gives a misleading impression of continuity.[73] We must therefore briefly examine how they are fundamentally different.

The doctrine of one world, one cosmos, of which only Earth, its center, is inhabited, was established by Plato's *Timaios* and Aristotle's *De caelo* and was able to dominate thinking about the universe for many centuries. But the idea of a plurality of worlds appears both before and alongside this doctrine. We find it in one form or another in antiquity in the writings of the pre-Socratic philosopher Anaximander (6th century B.C.), certain Pythagoreans (6th–4th centuries B.C.), and the Atomists Leucippus and Democritus and their followers Metrodorus of Chios (4th century B.C.), Epicurus (4th–3d centuries B.C.), and Lucretius (1st century B.C.).[74] According to modern classical schol-

[73]Even in Joseph Jérôme Lalande's *Astronomie*, 2d ed. (Paris, 1771), 3:452.
[74]Grant McColley, "The Seventeenth-Century Doctrine of a Plurality of Worlds," *Annals of Science* 1 (1936), 385–92; Milton K. Munitz, "One Universe or Many?" *Journal of the History of Ideas* 12 (1951), 232–39; Charles Mugler, *Deux thèmes de la cosmologie grecque: Devenir cyclique et pluralité des mondes* (Paris, 1953); Steven J. Dick, *Plurality of Worlds: The Origins of the Extraterrestrial Life Debate from Democritus to Kant* (Cambridge,

arship, Anaximander taught not a plurality of worlds in space but a succession of worlds in time, but some Pythagoreans are credited with believing that the visible heavenly bodies, the moon and stars, are simultaneously inhabited.[75] When the Atomists, in contrast, speak of a plurality of worlds, they regularly mean a multitude of universes that both succeed one another in time and exist simultaneously in space. They postulate, in other words, that the atoms present everywhere in space will have been tumbled together by chance and the mechanical laws of the universe to form a cosmos not just once, when our cosmos with all its visible celestial bodies was formed, but over and over again. These other universes, which need not be wholly identical with our own, cannot be perceived by us. Nevertheless, the existence on their earths, too, of living creatures, even of men, is assumed (Lucretius, *De rerum natura*, bk. 6, vv. 1048–89).

What distinguishes these cosmologic views from the post-Copernican conception of the plurality of planetary worlds in our solar system (and later on in other solar systems too) is the kind of reasoning on which they are based. They are not extrapolated from the results of empirical scientific investigation, and hence they make no claim to scientific validity. On the contrary, "without the slightest support from astronomical observation,"[76] they represent autonomous and self-sufficient metaphysical speculation deduced from ontological principles that are valid a priori and that themselves form the framework within which any possible observation of natural phenomena is undertaken.

This is equally true of the idea of plurality in the Christian Middle Ages. For Christian Aristotelians there were no living creatures except on Earth in the whole universe of geocentric spheres, unless it be the angels hovering in the heavenly regions between the moon and the primum mobile and assigned as "intelligences" to the stars circling the Earth. Not surprisingly, it is a medievalist, C. S. Lewis, who has reminded us of this notion—and used it, in adapted form, in his science fiction.[77] The existence of other *humankinds*, however, would only have been conceivable in the Middle Ages, if at all, somewhere outside the geocentric Ptolemaic-Aristotelian cosmos officially recognized by Christianity—in other words, at the center of another (possibly similar) cosmos. Whether such other *kosmoi* might exist was indeed

---

1982), chap. 1. On Anaximander see F. M. Cornford, "Innumerable Worlds in Presocratic Philosophy," *The Classical Quarterly* 28 (1934), 1–16.

[75]See also *Timaios*, 42D; Plutarch draws together the elements of this tradition in *De facie in orbe lunae*.

[76]Mugler, *Deux thèmes*, p. 179.

[77]*The Discarded Image* (Cambridge, 1964), pp. 113–21.

a recurrent subject of theological disputation from the thirteenth to the fifteenth century. Aristotle had given a negative verdict: in his view of physics, only a single center was possible. The medieval opposition to Aristotle (and to his theological champion, Thomas Aquinas) makes play with the idea that this need not be so, not of course on the basis of new discoveries in physics, nor on grounds of scientific theory, but for theological reasons. The Aristotelian assertion that only one world is possible limits the omnipotence and freedom of the Creator; one cannot have an adequate conception of his power, so runs the argument of several of the Scholastics, unless one concedes that he is capable of creating more than one world—in fact, of creating many, perhaps an infinite number of similar, or indeed different, universes in space and time, each with an inhabited Earth at its center. Saint Bonaventura, François de Meyronnes, William of Ockham, Jean Buridan, Nicole Oresme, Guillaume Vorilong, and many others adopt this line of reasoning.[78] This principle of the Creator's omnipotence and freedom of action, which makes the possibility of many created worlds the very essence of the divine, even received official sanction in 1277 when the bishop of Paris, Étienne Tempier, declared on papal authority that the contrary view was to be regarded as heretical. Even so, this by no means signified that the actual existence of more than one world created by God became part of the church's doctrine. The whole controversy was concerned ultimately not with the plurality of worlds as a matter of cosmology, but with God's infinite *potentia*, and for this the mere *possibility* of many worlds sufficed. That God had *not* in fact created the many worlds it lay in his power to create was the unanimous view even of the anti-Aristotelians among the Scholastics, and so after all they were in effective agreement with *De caelo* and accepted Genesis as a *complete* account of the Creation. Only Vorilong (d. 1464) found the actual existence of such "possible" worlds not wholly unthinkable and so became the first to ask himself the question characteristic of post-Copernican theology: had extraterrestrial humankinds suffered the Fall, and if so, had Christ redeemed them?[79]

A much bolder step was taken in 1440 by Nicholas of Cusa in his *De Docta Ignorantia* (bk. 2, chap. 12). He went far beyond the mere

[78]McColley, "Seventeenth-Century Doctrine," pp. 392–406; Munitz, "One Universe or Many?" pp. 239–42; Dick, *Plurality of Worlds*, chap. 2; Grant McColley and H. W. Miller, "Saint Bonaventure, Francis Mayron, William Vorilong, and the Doctrine of a Plurality of Worlds," *Speculum* 12 (1937), 386–89; Pierre Duhem, *Le système du monde*, vol. 9 (Paris, 1958), pp. 363–430.
[79]McColley and Miller, "Saint Bonaventure," pp. 388–89.

conceptual juggling of the thirteenth- and fourteenth-century theologians, and even further than Vorilong himself. For Cusa the actual existence of humankinds in extraterrestrial worlds is an absolute certainty and moreover these inhabited worlds are not merely speculative Aristotelian *kosmoi* but, as in post-Copernician discussions of the plurality of worlds, the visible heavenly bodies, the sun, moon, and stars. In addition, Nicholas abandons the Aristotelian hierarchy of these bodies, as Copernicus later does on a different basis, and conceives the idea, to which Copernicanism subsequently lends scientific plausibility, that they are all composed of the same elements. This further undermines Aristotelian and Christian cosmology, which represented the Earth as the "meanest and lowliest" of things because it supposedly consisted of matter different in kind from that of the other bodies.[80] In contrast to post-Copernican thinking, however, the grounds given in *De Docta Ignorantia* for the assumption of a plurality of inhabited worlds are not those of physics—which in this work is still speculative—but the purely metaphysical speculation that God in his omnipotence could not have left space void of life (p. 100). Unlike his thirteenth- and fourteenth-century predecessors (though still, like them, reasoning theologically), Cusa is thus using Plotinus's idea that whatever lies in God's power must also have been realized by him. (Arthur O. Lovejoy coined the term *plenitude* for this mode of thought and showed that the only way in which medieval and Renaissance Christian thinkers could conceivably accept the idea of a plurality of worlds *in the Aristotelian cosmos* was on the basis of this principle.)[81]

In this essentially metaphysical train of reasoning, however, it is the manner of Cusa's speculation about the nature of the extraterrestrial humankinds that is truly remarkable. Unlike Vorilong and the post-Copernican theologians, and despite being a cleric and a future cardinal himself, Cusa shows no interest in dogmatic quibbling about man's Fall and redemption. He is concerned with the status of man in terms of his degree of *perfectio* and *nobilitas*. In this he unmistakably foreshadows the self-confidence of Renaissance man emancipating himself from theology. Just as he dismantles the hierarchical structure of the Aristotelian-Christian cosmos, so too he discards the corres-

---

[80]*De Docta Ignorantia*, bk. 2, ed. Paul Wilpert (Hamburg, 1967), p. 94; see p. 96 on the uniformity of substance of all the planets, and pp. 96 and 92 on the absence of a center to the cosmos and the independent motion of the Earth.

[81]*The Great Chain of Being* (1936; New York: Harper Torchbooks, 1960), pp. 50–55; on Nicholas of Cusa, p. 114; see also Alexandre Koyré, *From the Closed World to the Infinite Universe* (1957; New York: Harper Torchbooks, 1958), pp. 6–24. On the Christian formulation of the idea of "plenitude" see Lovejoy, chap. 3.

ponding hierarchy of created beings, in which medieval man had been allotted a relatively lowly position (i.e., below the angels). And just as the Earth is a "noble star" (p. 98), or more precisely cannot be proved to be less noble than other heavenly bodies (p. 100), so man too is not an unworthy, sinful, imperfect creature, or at least is not demonstrably "of a meaner order [*ignobilior*] than the inhabitants of the region of the sun and other stars" (p. 100). Though inherently double edged and even possibly meaningless, this formulation moves Cusa to enthusiastic praise of mankind as created on our planet: "It appears that there could be no nobler and more perfect form of intellectual nature than that which dwells here on this earth and within its region, even if other stars may have inhabitants of a different sort [*alterius generis*]. For man strives after no other nature, but only after perfect fulfillment of his own" (p. 100). Strictly speaking, such glorification of earthly man does not actually mean that among the many supposed humankinds in the cosmos, he is credited with the highest or even a relatively high degree of perfection; for Cusa explicitly maintains that these various humankinds are not comparable with one another, just as the different parts of the human body are not comparable. In any case, he says, the inhabitants of other regions ultimately remain "totally unknown" to us (p. 102). Yet the revolutionary force of his argument remains undiminished: he is no longer prepared automatically to accept the church's doctrine which, while recognizing the distinction conferred by the Savior's human incarnation on this planet, nevertheless consigns "fallen" mankind to a lowly position in the whole of Creation. At the same time, the analogy between the different humankinds in the universe and the parts of the human body still leaves open the possibility that earthly man is in some sense unique, which does accord with Christian teaching.

As he performs this delicate philosophical-cum-theological balancing act, and cautiously covers himself by pointing out that of course he does not claim to speak from actual knowledge since there is no possibility of making comparisons (*improportionabiliter*, p. 102), it is astonishing to find that he nevertheless cannot resist speculating on the nature of the inhabitants of, for instance, the sun and the moon (he does this on the basis of the physical qualities of these bodies, "fiery" in the one case, "watery" and "airy" in the other)—and moreover that this train of thought once again relegates man on Earth (*materialis*) to a humble position, at least in one respect. For Nicholas assumes that the sun has "more sunlike, splendid, and illumined intellectual inhabitants" (*magis solares, claros et illuminatos intellectuales habitatores*) who are "more spiritual [*spiritualiores*], too, than the moon's," while the Earth's inhabitants are "more material and gross" (*magis materiales et grossi*) than those of the other two. Earthly men are thus inferior to the inhabitants of both the sun and the moon, but only in a sense that does not altogether cancel out Cusa's Renaissance conviction of man's intrinsic worth. For the distinction between earthly man and the others is also presented in this casually offered speculation in terms of *actus* and *potentia*. The sun's inhabitants are "multum in actu et parum in potentia," while we, at the other end of the scale, are "parum in actu" and yet "magis in potentia" (p. 102): their actual endowments are superior, but we are *potentially* greater.

We instantly recognize how revolutionary and indeed heretical Cusa's view of man is, for all his cautious reservations, when we compare it with a widely read didactic poem that embodies similar ideas and probably owes something to his influence, the *Zodiacus Vitae* (1534?) of Marcellus Palingenius Stellatus (Pier Angelo Manzoli).[82] Palingenius's "worlds" too are those of the heavenly bodies, not those of merely conceivable Aristotelian universes; they are real, not just possible; as in Cusa, they are inhabited, and again this is attributed to God's infinite plenitude. But Palingenius differs from Cusa in portraying mankind on Earth as unequivocally and irrevocably imperfect, while "the others" are just as certainly created perfect. This is manifestly a concession to the Christian view of the human condition in a cosmology otherwise heretical in tenor, specifically in its assertion of the existence of a plurality of worlds envisaged, for all their notional perfection, in concretely human terms.

Neither Palingenius nor Cusa based their cosmologic ideas on scientific theory derived from empirical investigation, which might have elevated them to the status of a truth challenging that of the theologians. Their ideas remained no more than metaphysical speculations—and therefore innocuous. The church saw no need to intervene, even though the ruling of 1277 had left unaltered the doctrine laid down by the great authorities of the faith—Saint Augustine, Albertus Magnus, Thomas Aquinas, and others—that the plurality of worlds (as a reality) was a pagan error that not only conflicted with the biblical account of the Creation but also cast doubt on God's wisdom and justice (Aquinas held that different worlds cannot all be perfect, and identical ones would be pointless).[83]

The reason why Giordano Bruno, whose ideas were similar to and

---

[82]See Koyré, *From the Closed World*, pp. 24–27; Lovejoy, *Great Chain*, pp. 115–16.
[83]See McColley, "Seventeenth-Century Doctrine," pp. 393–98; on Thomas Aquinas, p. 398.

in some measure derived from those of Nicholas of Cusa, paid for this heresy (as well as for others) at the stake, while Cusa did not, is that in the interim, Copernicus's *De Revolutionibus Orbium Coelestium* (1543) had given the idea of the plurality of worlds a totally new philosophical status—that of potential fact and secular truth. Only then did the notion that "we are not alone" begin to arouse the kind of public interest that has reached such a climax of intensity in our own day. Estimates of the frequency of the occurrence of intelligent life in the universe may differ widely, but there is surely scarcely a scientist alive today who believes that there is only a single instance of the emergence of higher life-forms in the entire universe (as distinct from just the Milky Way). For some, indeed, the empirical verification of the plurality of worlds is only a matter of time.[84]

[84]See Berendzen, *Life beyond Earth and the Mind of Man,* p. 2.

CHAPTER 2

The Renaissance:
Science Falls from Grace

1. The Scientific Validation of the "Plurality"
Idea by Copernicanism

The plurality of worlds is one of many concepts dating from antiquity that were revitalized by the Renaissance. But in the process this inherited concept underwent a radical transformation that made it irresistible as a literary subject. As already indicated, the old idea resurfaced in the climate of the Scientific Revolution, with the momentous result that what was previously a mere speculation acquired a more solid basis that gave it far higher credibility, so much so that many thinkers came to believe in it as a reality. In short, the plurality concept became linked with the new Copernican cosmology. Again and again we find the keenest minds of the sixteenth and seventeenth centuries pursuing the following train of thought: Copernicus, in *De Revolutionibus Orbium Coelestium* (1543), showed that both the Earth and the planets, which in the Ptolemaic system circled the Earth, in fact all revolved around the sun; and not the least of the implications was that the uniqueness of the Earth, hitherto taken for granted, was an illusion. The Aristotelian-Christian model of the universe divided it into the imperfect, changeable region of the Earth "under the moon," and the ethereal heaven of unchanging perfection in which all the other celestial bodies, from the moon upward, had their motion. Copernicus removed this division by raising the Earth's status to that of a planet: as Kepler said, he lifted the Earth up "into heaven."[1] But

---

[1]Johannes Kepler, *Gesammelte Werke,* ed. Walther von Dyck et al. (Munich, 1937 ff.), 2:224: "in coelum"; see *De Revolutionibus,* bk. 1, chap. 5.

if the whole universe was, in essence, similarly constituted, and the Earth was a planet like the others, then the way was open for the imagination, supported by scientific knowledge and critical judgment, to speculate that perhaps all the planets (and the Earth's planet, the moon) might be Earths: worlds possessing mountains, valleys, seas, rivers, plants, and animals—and why not human beings too, or at any rate, beings akin to humans? Might not these beings even be superior to ourselves, more perfect, happier, and dearer to God?

Even in the seventeenth century this idea must have come as a bolt from the blue to both friends and foes of Copernicanism. A harmless jeu d'esprit had suddenly become a serious possibility, arousing anxiety in some quarters, joy in others, excitement in all. It caused anxiety, for example, to the Jesuit mathematician Pierre de Cazre, who on 3 November 1642 wrote to Pierre Gassendi in the following terms: If, as Copernicus says, the Earth is a planet, then one may easily imagine that the other planets are also inhabited and, moreover, that those larger and more perfect than the Earth are inhabited by higher, "more perfect" beings (*praestantiores*) than those on Earth. But this would cast doubt on the biblical account of the Creation, which enjoins us to believe that the stars were created solely for the sake of man on Earth, in order to give him light and a means of measuring time. "The whole Christian faith" is therefore at stake. De Cazre urged Gassendi, whose thoughts tended in that direction, to recognize how "dangerous" (*periculose*) it was to make public the "monstrous conclusions, both unfitting and absurd, that follow from Copernicus's opinion" (*portentosa, quae ex Copernici opinatione sequuntur incommoda, atque absurda*).[2] Elsewhere, however, the idea was received with a sense of liberation, not only by Giordano Bruno, an avowed supporter of Copernicus, but also for instance by a scholar as sober as Christian Huygens, who expressed delight and wonder (*suavem admirationem*). His treatise on the inhabited universe, *Kosmotheoros* (1698), opens with the words "If one agrees with Copernicus's view that the Earth we live on is one of the planets that revolve around the sun and receive all their light from it, one cannot help sometimes thinking that it is reasonable to suppose that those other planets, like our own globe, may not be without institutions and embellishments and, perhaps, inhabitants."[3]

---

[2]Gassendi, *Opera Omnia* (Florence, 1727), 6:416. The date has been corrected in accordance with Henri Busson, *La pensée religieuse française de Charron à Pascal* (Paris, 1933), p. 291, n. 2.

[3]"Fieri vix potest, . . . si quis cum Copernico sentiat, Terramque, quam incolimus, e Planetorum numero unum esse existimet, qui circa solem circumferentur, ab eoque

The heliocentric view of the universe, accepted by the leading sixteenth- and seventeenth-century scientists, gave the idea of the plurality of worlds a new, "realistic" status; the significance of this development can hardly be overestimated. Yet it has received relatively little attention, and in more recent times there have even been denials of a link between Copernicanism and the revival of interest in the plurality idea (more on this later). Other consequences of the Copernican revolution have been seen as more disquieting and more radical in their effect.

Thus for instance Goethe—who wrote that "of all discoveries and convictions none could have had a more powerful effect on the human mind than the teaching of Copernicus"—considered that what was revolutionary about the sun-centered view of the world was that it deprived man of the "immense privilege of being at the center of the universe"; as a result, no less a thing than the "belief in a poetical-religious faith" dissolved into thin air.[4] Nietzsche was later to add that the loss of geocentrism was the starting point of modern nihilism; that "ever since Copernicus," man had been slipping "away from the center into the unknown" (*aus dem Zentrum ins x*).[5] Goethe and Nietzsche both have Christian cosmology in mind; in fact, however, the centrality of the Earth and of man was not especially significant in orthodox Christian teaching in pre-Copernican times. It played a more prominent role in, for instance, the syncretic humanism (regarded with some suspicion by Christian theologians) of Giovanni Pico della Mirandola. In his view, God placed man at the center of the world in order to provide an outward symbol of his status as the most highly favored and most fortunate of all creatures, because the highest possibilities were open to him. Man is represented as a being envied not only by the animals but also by the higher supraterrestrial beings that people Pico's hierarchically ordered universe.[6] True, the motif of the symbolic placing of man at the center of the universe does crop up from time to time in Christian polemics against Copernicus,[7] in a

---

lucem omnem accipiant; quin interdum cogitet haud a ratione alienum esse ut, quemadmodum noster hic Globus, ita caeteri quoque isti, cultu ornatuque, ac fortasse habitatoribus non vacent" (The Hague, 1698), pp. 3–4. The preceding quotation is from ibid., p. 4.

[4] *Werke,* ed. Erich Trunz, vol. 14 (Hamburg, 1960), p. 81.

[5] *Werke in drei Bänden,* ed. Karl Schlechta, vol. 3 (Munich, 1966), p. 882. For a modern instance of this view endorsed by such illustrious names, see Hugh Powell, *German Life and Letters,* n.s., 5 (1952), 275.

[6] *De Hominis Dignitate* (1496; written 1486), in *Opera Omnia* (Basel, n.d. [1572]), 1:314: "medium mundi"; "felicissimum"; "non brutis modo, sed astris, sed ultramundanis mentibus."

[7] See John Dillenberger, *Protestant Thought and Natural Science* (New York, 1960), p.

manner analogous to those world maps that placed Jerusalem at the center of the world and were long since out of date even then. But it is difficult to see man's displacement from this central position as the most unsettling effect of Copernicanism, since, after all, Christian theologians and Aristotelians by no means saw the Earth as occupying a favored or privileged place in the structure of the universe. Not only was the Earth the home of mutability and excluded from the "heaven" where the planets had their courses; it was also made up of the meanest, most ignoble material, "excrement" as Giordano Bruno drastically phrases the popular assumption; to Montaigne it was "the mire and dung of the world"; Pico della Mirandola himself speaks of it as a dung heap.[8] Also, the Earth was farthest from the divine empyrean beyond the heavenly spheres, and closest to hell, which was situated in the Earth's interior. Earth had special status not because of its position but because it was inhabited by man, who, though sinful, was created in God's image. The Earth was the setting for God's work of salvation; it was here that the all-important drama of the Fall of Man and his redemption by God's incarnate Son was enacted—events of which the stars were no more than distant spectators. Man was in this sense the apple of God's eye, the pinnacle of Creation, indeed the very purpose of Creation; and all other creatures, great and small, were subservient to him. This could still be true, even when his abode during his brief and relatively unimportant temporal existence was no longer the center of that structure of spheres which formed the cosmos. (Strictly speaking, in any case, the precise center was occupied by hell.)

Being at the center in the physical and in the religious sense need not go together, as has been amply shown by post-Copernican theology. Theologians have had little difficulty, either in the past or more recently, in coming to terms with the heliocentric view of the universe. Man's position is declared central in a nonspatial sense, independent of the mere outward forms of nature. As early as 1611, John Donne, in his satire on the Jesuits, *Ignatius His Conclave*, places this point of view in the mouth of none other than the zealot Saint Ignatius of Loyola, cast in the role of spokesman for hell: he declares to the supposed heretic, Copernicus, that in matters of faith it is

---

26; Hans Blumenberg, *Die kopernikanische Wende* (Frankfurt, 1965), passim, e.g. pp. 134–35.

[8]Giordano Bruno, *The Ash Wednesday Supper*, ed. and trans. Edward A. Gosselin and Lawrence S. Lerner (Hamden, Conn., 1977), p. 90; Montaigne, *Essais*, ed. Maurice Rat (Paris: Garnier, 1962), 1:496; Pico, *De Hominis Dignitate* 1:314: "excrementariae inferioris mundi partes."

immaterial whether the sun circles the Earth or the Earth the sun. Similarly, the Archangel in Milton's *Paradise Lost* declares, "Whether Heav'n move or earth, / Imports not" (bk. 8, lines 70–71). The question was of no importance because either way the certainty remained that God had created only one human race. This cardinal dogma, on which depended man's whole conception of himself and with it his self-esteem, was in no way threatened by Copernicus's theory that man was no longer at the center of the universe. On the contrary, his self-esteem was if anything enhanced, a development viewed with pious concern by the anti-Copernicans and also by Montaigne but greeted with delight by the Copernicans. For, as Kepler noted, the Earth and its inhabitants found themselves transported from the abode of corruption into the ethereal region of the stars.[9]

That other implication inherent in Copernicanism, the concept of the plurality of worlds, posed a far more dangerous threat. Man had to come to terms with a vastly expanded universe in which the Earth appeared as a mere drop in the ocean (Copernicus's calculation of the diameter of the cosmos was two thousand times the previously accepted estimate) and above all with the idea of a multitude of human-kinds on other planets, within our own solar system and in other analogous systems. This, and not the distance of the Earth from the center of the universe, was the most devastating blow Copernicus dealt to man's view of himself.

Consequently, when in the seventeenth and early eighteenth centuries Copernicus's teaching ceased to be an esoteric subject of scholarly inquiry and became more widely known, the belief that other planet-worlds were inhabited like the Earth was frowned upon by the church and became the "new heresy."[10] It is still sometimes claimed that Copernicanism caused no upheaval,[11] but this is very far from the truth. It exerted its subversive influence above all through the plurality idea. Wherever that idea took root, the Middle Ages were over and the modern era had begun. In 1640, Gabriel Naudé, physician to Louis XIII and librarian to Jules Mazarin, refers to the doctrine of plurality

[9]See n. 1 above; Montaigne, *Essais* 1:495–96; in general: Blumenberg, *Kopernikanische Wende*, p. 123; Arthur O. Lovejoy, *The Great Chain of Being* (1936; New York: Harper Torchbooks, 1960), p. 102; Alexandre Koyré, *From the Closed World to the Infinite Universe* (1957; New York: Harper Torchbooks, 1958), p. 43. Human self-esteem was also enhanced by the fact that human reason was capable of such revolutionary insights; see for instance Kepler's letter to Maestlin, 9 April 1597, in *Johannes Kepler in seinen Briefen*, ed. Max Caspar and Walther von Dyck (Munich, 1930), 1:44–45 (*Ges. Werke* 13:113).

[10]Marjorie Nicolson, *Science and Imagination* (Hamden, Conn., 1976), p. 55.

[11]Siegfried Korninger, *Die Naturauffassung in der englischen Dichtung des 17. Jahrhunderts* (Vienna, 1956), p. 30; see also pp. 11, 22. See n. 19 below and chap. 3, n. 3.

not only as a sin of thought but as the most revolutionary of heresies. Significantly, he makes particular reference to the "new science":

> I am afraid that those old theological heresies are nothing compared to the new ones the astronomers are seeking to introduce with their worlds, or rather lunar and celestial earths. For the consequence of these heresies will be far more dangerous than those of the earlier ones and will bring about many stranger upheavals [bien de plus estranges revolutions]. By heaven, if anyone had told Lucian in ancient times that the tales he presented to us as wild inventions, which he declared were not and could not be true, were absolutely truthful, what do you think he would have said? This reminds me of the antipodeans, whom two or three hundred years ago no one could believe in without being declared a heretic. I do not know whether you have noticed in the commentaries by Cichus Asculanus on the Sphere of Sacrobosco that a familiar Spirit the author claimed to have, when asked by him what the moon was, replied in these very words: Ut Terra, terra est [It is like the Earth, it is an Earth]. I am not concerned with the Spirit, which I do not believe ever said such things, but with the philosopher, who lived more than two hundred years ago.[12]

As early as 1612 the anti-Copernican Jesuit Julius Caesar La Galla had made similar comments in his De Phaenomenis in Orbe Lunae. That was just twelve years after Bruno had been put to death for his belief in plurality (and related ideas) and four years before De Revolutionibus was banned by the Roman Catholic church.

Copernicus was also attacked by the Protestants in Germany, France, and England, mainly for the same reasons.[13] Copernicus questioned the authority of Aristotle as interpreted by the Christian church; he contradicted certain biblical passages that spoke of the motion of the sun (Jos. 10:12–13) and the immobility of the Earth (Job 26:7); and he chose to derive his knowledge from a source—Nature interpreted by reason—that seemed to challenge the supremacy of revealed truth. For Protestants too, however, the doctrine that the Earth moved was less disturbing than the plurality of worlds, which followed from the

---

[12]In a letter to Ismael Boulliau dated 15 August 1640, first published in René Pintard, Le libertinage dans la première moitié du XVIIe siècle (Paris, 1943), pp. 473–74, and quoted from there.

[13]For a detailed historical account see Dillenberger, Protestant Thought. S. F. Mason, "The Scientific Revolution and the Protestant Reformation" (pt. 1), Annals of Science 9 (1953), 64–87, highlights an affinity between Protestantism and "new science" (which does not, however, directly concern the theme of plurality), namely, their shared opposition to the traditional hierarchical view of cosmology.

Earth's elevation to the status of a planet. They too, like the Jesuit de Cazre, saw this idea as striking at the very foundations of faith. After all, the whole of Christian teaching on salvation rested ultimately on the Creation story, which spoke of only one humankind, descended from Adam and Eve, one Garden of Eden, and one Fall. It also declared that the stars were created, like everything else, for the benefit of this one humankind, and for no other purpose. There was not a single word about inhabited planets; therefore belief in other human beings on the stars, or even closer at hand, on the moon, must be at variance with true doctrine. This was the line still being adopted in 1646 by Alexander Ross in his treatise *The New Planet No Planet*, directed against the Copernican John Wilkins. The French Protestant Lambert Daneau had used a similar argument as far back as 1576, in his *Physica Christiana*:[14] since the Bible was the only authentic source of knowledge about nature, a belief in the plurality of worlds, which was not supported by Genesis, must be a false doctrine.

There were also of course the christological objections (see chap. 1, sec. 4 above), raised particularly often by the Protestants. These were most powerfully expressed by Philipp Melanchthon, whose arguments continually recur in the debate about plurality.[15] In his *Initia Doctrinae Physicae* (1550), in which he does not refer to Copernicus by name, although he was already familiar with his work,[16] he argues (fols. 43–44) that the assumption of a plurality of worlds is monstrous and absurd (*portentosus*) not only for reasons of physics (i.e., Aristotelian physics) but above all on dogmatic grounds. Nothing proves more conclusively that we alone are the objects of the Creator's care and love than that he sent his only Son to our world to redeem us from original sin. There is no authority for thinking that Christ appeared, was crucified, and rose from the dead in other worlds too, and it is indeed inconceivable. It is equally "unthinkable" (*nec cogitandum*), however, that human beings in other worlds can attain salvation and be awakened to eternal life without being redeemed by the Son of God, in other words, that they can be perfect by nature (*naturaliter*).

[14]*The Wonderfull Woorkmanship of the World* (London, 1578), chap. 12, esp. fol. 26v. If there were more than one world, God would have told us of this in the Bible. To assume that other worlds had come into being of their own accord or had been created by another god was, of course, blasphemous.

[15]See Paolo Rossi, "Nobility of Man and Plurality of Worlds," in *Science, Medicine and Society in the Renaissance: Essays to Honor Walter Pagel*, ed. Allen G. Debus (New York, 1972), 2:145.

[16]See Robert S. Westman, "The Wittenberg Interpretation of the Copernican Theory," in *The Nature of Scientific Discovery*, ed. Owen Gingerich (Washington, 1975), pp. 393–429.

So runs Melanchthon's argument, which one would hardly expect to find at all in the context of a textbook on physics. The reason for this theological offensive directed ostensibly against Democritus and his hypothesis of a multiplicity of *kosmoi* can only be that a similar cosmology, this time implying a multiplicity of planet-worlds, was currently the subject of scandalized debate and seemed to be posing a more serious threat than ever before. Melanchthon does apologize at the end of the section, arguing rather lamely that even if his reflections do not strictly belong to the realm of physics, they are still not superfluous; for if anyone who imagined other inhabited worlds were also to entertain ideas of other religions and of "another human nature" (*alia hominum natura*), these arguments might dissuade him.

These words also reflect historical events that were contributing to the unease of theologians. It was the age of discovery and conquest, and thus of encounters with "heathens" in far-off continents beyond the seas. Christianity had to face the challenge posed by the religions of other peoples that were not necessarily uncivilized and unworthy of emulation. Hence, it is not surprising that the controversy over "plurality" soon became in some ways analogous to the argument over the "antipodeans," and this parallelism continued up to Bernard Le Bovier de Fontenelle and beyond. The *existence* of antipodeans, once categorically denied by Saint Augustine, was now no longer an issue. The question was whether the Indians discovered by Columbus were to be regarded as human beings, that is, descendants of Adam and Eve; or as animals; or as human beings untouched by original sin, descendants of an *Adam americanus*. And just as the church's opposition failed to silence the controversy over plurality, so the papal bull of 1537, which declared the Indians to be human and capable of being received into the Catholic faith, could not lay to rest the dispute about polygenism.[17] In fact these two areas of controversy became to some extent linked together as two "new worlds" of heretical thought. We have seen how Naudé mentioned both in quick succession, and we shall shortly see Bruno doing the same. Again and again the two questions crop up side by side, especially once the Copernican Galileo comes to be seen as the "new Columbus." Both subjects raise the problem of defining "human nature," about which Melanchthon, quite understandably in the intellectual climate of his time, was so

[17]Otto Zöckler, *Geschichte der Beziehungen zwischen Theologie und Naturwissenschaft,* vol. 1 (Gütersloh, 1877), pp. 542–48; Edward Dudley and Maximillian E. Novak, eds., *The Wild Man Within* (Pittsburgh, 1972), esp. pp. 262–72.

concerned. But the challenge posed by the plurality of worlds was the greater of the two: even the tamest fictional account of travel in space demanded a more radical reappraisal of man's nature than the most exotic account of a journey on Earth.

As for the man who had initiated these unsettling ideas, what was his own attitude? Copernicus, canon of Frauenburg, must have foreseen and feared that doubts would be cast on the orthodoxy of his beliefs; otherwise he would hardly have taken the precaution of dedicating his work to Pope Paul III, the man who had declared the antipodeans to be members of the human race. But in fact Copernicus himself was not directly guilty of the "new heresy": his scientific speculation did not extend to the assumption of a plurality of planetary worlds. Nevertheless, in the sixteenth and seventeenth centuries, especially in the wake of Galileo's observations, a close link was perceived between the teaching of Copernicus, which demolished the geocentric view of the universe, and the concept of a plurality of worlds, which destroyed anthropocentrism. As we shall see, this perception was shared by such very different thinkers as Kepler, Tommaso Campanella, Robert Burton, John Wilkins, Pierre Borel, Hans Jakob Christoffel von Grimmelshausen, Savinien Cyrano de Bergerac, Christian Huygens, and Fontenelle. They and many others believed that it was precisely the Copernican world view that provided a logical basis for all the speculations—often worked out in considerable detail—about the human or near-human inhabitants of other planets in the solar system, who in one sense or another were our "rivals." Conversely, it has been convincingly argued that John Milton had misgivings about the Copernican model of the universe precisely because he felt that it implied the existence of human beings on other planets.[18]

So the view that the Copernican theory caused not the slightest upheaval in sixteenth- and seventeenth-century thinking, put forward even recently by an expert in the field, cannot be upheld.[19] On the contrary, as knowledge of the Copernican theory gradually spread, anyone who "did not wish to be a stranger in the world" (to quote G.

[18]Grant McColley, "The Theory of a Plurality of Worlds as a Factor in Milton's Attitude toward the Copernican Hypothesis," *Modern Language Notes* 47 (1932), 319–25. Robert Burton's initial reaction to the Copernican theory was similar to Milton's; see Robert M. Browne, "Robert Burton and the New Cosmology," *Modern Language Quarterly* 13 (1952), 145.

[19]Marjorie Nicolson, "The Discovery of Space," *Medieval and Renaissance Studies* 1 (1965), 41: "So far as imagination was concerned, the Copernican theory had no effect at all." See n. 11 above.

E. Lessing, who devoted an early poem to the plurality of worlds)[20] was bound to react. He could not but give some thought to the inhabited worlds on other planets in our solar system as well as in those systems that, since Bruno and Galileo, were thought to exist around other stars; he had to consider the nature of mankind in "this world" in relation to those other beings. How the heliocentric model of the world must have appealed above all to the literary imagination is vividly illustrated, centuries later, by the poetic dialogue "Copernico" by Giacomo Leopardi—who, incidentally, also wrote a history of astronomy. Like so many others, Leopardi feels that of all the ideas that follow from the Copernican theory, the concept of plurality has the most exciting, and at the same time the most alarming, implications for our view of humankind. In Leopardi's dialogue, Copernicus addresses the sun as follows:

> So that, indeed, the Earth has always believed herself to be the Empress of the universe; and truly, matters being as they have been till now, one cannot say that she was wrong; indeed I shall not deny that these ideas had a sound foundation. But what shall I say about men? For believing ourselves to be (as we always shall believe) the first and most important of all terrestrial creatures, each one of us, even if dressed in rags and with nothing but a dry crust of bread to gnaw, has yet esteemed himself an emperor—not only of Constantinople or Germany, or of half the Earth, as the Roman emperors were, but an Emperor of the Universe, ruling the sun, the planets, and all the visible and invisible stars; and the final cause of the stars, the planets . . . and everything else. . . . a great revolution even in metaphysics. . . . And the result will be that men, if indeed they are still able or willing to talk sanely about it, will discover that they themselves are something quite different from what they had been until now, or imagined themselves to be. . . . But consider . . . what may reasonably be expected to happen to the other planets. For when they see the Earth doing precisely what they do, just like one of themselves, and become one of them, they will be dissatisfied with their own bare simplicity and lack of adornment, and the mournful desolation they have known until now, while the Earth alone is so richly adorned. They too will want their rivers, seas, mountains, plants, and also animals and men, for they will see no reason why they should be at all inferior to the Earth. And so there will be another great revolution in the universe: and an infinite number of new races

[20]*Sämtliche Schriften*, 3d ed., ed. K. Lachmann and F. Muncker, vol. 1 (Stuttgart, 1886), p. 244.

and peoples will spring up in a moment on every side, like mushrooms. . . . For the stars, when they see that you have seated yourself, and not only on a stool, but on a throne, and that you are surrounded by this fine court and population of planets, will not only wish to seat themselves too and take their rest, but will also want to reign; and in order to reign one must have subjects, so they too will wish to have their planets, as you have, each one of them his own. And these new planets will have to be inhabited and adorned like the Earth. And it is unnecessary to mention the poor human race, which has already become wholly insignificant with regard to this system alone. What will become of it, when so many thousand new worlds appear, so that even the smallest star in the Milky Way does not lack its own? . . . I would rather not be roasted alive on this account, like the Phoenix.[21]

Many other examples could be given from among the literary and philosophical treatments of the theme; some of these are discussed fully in the ensuing chapters. Here it may suffice to cite just one more source, this time a scientific work. For the experimental physicist David Brewster, one of the most vigorous later supporters of the plurality of worlds, it is, once again, the link with heliocentrism as propounded by Copernicus that makes the plurality idea worthy of serious consideration. In 1854 he published his book *More Worlds Than One: The Creed of the Philosopher and the Hope of the Christian* (London) in response to William Whewell's essay *Of the Plurality of Worlds*, which had appeared, and excited much comment, in the previous year. Brewster's book created a considerable stir, so that the subject became a highly topical issue. Brewster is as strongly convinced as the literary authors already mentioned that however much the idea of a plurality of worlds may have appealed to the human imagination from earliest times, it was only the discovery of the shape, size, and motions of the Earth that forced us to infer by analogy "that these planets must be inhabited like our own" (New York 1854, pp. 7–8).

Thus, as a result of Copernicus's cosmology, the plurality idea entered on a new phase in which the dreams or nightmares it inspired were more convincing than ever before. This does not mean, however, that when the subject was discussed in works of theology, philosophy, or literature, the arguments used in ancient and medieval speculations about plurality were wholly forgotten. What it does mean, in fact, is that these older variations on the idea could now be developed to their

---

[21]*Selected Prose and Poetry*, ed., trans., and introduced by Iris Origo and John Heath-Stubbs (London, 1966), pp. 171–74.

full potential. They ceased to be merely an esoteric heresy for the discerning few, as it were, and began to pose a genuine threat because they now had the support of a verifiable scientific theory about the composition and structure of the universe. A brief look at particular arguments gives a clearer picture of this combination of continuity and change.

In his influential book *The Great Chain of Being* (1936), a classic among English and American studies in the history of ideas, Arthur O. Lovejoy gave currency to the view that even after Copernicus, plurality continued to be derived from the neo-Platonic principle of "plenitude," just as it had been in earlier writers such as Nicholas of Cusa. According to this principle, the concept of God's omnipotence necessarily implied that that omnipotence must be continuously active everywhere and at all times, as a creative force knowing no bounds. Potentiality must become reality. Lovejoy considers that the Copernican revolution made no significant contribution to the continuing speculation about the plurality of worlds during the Renaissance. This view, which is disputed by recent work in the history of science,[22] fails to recognize how decisively the concept of "more than one world" was transformed once the traditional arguments based on the concept of "plenitude" were subjected to the influence of Copernican ideas. What were the sorts of transformations that took place?

First of all, the doctrine of "plenitude" as understood by Lovejoy related (up to Nicholas of Cusa, at any rate) to the single cosmos of concentric spheres belonging to Aristotelian-Ptolemaic-Christian tradition, and it referred to the Great Chain of Being, the hierarchy reaching from inorganic matter through plants, animals, man, and angels, up to God. In this form it undoubtedly exerted an immense influence on the way the world was visualized in ancient times and in the Middle Ages. What was novel in the post-Copernican age is that the many life-forms in nature were no longer arranged in a vertical hierarchy—like rungs on a ladder—within the geocentric system of spheres; instead they were spread out in all directions within the heliocentric model of the universe. In this model, since there were planets basically similar to Earth in our solar system (and possibly in other systems too), these planets might perhaps also be the home

---

[22]Rossi, "Nobility of Man"; Steven J. Dick, *Plurality of Worlds: Origins of the Extraterrestrial Life Debate from Democritus to Kant* (Cambridge, 1982); summary in *Journal of the History of Ideas* 41 (1980), 3–27; Dick's work has served as a useful compendium of sources for some sections of this book, though his interpretation of the sources is sometimes different from mine and, in my view, untenable. See also Lovejoy, *Great Chain of Being*, chaps. 2, 4, esp. pp. 99, 108, 110.

of those life-forms that seemed to be missing from the hierarchy—for instance creatures occupying the places between animals and man, and between man and the angels. The absence of creatures representing these logically necessary stages was one of the problems that beset the protagonists of the Great Chain of Being. But leaving that aside, the old belief in the angels who kept the spheres of the Aristotelian cosmos in motion—or in the transfigured souls dwelling in the higher regions of the Aristotelian-Christian cosmos—was plainly a very different matter from a belief in human or quasi-human races scattered through a universe whose hierarchical arrangement—symbolically conveying religious meaning—had proved to be only an optical illusion. Lovejoy tries to conflate the image of the Great Chain of Being and its vertical continuity with the Renaissance conception of a plurality of worlds, but in fact they are totally different. And it was Copernicanism that made them different. Giordano Bruno, an avowed Copernican and the most eloquent advocate of the doctrine of plurality in the sixteenth century, was already able to recognize this: he dismissed "that beautiful order and ladder of nature" as a "charming dream, an old wives' tale."[23] At the very least it has to be said that the idea of the Great Chain of Being was so radically altered by Copernicanism that it becomes pointless to go on using the image in the context of the new perspective. The theologically based hierarchy in which man had a fixed place is replaced by a plurality of planetary worlds (and of humankinds?) that is theologically highly problematic and that forces man to ask what his status really is. This must be a major reason why the idea of plurality, which had led rather a shadowy existence in the Middle Ages and had been taken rather lightly as an intellectual game, now began to flourish so vigorously and to arouse such very widespread interest.

A second way in which Copernicanism transformed the idea of plurality has already been outlined and can now be examined more closely. As long as the concept remained based on the neo-Platonic doctrine of plenitude, it was merely one purely metaphysical speculation among others—including contrary theories—and had no verifiable basis in the natural sciences. The same concept gained quite a different order of credibility, however, once Copernicus had argued

[23]*On the Infinite Universe and Worlds*, trans. Dorothea Waley Singer, in Singer, *Giordano Bruno: His Life and Thought* (New York, 1950), p. 239. On the radical transformation of the Great Chain of Being in the context of the universe as seen by the "new science," see also F. E. L. Priestley, "Pope and the Great Chain of Being," in *Essays in English Literature from the Renaissance to the Victorian Age*, ed. Millar MacLure and F. W. Watt (Toronto, 1964), pp. 213–28; Bernhard Fabian, "Pope und die Goldene Kette Homers," *Anglia* 82 (1964), 150–71.

persuasively that the Earth and the other planets were qualitatively and functionally alike, basing his argument on scientific conclusions drawn from astronomical observation. And this was an age increasingly willing to accept the truth of Nature as against the truth of Revelation. What had been a pure speculation was now a scientific inference by analogy. That the plurality idea developed at this time into a major challenge to man's self-esteem came about not independently of, or even despite, Copernicus, as Lovejoy, with the limited vision of his "history of ideas" methodology, inevitably concluded: it happened *because* of Copernicus, "per consequens," as Robert Burton said in 1621.[24]

This new stage in the development of the plurality idea, in which it is presented as supported by scientific evidence, needs to be distinguished not only from the doctrine of plenitude but also from the Scholastics' speculations about a plurality of Aristotelian *kosmoi*. Oddly enough, Lovejoy, who concentrates his attention upon the Chain of Being within the one cosmos that was the medieval image of the universe, does not mention these speculations at all. They are the speculations, briefly referred to earlier, that were unleashed by the decretal published in 1277 by Étienne Tempier, bishop of Paris, with the authorization of the pope. They too were based on the argument of God's omnipotence (just as the doctrine of plenitude was), but they also stressed his free will. To propound as a truth that God *could* have created only one universe was incompatible with his possession of these two qualities—it was impudent to set limits to his power and free will in that way. Metaphysics and natural philosophy, it was said, could not judge definitively on such matters. It was precisely on the basis of Christian beliefs that the possibility of many inhabited worlds could not be doubted. But only the possibility: unlike the Christian supporters of the plenitude idea, not one of the Scholastic thinkers assumed that more than one world existed in reality.[25] The uniqueness of the world remained unshakable as a revealed religious truth. The idea that there could be more than one humankind was thus merely a hypothesis, in no way binding—more in the nature of a fiction.

Copernicus's cosmology itself was first presented as such a hypothesis, with no claim that it represented reality or truth. It was

[24]See Lovejoy, *Great Chain of Being*, p. 108, for his interpretation based on the "history of ideas" approach. Burton, *The Anatomy of Melancholy*, ed. Holbrook Jackson (London: Everyman's Library, 1968), 2:53.

[25]For this reason I find it unfortunate that McColley sees the judgment of 1277 as being based on the doctrine of plenitude; see Grant McColley, "The Seventeenth-Century Doctrine of a Plurality of Worlds," *Annals of Science* 1 (1936), 399, 429.

introduced to the public in this way by the Lutheran theologian Andreas Osiander, who wrote a preface to *De Revolutionibus*. The work was unobjectionable, he said, because it set up mere hypotheses about Nature; they were logical in themselves but were not supposed to be true or even probable; it was not suggested that they were a representation of reality. Truth could be arrived at only by divine inspiration or revelation, not by a study of nature. Osiander's foreword was, however, printed without the author's knowledge. Copernicus himself took a very different view of the relation his cosmology bore to truth, as is at once apparent from the dedication to Pope Paul III with which he introduced *De Revolutionibus*. He believed that his theory was a correct interpretation of physical phenomena, a fair representation of the way the universe actually worked. The claim he made for his theory was that it conveyed verifiable truth about the real world, and this is how his successors Bruno, Kepler, and Galileo read it.[26] As already mentioned, Copernicus, scrupulously scientific in his approach, never refers to the plurality of worlds either in the dedication or elsewhere. But it is obvious that that concept automatically acquires greater plausibility when it is linked with Copernican cosmology rather than with the Scholastic judgment of 1277. Its association with Copernicus's theory gives it a stronger claim to truth, even though, strictly speaking, the truth of that theory itself was not proved until Newton published his analysis of the mechanics of the planetary system in 1687.[27] Thus the Scholastics' speculation on plurality, too, was transformed. The idea is now related to a plurality of planetary worlds, not *kosmoi*; but above all, and in a manner comparable to the transformation of the speculation based on the notion of plenitude, it changed from being more or less a flight of fancy, with no serious implications, to a scientifically justifiable extrapolation. And not surprisingly, it was an astronomer of the Copernican school—Kepler—who formulated the new theory of plurality soon afterward.

The change has a direct bearing on "space literature." Any literary account of a journey in space would formerly have been either a pure

[26]On the difference between Osiander's and Copernicus's views see Blumenberg, *Kopernikanische Wende*, pp. 41–99; Edward Grant, "Late Medieval Thought, Copernicus, and the Scientific Revolution," *Journal of the History of Ideas* 23 (1962), 197–220. See also n. 27.

[27]A. R. Hall, *The Scientific Revolution, 1500–1800* (London, 1954), p. 102. On the spread of Copernicanism and rival cosmologies see, among others, Dorothy Stimson, *The Gradual Acceptance of the Copernican Theory of the Universe* (1917; Gloucester, Mass., 1972); Lynn Thorndike, *A History of Magic and Experimental Science*, vol. 6 (New York, 1941); S. K. Heninger, *The Cosmographical Glass: Renaissance Diagrams of the Universe* (San Marino, Calif., 1977).

jeu d'esprit or a work in the conventional form of the utopia or satire with no original content of its own. Now such a book could be read as a search for potential truth, as a serious and, in principle, verifiable anticipation of scientific knowledge. This seriousness distinguishes space literature from the genre of visits to the Earth's center and the "people" living there. (Grimmelshausen, Ludvig Holberg, and Charles de Fieux de Mouhy were among those who wrote in this genre, which continued to flourish for a long time.)[28] From now on there was to be an essential difference between the latter genre and most, if not all, literature about space travel.

The Copernican revolution revived a third variation of the plurality idea, namely the Pythagorean speculation on the subject, and again gave it a wholly new aspect. Copernicus himself stressed in his dedication to the pope that the arguments for the Earth's movement around the sun in *De Revolutionibus* revived ideas of the Pythagoreans, and in the sixteenth and seventeenth centuries his teaching was widely, and justifiably, seen as a mathematical formalization and confirmation of Pythagorean cosmology. This view is expressed in the very title of Thomas Digges's translation of part of *De Revolutionibus: A Perfit Description of the Caelestiall Orbes According to the Most Aunciente Doctrine of the Pythagoreans Lately Revived by Copernicus and by Geometrical Demonstrations Approved* (1576). The title also reveals the difference between Copernicus and the Pythagoreans: semipoetic philosophical speculation has become scientific truth, confirmed ("approved") by mathematics and physical observation. Linked with heliocentrism in the thinking of the Pythagoreans is the idea that just as the Earth is a star, so the stars are inhabited worlds, and it is fair to say that the claim of Copernicus's cosmic model to scientific validity to some extent validates this idea too, as a philosophical corollary to it. In his foreword, Digges, presenting Copernicus to a wider public, has no hesitation—unlike his author—in alluding to the plurality concept of antiquity and does not play down its significance by suggesting that it is a mere fiction. Suggestively, he quotes a passage from Palingenius's *Zodiacus Vitae* (1534?): "Many believe that individual stars can also be called worlds, and they call the Earth the dark star" (i.e., a planet that does not give out light).[29]

---

[28]Régis Messac, "Voyages modernes au centre de la terre," *Revue de littérature comparée* 9 (1929), 74–104.

[29]From the reprint by Francis R. Johnson and Sanford V. Larkey in *Huntington Library Bulletin* 5 (April 1934), 81. *Zodiacus Vitae*, bk. 7, lines 497–98: "Singula nonnulli credunt quoque sydera posse / Dici Orbes, TERRAMque apellant sydus opacum [.]" Palingenius of course still thinks in terms of the Ptolemaic universe.

In the writings of Montaigne, Bruno, and others, not only the Pythagoreans but also the Atomists experienced a revival in the light of Copernicanism. Even the literary fantasies of Lucian, never intended as contributions to knowledge about the real nature of the universe, acquired a new degree of persuasiveness (which could be either exhilarating or alarming) when referred to by, say, Bruno,[30] or Naudé in the letter already quoted. And that Plutarch's speculations about the moon's being inhabited also began to carry more weight in the light of Copernicus's discoveries is shown most clearly by Kepler, who made respectful mention of them a number of times and even translated them himself into Latin to be printed as an appendix to his novel about the moon (see also sec. 7).

Clearly, it was only in the wake of the Copernican revolution that the old concept of plurality of worlds could become a significant factor in the history of thought. Equally, it was that revolution which for the first time made the associated question of the nature of man a matter of urgent debate. As long as the Christian had been able to see himself as the pinnacle of Creation, his self-confidence was assured, even if he had to admit to being sinful and in need of salvation. As soon, however, as the plurality of worlds (which, to recapitulate, now means a plurality not of Aristotelian *kosmoi* but of planetary worlds in our solar system and perhaps also in other star systems) becomes a realistic possibility—no longer just a possibility without reality as entertained by medieval scholars or the merely speculative postulation of the Atomists—man has to confront a counterpart that forces him to reassess himself on the basis of quite different premises. Is this the death knell of the anthropocentric beliefs that went together with Christianity's geocentric view of the universe? Donne's endlessly quoted line "New philosophy calls all in doubt" ("The First Anniversary," line 205) may not have applied to everyone. But the possibility that "we are not alone" must surely have undermined not only the self-confidence of the Christian as the only child of God but also the confidence of Renaissance man, who, like Pico, saw himself as the most fortunate of creatures (*felicissimum animal*). "What a piece of work is man! how noble in reason! how infinite in faculty! . . . in action how like an angel! in apprehension how like a god!" And yet he is also "this quintessence of dust" (*Hamlet*, act 2, sc. 2)?

The question What is man? acquires a totally new dimension and becomes more urgent than ever before. It becomes an inquiry, taking different forms over the centuries, into man's status. Is he still the

[30]Bruno, *Ash Wednesday Supper*, pp. 155, 157.

most perfect of all creatures, the most intelligent, the dearest to God, the most fortunate? Whether he is or not, if we assume the existence of a *variety* of worlds, how is that consistent with God's justice? Bruno thought (see sec. 3) that God demonstrates his own perfection all the more gloriously by having created other, numerous races of human beings, possibly including some superior to ourselves. Or need one no longer assume a Creator at all? As early as 1610, Kepler asks: If other heavenly bodies are similar to our Earth, who then "occupies the better portion of the universe?" Are we still "the noblest of rational creatures"? How can it still be true that everything was created for the sake of man, and "how can we be the masters of God's handiwork"? (see this chap., sec. 7). In 1621, Robert Burton, in his widely read *Anatomy of Melancholy*, takes up Kepler's words and formulates the question still more pointedly: "Are we or are they Lords of the World?" (2:55). And in 1897, H. G. Wells is still asking the same question, placing it as a motto at the head of his classic science fiction novel *The War of the Worlds*. As Montaigne, among others, critically asked, did the new cosmology really elevate man? Did it not rather demote him? Perhaps, as Kepler hoped, a way could still be found for man to retain the central position theology had up to now accorded him? This question is a leitmotif that runs through successive post-Copernican treatments of the theme, in theology and philosophy as well as in literature.

The extent of this debate remained limited, however, in the decades immediately after *De Revolutionibus*. The lay public was not sufficiently convinced by Copernicus's theory until Galileo had actually *seen* through his telescope in 1609/10 that the surface of the moon was extremely similar to that of the Earth; that Jupiter was circled not by one moon (like the Earth) but by four, which could only shine on Jupiter's inhabitants; and that the stars of the Milky Way were suns. Only after Galileo had popularized these findings in his *Dialogo dei due massimi sistemi del mondo* (1632) did the consequences and implications of Copernicus's theory achieve their full impact. Astronomy became a fashionable science, and the remainder of the seventeenth century saw the appearance of more and more philosophical and literary accounts of the nature of rational beings on other planets, extrapolated from the scientific facts.

Nevertheless, a poet's imagination could be fired by things only thought possible, without waiting for them actually to be seen; Edmund Spenser's was, less than two generations after *De Revolutionibus* appeared. In the proemium to the second book of *The Faerie Queene* (1590), he speaks of the new lands of North and South

America that, though they have only become known in his lifetime, have always existed. He challengingly makes the analogy between the dispute about the antipodeans and the idea of plurality in the cosmos:

Why then should witlesse man so much misweene,
That nothing is, but that which he hath seene?
What if within the moones fayre shining spheare,
What if in every other starre unseene,
Of other worldes he happily should heare?
He wonder would much more; yet such to some appeare.

"What if" . . . ? With all due caution, others had already begun to explore this idea.

2. Cautious Explorations: Montaigne and Benedetti

Before Galileo's discoveries became known, accounts of imaginary space travel influenced by Copernicus's ideas went into very little detail. Authors were more concerned with *whether* extraterrestrial worlds might exist than with the nature and habits of the inhabitants, except in the broadest outline. Yet even at this time, the meeting between Copernicanism and literary imagination—in its widest sense—gave rise to a few variations that stand out as especially interesting. Bruno and Kepler are the chief figures in the conquest of this new territory, but they are not the only ones.

Before going further, it should be mentioned that this period between Copernicus and Galileo naturally still also saw the publication of literary works, specifically works describing space journeys, that do not acknowledge the new cosmology, or at any rate do not take it seriously into account. The problem of whether the planets or the moon are inhabited is treated, when raised at all in these works, in one of two ways. The idea may be denounced as absurd or at least improbable, as for instance in Jean Edouard du Monin's *L'uranologie, ou le ciel* (1583); it is no coincidence that this author simultaneously rejects Copernicus's theory. Similarly, Joseph du Chesne, in *La morocosmie, ou de la folie, vanité, et inconstance du monde* (1583), treats both plurality and Copernicanism as erroneous ideas.[31] Alternatively, the

[31]See Beverly S. Ridgely, "The Cosmic Voyage in French Sixteenth-Century Learned Poetry," *Studies in the Renaissance* 10 (1963), 136–62. Space journeys in a geocentric universe are also discussed in Carrie Esther Hammil, "The Celestial Journey and

notion of inhabited planets is simply a literary convention, with no claim to scientific or philosophical validity. The convention is used, in the tradition of Lucian, as a vehicle for contemporary satire, allegory, or the depiction of a utopia. An example of this is Barnaby Rich's *A Right Excellent and Pleasaunt Dialogue, betwene Mercury and an English Souldier* (1574), in which an English soldier is transported by Mercury first to Mars, where he finds a warlike society suspiciously reminiscent of England, and then to an ideal, peace-loving world on Venus. The author's didactic intention is plain.[32] In a similar way, the political, social, and religious situation in France is satirized in the anonymous account of a cosmic voyage, *Le supplément du Catholicon, ou nouvelles des régions de la lune* (1595). Everything on the moon, from human nature in general to specific features of political life, is *comme chez nous*.[33] To cite one final example, *The Man in the Moone, Telling Strange Fortunes; or, The English Fortune-teller* (1609) is a series of character portraits that pointedly satirize English society of the day. There is barely a trace of astronomy, either old or "new." Yet at the time of its publication, the question of whether the moon and planets were inhabited was being treated with increasing seriousness. Kepler's novel about the moon is of the same year, and as far back as 1588, candidates for master's degrees at Oxford had had to debate the question of "whether there are many worlds." The author of *The Man in the Moone* chooses completely to ignore these developments.[34]

Not so a lively and critical thinker like Montaigne. In the context of our topic, Montaigne interestingly represents a transitional phase: he accepts the notion that the planets may be inhabited with a shrug of the shoulders, as it were, without finally committing himself. He is usually considered the first French literary author—perhaps he is the first of all nonscientific writers—to pay the Copernican system the sort of attention "one accords to hypotheses of this kind."[35] In the famous "Apologie de Raimond Sebond," the twelfth chapter of the second book of his *Essais*, he mentions that for three thousand years

---

the Harmony of the Spheres in English Literature, 1300–1700" (Ph.D. diss., Texas Christian University, 1972).

[32]For more detail see Michael Winter, *Compendium Utopiarum*, vol. 1 (Stuttgart, 1978), pp. 40–41.

[33]See Beverly S. Ridgely, "A Sixteenth-Century French Cosmic Voyage: *Nouvelles des régions de la lune*," *Studies in the Renaissance* 4 (1957), 169–89.

[34]For more detail see Winter, *Compendium Utopiarum*, p. 48. On the Oxford M.A. examination question see Francis R. Johnson, *Astronomical Thought in Renaissance England* (Baltimore, 1937), p. 181.

[35]Jean Plattard, "Le système de Copernic dans la littérature française au XVIᵉ siècle," *Revue du seizième siècle* 1 (1913), 234.

everyone had believed that the Earth was immobile, until suddenly the Stoic Cleanthes, or else Nicetas of Syracuse, declared that the Earth revolved around the sun (and on its own axis). Montaigne adds: "And, in our days [de nostre temps], Copernicus has so well grounded this theory, that he very lawfully uses it for all astronomical conclusions."[36] This reference, though brief, is very precise and indicates a degree of knowledge of De Revolutionibus. Copernicus himself, in the dedication, had sought to justify his theory precisely by claiming that the heliocentric cosmology could be used to explain those astronomical phenomena that presented insuperable problems in the context of the Ptolemaic system. But, having conceded this point, what is Montaigne's conclusion? It is that we need not worry about which system is the correct one. On principle skeptical of the validity of knowledge, he goes so far as to deny any truth. Who knows whether in the next thousand years a new theory will not come along and overthrow both the Ptolemaic and the Copernican systems? The history of philosophy and of science is a history of supposed truths being disproved: why should this not continue to happen in the future, given that human reason remains the same? Of course there is an innate fallacy in this Pyrrhonian position: if nothing can be believed, then the validity of one's own doubt becomes doubtful. The converse is that anything is believable; and in the passage quoted above, Montaigne did unmistakably acknowledge the irrefutable internal logic of Copernicus's reasoning, however much he considered that all theories were to be doubted on principle: he credited Copernicus's system with at least a quality of mathematical truth not of the same order as other, more problematic kinds. At the very least, then, Montaigne placed Copernicus's theory in a sort of philosophical limbo. It might not be true, but it was not false either: it was potentially both and, at that particular time, rather convincing than otherwise.

It is as well to bear in mind this ambivalence toward Copernicus when, elsewhere in the same chapter, the plurality of worlds is subjected to the twists and turns of Montaigne's logic. The general context is again that of the doubtfulness of all knowledge, but seen now in relation to the unfathomable omnipotence of God. God acts according to principles our reason cannot encompass. Therefore it is an arrogant overestimation of man's range of vision to believe that in creating our

[36]The Essays of Montaigne, trans. E. J. Trechmann (London, 1927), 2:15; Essais, ed. Rat, 1:640. For the quotations that follow, two page references are given: the first is to Trechmann's translation, vol. 1; the second, to the Rat Essais, vol. 1. The two major editions of the Essais during Montaigne's lifetime are of 1580 and 1588. The text quoted is that of the author's subsequent final revision.

world God exhausted his powers. "Why, indeed, omnipotent as he is, should he have restricted his powers within a certain measure? In whose favour do you suppose he has renounced his privilege? Of nothing can thy reason convince thee with better grounds and more likelihood than of the plurality of worlds." This is followed by a quotation from Lucretius saying that there is not only one Earth, one moon, one sun, and so on, but countless numbers of them (pp. 522–23/583–84). In Montaigne's context this might seem to mean that he is as little convinced of the likelihood of a plurality of worlds as of the opposite—all the more so, as elsewhere in the twelfth chapter he exclaims: "Is it not a delusion of human vanity to make the moon a celestial earth, and to imagine that there are mountains and valleys upon it, as did Anaxagoras; to set up human habitations and dwellings" (pp. 443/495–96). But in fact this is not Montaigne's view. For in the just-quoted passage on the "plurality of worlds," he continues in a way that suggests the opposite of the doubt he appeared to be expressing. He says that reason errs in considering the plurality of worlds to be improbable, and this assertion would accord well with the opening words "Why, indeed . . . ." The most famous thinkers of the past, he now adds, believed in plurality, and not they alone but also "some of our own time [aucuns des nostres mesmes], compelled by the evidence of human reason" (pp. 523/584). It cannot be objected that it is Lucretius, rather than Copernicus and the modern astronomers, who turns Montaigne's thoughts in this direction.[37] "Nostres mesmes" seems too closely to anticipate "de nostre temps, Copernicus" for this association to be discounted. And the further Montaigne pursues the idea, the more he seems, almost unwillingly, to admit its plausibility. Nothing in the world we know is single and unique. "Wherefore it seems unlikely that God should have created this work [our Earth] without a fellow, and that the matter of this form should have been exhausted upon this single individual" (pp. 523/584). Here he harks back to the actual wording of his criticism, a page earlier, of those rationalists who doubted the plurality of worlds. And indeed in the very next paragraph he uses the idea that we have a counterpart elsewhere in the universe as a weapon against the "rational" opponents of the idea:

> Now if there be many worlds, as Democritus, Epicurus and almost all the philosophers have believed, how do we know if the principles and laws of this one in like manner apply to the others? They have

[37]Plattard, "Système de Copernic," p. 237.

perhaps a different appearance and a different constitution. Epicurus imagines them either like or unlike.

In this world we see endless differences and variations, due merely to distance in place. . . .

Besides, how many things there are within our knowledge that impugn those fine rules that we have cut out for and prescribed to Nature! And shall we attempt to bind even God to them? (Pp. 523–24/584–85)

So here we have the plurality of worlds used as an argument against the arrogance of reason! This may seem to echo the reasoning of Étienne Tempier, insisting that God's omnipotence and free will cannot be restricted to a scale within the compass of human reason. But there is an obvious difference: Montaigne does *not* add that such worlds have no reality. He refers, of course, not to the bishop of Paris but to the Atomists of antiquity; but he cites them specifically in relation to the modern perspective that Copernicus, a man "of our own time," had revealed. And the thrust of the whole argument—perhaps against Montaigne's original intention—is to undermine the complacency of those rationalists who held that the existence of a plurality of worlds was contrary to reason.

Did Montaigne, then, believe that there were other inhabited worlds? Some have thought so, and admired his courage.[38] But for a truer assessment of his position we need to bear in mind the spirit of yes-and-no that characterizes so many of Montaigne's utterances. He seems to have felt more keenly than most that he was playing with fire; hence the tentative manner of his approach. Montaigne should therefore be seen as only a transitional figure reflecting a time when people were only just beginning cautiously to link the Copernican theory with the idea that man was not unique—but as a most stimulating one nevertheless.

There was no such equivocation in the mind of the Venetian mathematician Giovanni Battista Benedetti, who made his own rather eccentric contribution to the debate in the 1580s. Criticism of Aristotle's cosmology is a thread running right through his *Diversarum Speculationum Mathematicarum et Physicarum Liber* (1585), and in one passage he invokes Copernicus as an ally in this campaign. Attacking Aristotle and Christian metaphysics, he argues, with astonishing bravura, that the heaven and stars cannot exist solely for the benefit of the Earth's

---

[38]Ibid. See also Camille Flammarion, *Les mondes imaginaires et les mondes réels*, 12th ed. (1865; Paris, 1874), p. 298.

inhabitants. Copernicus has shown, he says, that the moon revolves around the Earth; what if Saturn, Jupiter, Mars, Mercury, and Venus are also moons circling other earths (*corpora huic terrae similia*) invisible to us? "Who knows . . .?" he asks (pp. 255–56).

Leaving aside the eccentricity of the analogical inference from Copernicus's theory (which replaces the more usual inference that if the Earth is a planet, then the planets are Earths), it is remarkable that the author can accept as a matter of course that *real* worlds, Earths in space, are conceivable. And nothing either here or elsewhere in the *Speculationum Liber* leads us to suspect that Benedetti had any serious doubts about the anti-Aristotelian theory that is the basis for this extrapolation about the plurality of worlds. In fact he devotes a whole further chapter of his book to this context (pp. 195–96). In it he uses his idea that there might be other worlds with the planets as their moons as an argument against the Aristotelian view that there is only one world: "Aristotle did not satisfactorily refute the view of those who believed that many worlds existed." Aristotle's main argument against a plurality of inhabited worlds was that all heavy bodies, all assumed other worlds with their ring of fire, would be drawn to the *one* natural and absolute center, the center of the Earth, with catastrophic results. Benedetti considers this to be not worth the trouble of refuting. Putting the opposite case, he bases his argument on the heliocentric system as a matter of course. In this passage he does not invoke Copernicus, but he does refer to his predecessor, Aristarchus of Samos ("if the opinion of the most learned Aristarchus is true . . ."), whom he named in the same breath as Copernicus in the passage discussed above. Just as Copernicus, in his dedication to the pope, had expressly claimed to be guided by reason, so Benedetti also claims that his argument will be seen to be wholly consistent with reason. Benedetti implicitly rejects the Aristotelian division of the universe into an uninhabited realm of perfection and an inhabited one of imperfection, and he pioneers the idea of plurality in its modern form (a plurality not of Aristotelian *kosmoi*, but of planet worlds, which he assumes to be in our solar system). In this passage too he speaks of Saturn, Jupiter, and other planets circling around "Earths" (*terrae*) we cannot see and whose "conditiones," he says, are similar to those of our Earth. Does he mean that they are similar in that they are inhabited, perhaps even by rational beings? Having gone so far and prompted this question, Benedetti suddenly holds back; even the "conditiones"—conditions of life?—are not enlarged upon. Certainly his idea about the planets' being moons is a fantastic one, incompatible with the very heliocentric theory he himself champions. He seems

prepared to follow Aristarchus and Copernicus only as far as is necessary for his campaign against traditional cosmology. He accepts both
the alternative cosmos established by the new science and the view,
set out in *De Revolutionibus*, that there are no qualitative differences
between one part of the universe and another. But he then indulges in
a wild flight of fancy that makes no sense in the light of the heliocentric
theory. Even so, his own exercise in analogical reasoning—in principle
a scientific procedure—leads him, too, to assume the existence of
worlds that previously could only have been the stuff of dreams.
For all its eccentricity, his conclusion represents a remarkably bold
attempt, at that early date, to let Copernican thinking breathe new life
into the idea of plurality.

3. *"New Science" and the Visionary: Bruno*

At about the same time, Giordano Bruno was also undertaking
the revitalization of the plurality idea in the light of the new anti-
Aristotelian cosmology, but on a far larger scale and with far-reaching
implications. He brought incomparably stronger conviction and
greater literary eloquence to the task, so that when the idea burst upon
public awareness and became a matter of scandal, it was his doing—
the result of his writings and of his death at the stake. For at his
sensational trial by the Inquisition, Bruno's belief in a plurality of
inhabited worlds may well have been the main charge brought against
him, whether or not the official verdict, now lost, specified it as such.
It certainly figured prominently in the hearings.[39] And though the
notion that Bruno was a martyr to science is no longer as commonly
held as it once was (plurality was not the only objectionable idea he
advocated), his heterodoxy in this respect is salient in our context. But
we need to define more precisely the link between the Copernican
theory and Bruno's belief in plurality. His conviction that not only our
solar system but innumerable other star systems are inhabited does
not derive directly or solely from his support for Copernicus's doctrine. Bruno was equally familiar with the pluralistic cosmology of the
Atomists, above all that of Lucretius, whom he frequently quotes, and

---

[39]According to Blumenberg in his edition of *La cena de le ceneri* (*The Ash Wednesday
Supper*) (Bruno, *Das Aschermittwochsmahl*, trans. into German by Ferdinand Fellmann,
with an introduction by Hans Blumenberg [Frankfurt, 1969]), pp. 47, 50. Here Blumenberg is implicitly responding to the claim made by Frances A. Yates that Bruno was
condemned not for his belief in plurality but for his Hermetic beliefs: *Giordano Bruno
and the Hermetic Tradition* (London, 1964), p. 355.

also with the neo-Platonic arguments relating to plenitude. But, as already stated, it as only in the context of Copernicanism that these traditional ideas, mere intellectual exercises with no claim to truth, could attain the status of a hypothesis that might actually correspond to reality. A modern interpretation that has received much attention maintains that Bruno's support for Copernicanism must in its turn be seen in a wider context, that of the Hermetic beliefs attributed to him. According to this interpretation, Bruno used the central position of the sun as a religious hieroglyph in accordance with the religious mysticism and gnosis of Hermes Trismegistus. These teachings also offer a parallel, though by no means the only one, to Bruno's belief in the magical animation of the planets, which he calls "great living creatures" (grandi animali). There may be something in this, but it is irrelevant to our purposes here, since Hermetic philosophy does not include a belief in the plurality of worlds.[40]

At all events, Copernicanism was Bruno's starting point: it was an early, formative influence and introduced him to ideas that were to form the basis of his philosophy of plurality.[41] It is generally agreed that during his novitiate in Naples he already knew and accepted Copernicus's teaching. In his wanderings through Europe after fleeing from his monastic order, he continually spread Copernicus's heliocentric doctrine. He made an especially strong impact with his enthusiastic lectures on the Copernican theory at the University of Oxford in 1583. True, they were soon banned; but it was during that stay in England in 1583/84 under the protection of the French ambassador to London that he wrote his important Italian dialogues, which vividly convey his enthusiasm for Copernicanism. A distant echo still sounds as late as 1591, in the didactic astronomical poem De Immenso. It contains a hymn celebrating Copernicus as the bold spirit who opened up the "veri fontes"—the true springs, or sources of truth (bk. 3, chap. 9).

To be sure, Bruno's acceptance of Copernicanism and his associated rejection of the Aristotelian cosmic model is only the first step in the shaping of his own cosmology, but it is the decisive one. For while he goes further than Copernicus, he remains bound by Copernicus's

[40]Yates, Giordano Bruno, for example, pp. 155, 168, 238, 246, and passim. On the absence of a belief in plurality in Hermeticism see p. 245.

[41]See for instance McColley, "Seventeenth-Century Doctrine," p. 414; Thomas S. Kuhn, The Copernican Revolution (1957; Cambridge, Mass., 1976), p. 235; Singer, Giordano Bruno, e.g., p. 50; Antoinette Mann Paterson, The Infinite Worlds of Giordano Bruno (Springfield, Ill., 1970), pp. 9, 12; Waldemar Voisé, "Giordano Bruno, sur la morale dans un monde infini et sur la passion de connaître chez Copernic," Revue de synthèse 94 (1973), 19, 27; Blumenberg, Aschermittwochsmahl, pp. 15, 31–32, 38; Alfonso Ingegno, Cosmologia e filosofia nel pensiero di Giordano Bruno (Florence, 1978), chap. 2.

general conception. Copernicus had destroyed the hierarchical series of spheres with the Earth at its center, but he had retained the traditional sphere with the fixed stars as the outermost limit of a finite universe; Bruno discarded this sphere too. (So, incidentally, did that other follower of Copernicus, Thomas Digges, whose 1576 interpretation of Copernicus, Bruno may well have read while he was in England.) But when Bruno goes on to declare that the fixed stars are suns, his assumptions about their physical properties and relationships are still inspired by the Copernican model: he supposes them, by analogy with our sun, to be circled by planets. For him, moreover, these planets are worlds, every one of which is inhabited by living beings. The universe, harboring life on countless planets, is infinite, with no center and no outer limit. With truly religious fervor, Bruno proclaimed this grandiose vision of the cosmos again and again, in *De Immenso* and above all in the dialogues *La cena de le ceneri* (1584) and *De l'infinito universo e mondi* (1584). Physical cosmology imperceptibly slides over into rapturous metaphysics; thus in *De l'infinito universo*:

> Thus is the excellence of God magnified and the greatness of his kingdom made manifest; he is glorified not in one, but in countless suns; not in a single earth, a single world, but in a thousand thousand, I say in an infinity of worlds.
>
> Thus not in vain is that power of the intellect which ever seeketh, yea, and achieveth the addition of space to space, mass to mass, unity to unity, number to number, by the science which dischargeth us from the fetters of a most narrow kingdom and promoteth us to the freedom of a truly august realm, which freeth us from an imagined poverty and straitness to the possession of the myriad riches of so vast a space, of so worthy a field, of so many most cultivated worlds.[42]

"By the science . . .": it is interesting to see Bruno himself pointing to Copernicus as his predecessor and to Aristotle as his adversary. To begin with Aristotle, in all three of the works mentioned, Bruno refutes each of Aristotle's objections to the assumption that "beyond this world there lieth another" (p. 328). But he also puts the discussion on a quite different footing by being the first to give a clear definition of the new conception of plurality, which has already been mentioned and which was to become the accepted basis for future discussions of

[42]*On the Infinite Universe and Worlds*, in Singer, *Giordano Bruno*, p. 246. Page references are to this text except where *Ash Wednesday Supper*, ed. Gosselin and Lerner, is named as the source. In quotations from these two works, angle brackets indicate part of the text of the published English translations.

the topic. Plurality applies now not to geocentric *kosmoi* but to worlds similar or identical to Earth—inhabited worlds situated on the planets and stars randomly scattered throughout the visible universe:

> Concerning this question, you know that his [Aristotle's] interpretation of the word *world* is different from ours. For we join world to world and star to star in this vast ethereal bosom, as is seemly and hath been understood by all those wise men who have believed in innumerable and infinite worlds. But he applied the name *world* to an aggregate of all those ranged elements and fantastic spheres reaching to the convex surface of that *primum mobile,* the perfect sphere which draweth the whole revolving with it at immense speeds around the centre near which we are placed. (P. 329)

The "wise men" Bruno invokes here are, as parallel passages confirm (pp. 240, 374), the Atomists. Of the influence of Lucretius there is no doubt. But it is clear that Bruno, like Montaigne, interprets the Atomists' concept of plurality as a plurality not of *kosmoi* but of planetary or stellar worlds. In other words, his understanding of the Atomists' theory is colored by his Copernicanism. Thomas S. Kuhn has recognized that Bruno's chief contribution to modern cosmology lies in his recognition of the affinity between Atomism and Copernicanism, which he took as the basis for his philosophy of infinity.[43] It would be idle to speculate on which of the two formative influences, Copernicus or Lucretius, either came first or was more significant. It was in combining the two that Bruno laid the foundation of his anti-Aristotelian theory of plurality in the context of the new science.

The very first of the Ash Wednesday dialogues sets out to show "how praiseworthy" Copernicus is as the exponent of this new knowledge (p. 69). But Bruno's praise is not uncritical:

> He was . . . a man who, in regard to innate intellect, was greatly superior to Ptolemy, Hipparchus, Eudoxus and all others who followed in their footsteps. This estate he attained by freeing himself from a number of false presuppositions of the common and vulgar philosophy, which I will not go so far as to term blindness. Yet, Copernicus did not go much further ⟨away from the common and vulgar philosophy⟩ because, being more a student of mathematics than of nature, he could not plumb and probe into matters to the extent that he could completely uproot unsuitable and empty principles. . . . [His achievement is that he provided mathematical

[43]Kuhn, *Copernican Revolution,* pp. 236–37.

proof of heliocentrism.] Who, then, will be so rude and discourteous toward the labours of this man as to forget how much he accomplished, and not to consider that he was ordained by the gods to be the dawn which must precede the rising of the sun of the ancient and true philosophy, for so many centuries entombed in the dark caverns of blind, spiteful, arrogant and envious ignorance? Who, marking what he could not do, would place him among the common herd who are moved and guided by, and throw themselves headlong after, the voice of a brutish and ignoble fancy sounding at their ears? Who would not rather count him among those who, with happy genius, have been able to raise themselves and stand erect, most faithfully guided by the eye of Divine Intelligence? (Pp. 86–87)

The whole "truth," however, was revealed only by Bruno himself with his conception of the infinite, inhabited universe, a conception in which natural philosophy combines with quasi-theological elements (pp. 89 ff.). He does seem to concede that his ideas are derived from those of Copernicus when he stresses, at the very opening of the first Ash Wednesday dialogue, that he "saw through neither the eyes of Copernicus nor those of Ptolemy, but through his own eyes" *when it came to* "judging and determining" (or drawing *conclusions*, p. 85). Though he readily acknowledges Copernicus as a predecessor, he does nevertheless claim in the same dialogues to have his own reasons for holding views that agree with, or are developments of, Copernicus's views:

But in truth it signified little for the Nolan [Bruno] that the aforesaid ⟨motion⟩ had been stated, taught and confirmed before him by Copernicus, Niceta Syracusus the Pythagorean, Philolaus, Heraclitus of Pontus, Hecphantus the Pythagorean, Plato in his *Timaeus* (where the author states this theory timidly and inconstantly, since he held it more by faith than by knowledge), and the divine Cusanus in the second book of his *On Learned Ignorance*, and others in all sorts of first-rate discourses. For he ⟨the Nolan⟩ holds ⟨the mobility of the earth⟩ on other, more solid grounds of his own. On this basis, not by authority but through keen perception and reason, he holds it just as certain as anything else of which he can have certainty. (P. 139)

What are his own "grounds"? Let us consider only his grounds for assuming a plurality of inhabited worlds. Writers on Bruno who have stressed that his reasons are not derived from Copernicus have instead identified them strongly with the doctrine of plenitude.[44] Certainly there is no mistaking the part that plenitude plays in Bruno's thought. In the dialogues *De l'infinito universo* we need look no further than the summary of the contents to be told that "divine power should not be otiose" and that it *would* be otiose if it produced only "a finite effect," which would not be consistent with God's "goodness and greatness" (pp. 234–35). "Why do you desire that center of divinity which can . . . extend infinitely to an infinite sphere, why do you desire that it should remain grudgingly sterile rather than extend itself. . . . Why should infinite amplitude be frustrated, the possibility of an infinity of worlds be defrauded?" (p. 260). For "if in the first efficient Cause there be infinite power, there is also action from which there resulteth a universe of infinite size and worlds infinite in number" (p. 265). This is the doctrine of plenitude in its purest form.

This does not mean, however, that the commentators referred to are justified in triumphantly pointing to such passages as proof that Bruno's belief in plurality rests on these neo-Platonic arguments and not on extrapolation from Copernicus's teaching, nor in concluding that his views have a metaphysical and not a scientific basis. True, Bruno is not so much a scientist, despite his extensive and detailed knowledge of astronomy, as he is a religious philosopher to whom enthusiasm comes more naturally than precision. As a type he is reminiscent of Nicholas of Cusa and Palingenius, both of whom he refers to on occasion in connection with plurality.[45] Nevertheless, his ideas, particularly those relating to the plurality of mankinds, are firmly set in the context of modern scientific thinking. He transfers the arguments based on plenitude from the one hierarchically ordered universe of Aristotelian-Christian thought to his own unbounded infinite universe of worlds, a conception derived from Copernican cosmology, and he can therefore lay claim to scientific as well as metaphysical plausibility. It takes only a superficial reading of his pertinent works to see how clearly they reflect the new science: the dialogues and even some chapters of the didactic poem *De Immenso* are larded with mathematical calculations and astronomical diagrams. Bruno's approach is physical as well as metaphysical, inductive as well as deductive.[46] And it is, after all, analogical reasoning—a scientific procedure—that convinces him that other stars have planetary systems: he assumes that analogous conditions apply to all earths

[44]Lovejoy, *Great Chain of Being*, pp. 116–17; Dick, *Plurality of Worlds*, pp. 61–69.

[45]He refers to Nicholas of Cusa for example in *On the Infinite Universe*, p. 307, and to Palingenius for example in *De Immenso*, bk. 8, chaps. 2, 4.

[46]See for example Singer, *Giordano Bruno*, p. 50; Paterson, *Infinite Worlds*, pp. 17–21.

and all suns.[47] He accepts the doctrine of plentitude, like the ideas of the Atomists, only in a version adapted to Copernicanism. And if this is what he means by having "grounds of his own," it is, in a sense, a fair enough claim.

But how does Bruno visualize the beings who inhabit the innumerable other worlds or earths? In his day, the question had to be considered in relation, above all, to Christian dogma. We may assume that Bruno was aware of this theological angle to the question; it probably accounts for a statement of his, which stands out because it is not logically necessary in the context, to the effect that the assumption of a plurality of mankinds on other heavenly bodies enhances the glory of God (see the quotation at n. 42 above). Not that this is merely a defensive ploy; to judge by the rest of his work, it is his honest opinion. But equally unmistakably, this opinion is a heretical one. The question, then, is how Bruno deals with the inevitable christological objections, raised, as we have seen, first by Vorilong and later notably by Melanchthon: are all these mankinds descended from Adam and Eve, and was Christ crucified in all these worlds? Or, alternatively, did these human beings have no need of redemption from original sin—in which case, their humanity must be of a different order from ours? Either supposition would have been heretical.

Bruno's attempts to tackle this problem, even in the most general terms, are suspiciously few and far between. Where he does refer to it, he stifles specific questions; before they have been fully formulated, he breaks into hymnic praise of God's greatness, as if it were self-evident that all worlds were united in an impeccably orthodox hosanna. We can nevertheless make out the general drift of his answer. He stresses the *unity* of the whole, infinite Creation. As a Copernican he rejects the Aristotelian division of the universe into the realm of the four elements on the one hand and the divine *quinta essentia*, the ether, on the other (a case he argues most eloquently in the dialogues *On the Infinite Universe*, pp. 311 ff.). He maintains that the physical composition of all heavenly bodies is essentially the same (p. 313). All of them are composed, like the Earth, of the four elements; only the relative quantities of the elements vary according to the kind of celestial body concerned (pp. 314, 371). But in fact he consistently distinguishes only two types, the suns and the planets: the "fiery worlds," which radiate light and warmth, and the "watery worlds," which receive it. Pursuing this further in the *Ash Wednesday Supper*, he says that there are no differences of "kind" within each of the two catego-

[47]*On the Infinite Universe*, p. 305; and see p. 304.

ries; therefore we must assume other earths constituted like ours, indeed "innumerable earths similar to ours," and likewise innumerable other "worlds" that exactly resemble the sun (pp. 153–55).

All these heavenly bodies, we are told, those that shed light and those that receive it, are inhabited, like our Earth. "Then the other worlds are inhabited like our own?" asks the simple, doubting character Burchio in *De l'infinito universo*, and Fracastorio replies:

> If not exactly as our own, and if not more nobly, at least no less inhabited and no less nobly. For it is impossible that a rational being fairly vigilant, can imagine that these innumerable worlds, manifest as like to our own or yet more magnificent, should be destitute of similar and even superior inhabitants; for all are either themselves suns or the sun doth diffuse to them no less than to us those most divine and fertilizing rays, which convince us of the joy that reigneth at their source and origin and bring fortune to those stationed around who thus participate in the diffused quality. (P. 323; cf. p. 240)

There are two questions Bruno cannot wholly bypass. First, are all those inhabitants of planets and suns, that are to be thought of in human terms, descendants of the biblical ancestors of mankind? Without putting the question too directly, so as not to seem to challenge dogma, Bruno tells us that they are not, just as in *De Immenso* his verdict on the question of the antipodeans is that all mankind on this Earth does *not* share the same origin.[48] A passage in *De l'infinito universo* reads:

> The ELEVENTH [argument] asserteth that Nature having multiplied by definition and division of matter, entereth on this act only by the method of generation, when the individual as parent produceth another individual. We reply that this is not universally true. For by the act of a single efficient cause there are produced from one mass many and diverse vessels of various forms and innumerable shapes. . . . If there should come to pass the destruction of a world followed by the renewal thereof, then the production therein of animals alike perfect and imperfect would occur without an original act of generation, by the mere force and innate vigour of Nature. (P. 376)

Bruno wisely avoids a specific discussion of Christian doctrine.

The second question arises from the answer to the first. It is the old question of whether, if we think in terms of polygenesis, the

---

[48]See Paul Henri Michel, *La cosmologie de Giordano Bruno* (Paris, 1962), pp. 258–60.

resulting groups of human beings on different planets are equal. This in turn leads on to the question of whether the Creator, working through Nature, has acted with justice. (For Bruno, spontaneous origin followed by an evolutionary process is not a real possibility, even though his vague use of terminology occasionally seems to imply it.) The last passage quoted speaks of perfect and imperfect beings (see also p. 309), while, paradoxically, the very next paragraph refers to the "perfection" of *every* world: "It . . . followeth not from the perfection of this [world] or of those that those or this be less perfect; for this world even as those others, and those others even as this, are made up of their parts, and each is a single whole by virtue of his members" (p. 376). He admits that there are differences between the groups of living beings because of the difference in the physical composition of their worlds: "Therefore, as in this most frigid body primarily cold and dark, there dwell animals which live by the heat and light of the sun, so in that most torrid and shining body there are beings which can vegetate by aid of the chill from surrounding cold bodies" (p. 309). This physical difference may yet be compatible with equality in the philosophical or theological sense. But this is precisely what Fracastorio seems to call in question when he admits it is unthinkable that the other heavenly bodies are not populated by "similar or even superior inhabitants." If the various beings are "all subject to a perfect Power" (p. 244), are they all equally fortunate and perfect? Or is it part of the Creator's plan that they should not be equal? In the *Ash Wednesday Supper*, Bruno makes Teofilo, who is usually his spokesman, say, "In this way, we know that if we were on the moon or on other stars, we would not be in a place very different from this—and maybe in a worse place, just as there may be other bodies quite as good and even better in themselves and in the greater happiness of their inhabitants" (p. 90).

What is most disturbing in all this is that Bruno barely hints at, and certainly does not discuss, the problem that in the view of Thomas Aquinas was so important it alone could invalidate the assumption of a plurality of worlds. Either the many worlds are equal, in which case plurality is superfluous and therefore pointless; or they are unequal, and this leaves God's justice open to doubt (see chap. 1 above, at n. 83). Can man on this Earth still be the apple of God's eye if he is not the most highly favored of all created beings? A possible solution offers itself if we remember that for Bruno the supreme characteristic of the universe is its homogeneity, so that each world is exemplary for, or representative of, every other, differing only as one individual differs from another.[49] But this is hardly satisfactory. The fact remains that Bruno obstinately refuses to confront this crucial issue. On the one hand he undermines the confidence of Christians and does not attempt to restore it, as Kepler later tried to do, by finding some way of reconciling belief in man's special status in the eyes of God with the new cosmology. Nor does he, on the other hand, do anything to aggravate their doubts and their fear of superior beings. Basically Bruno is more interested in his grand overall conception of the universe than in the humans who have to live in it and come to terms with the thought of that boundless expanse crammed with other occupants. Unlike him, his immediate successors, Nicholas Hill and Johannes Kepler, are not content to postulate a plurality of worlds, with its inevitably disorienting effect, and leave it at that. Their reflections lead directly to the issue of man's status, which Bruno had merely touched on in passing and then evaded.

4. Disquiet in the "School of Night": Hill

When the Inquisition ordered the public execution of Bruno in 1600, it could hardly have done itself a greater disservice. By making Bruno suffer a conspicuous and widely reported martyrdom for the new science (though not *only* for the new science), it accelerated the process by which its own philosophical position, where it conflicted with science, became increasingly indefensible. Right through the seventeenth and even the eighteenth century, Bruno's influence, though largely hidden, undermined the foundations of Christian dogma. When it did break the surface, as it were, when people dared to whisper his name ever more audibly, it was usually linked with an assertion of the belief that we were "not alone"; this was the idea with which he was associated. In the following decades we find the linkage, for example, in the writings of Edmund Spenser, Francis Godwin, Wilkins, Huygens, and above all the so-called School of Night, a group of philosophers, "new scientists," and writers whom Henry Percy, ninth duke of Northumberland, gathered around him. Famous figures like Thomas Hariot, Robert Hues, John Donne, Christopher Marlowe, Walter Raleigh, and many of England's other leading Renaissance thinkers were members of this circle. Like Bruno, whom the group adopted, through his works, as their mentor, they held beliefs that were a fusion of Copernican and Atomist ideas. Nicholas Hill, who

[49]Ibid., pp. 265–66.

has been called a mirror of this whole group of English anti-Aristotelians,[50] in his *Philosophia Epicurea, Democritiana, Theophrastica* (1601), attempted a Bruno-like combination of Copernicus and Democritus which forms the basis of his philosophy of the plurality of worlds. So closely does his treatment of the subject of plurality follow Bruno's that there are actual verbal echoes.[51] Nevertheless, Hill merits a certain amount of attention in his own right. Although he has suffered centuries of oblivion and has only been rediscovered in our own day as the "first 'modern' English Atomist," he was very highly regarded by his contemporaries in England and on the Continent, for instance by Burton, Wilkins, and Marin Mersenne. In the foreword to Campanella's *Apologia pro Galileo* (1622) his name appears alongside those of Nicholas of Cusa, Copernicus, Michael Maestlin, Galileo, Kepler, and of course Bruno.[52]

In the context of "extraterrestrial anthropology" it is interesting to note that Nicholas Hill goes further than Bruno and tackles, albeit very hesitantly, that touchstone of religious orthodoxy, the question of man's unique position as the pinnacle of Creation. If, according to the anti-Aristotelian view of the universe, everything is made up of the same substance (*materia*), then it follows, as Hill says in so many words, that by analogy (*secundum analogiam*), the other heavenly bodies must contain all the same things as the Earth (*omnia quae apud nos*), including living creatures and human beings (*homines*, § 278). But what kind of human beings—identical to ourselves, or even superior, or alternatively, perhaps humanoids like the Tierra del Fuegans (whom Darwin was still to consider half-animal), or like Caliban in *The Tempest*? (Ideas of this kind were very much in the air in Elizabethan England.)

For a start, Hill assumes a relationship between the size of the heavenly body and that of its human inhabitants, and accordingly he envisages pygmies on the moon (which Kepler was soon afterward to populate with giants) and giants on the sun (§ 278). To forestall objections from orthodox believers, Hill reassures the reader at the end of

---

[50]Robert Hugh Kargon, *Atomism in England from Hariot to Newton* (Oxford, 1966), p. 14; on the Northumberland circle in general see chaps. 2–4. See also n. 51.

[51]Daniel Massa, "Giordano Bruno's Ideas in Seventeenth-Century England," *Journal of the History of Ideas* 38 (1977), 227–42. On Bruno's influence see also Singer, *Giordano Bruno*, pp. 181–95.

[52]See Massa, "Giordano Bruno's Ideas," pp. 230 (quotation), 229, 239; also Grant McColley, "Nicholas Hill and the *Philosophia Epicurea*," *Annals of Science* 4 (1939), esp. 403–5. In the remainder of this section, references are to the (inconsistently numbered) "propositions" in Hill's *Philosophia Epicurea*. On the errors in Hill's numbering, which cause considerable inconvenience, see the collation by McColley, pp. 390–91.

his book that this assumption of a plurality of worlds, though lacking the authority of the Bible, does not detract from the oneness (*unitas*) of God (§ 508)—much as in the eyes of Bruno, God's perfection revealed itself precisely in the infinity of his Creation. But what of man on Earth: is there any sense in which he can still be seen as unique or of special consequence? As Kepler anxiously asked, are we still "the masters of God's handiwork?" Anticipating the raised eyebrows of his public, Hill spares no pains to offer reassurance: the assumption of an infinite number of mankinds in a boundless universe does not reduce man on our planet to an ant, to a "homunculus"; there is no reason for us to "envy" the inhabitants of other heavenly bodies their "humanity," but every reason to praise God (§ 482). We recall that in Pico della Mirandola, in pre-Copernican times, not only animals but even angels envied terrestrial man his privileged status. Now, against the background of Bruno's radical version of Copernicanism, it is difficult to make a convincing case for man's possessing a status even equal to that of his newly discovered counterpart elsewhere in the universe. For Hill's concluding arguments on this matter are not very impressive. He does not refer to any of the biblical or christological objections, which were by this time not unknown in England even if they were not yet in the forefront of the debate; and he ends his discussion— not so much concluding as breaking off—with a sophistic piece of numerological argument: the special preeminence of human nature is not impaired by the multiplicity of (nonidentical) mankinds, and the infinitely multiplied number of souls in the universe must be accounted an advantage (§ 482). And the status of man on Earth? It is not diminished, we are told (§ 482). Such anodyne statements are all that Nicholas Hill offers to counteract the anxiety caused by disorientation. Kepler was shortly afterward to attempt a more precise—mathematically precise—answer, but not before he had followed all sorts of overscrupulous and inconclusive lines of thought that, as long as the telescope had not been developed, brought him only to the threshold of a solution. This perplexity is typical of the effect that the idea of plurality as a consequence of Copernicanism had on scholars and scientists at that time. And as Kepler pondered on the plurality of humankinds in the universe, the shadow of Giordano Bruno hung over him, even more menacingly than it did over the School of Night.

5. *Reason Speaking True Words in Jest: Kepler*

On the face of it, there could hardly be a greater contrast than that between Bruno and Kepler—the one an impassioned mystic for all his scientific knowledge, the other a scientist wholly committed to empirical observation and mathematical calculation. And yet Kepler saw himself, in a sense, as in the tradition of Bruno as well as of Copernicus.[53] This is surprising; for Kepler, who all his life viewed himself as Copernicus's executor, remained almost compulsively faithful to Copernicus's belief that the universe was finite, with the sun at its center, and explicitly distanced himself from the view of "that unfortunate man, Giordano Bruno" on this point (1:253). It is surprising, however, only at first glance. For although Kepler is not prepared to go beyond Copernicus's one heliocentric system and assume, like Bruno, the existence of countless analogous solar systems, he does share with him the conviction that our solar system is inhabited beyond the Earth. He expresses this belief in a letter written on 30 November 1607 to the physician Johann Georg Brengger of Kaufbeuren. Astonishingly, for one so excessively cautious in matters of religious dogma, he makes no bones about naming the heretic Giordano Bruno as his precursor in this view: "For not only that unfortunate man Bruno, burnt at the stake in Rome, but also my friend Brahe agreed with this opinion that the stars are inhabited" (16:86). The next sentence clearly implies that for Kepler, as for Bruno, the assumption was based on the Copernican model of the universe in which the Earth is one of the planets: "I am the more ready to adopt this view because I believe, with Aristarchus, that the Earth moves just like the planets." Not only with Aristarchus, of course, but also with his avowed disciple, Copernicus,[54] as Brengger duly pointed out to Kepler, bluntly and critically (16:116). Kepler is thus one of the most plausible counterexamples to Lovejoy's influential claim that the idea of plurality in the Renaissance was not a development of Copernicanism. (It is noteworthy that Lovejoy, in *The Great Chain of Being*, refers [p. 121] to Kepler as a believer in plurality, without indicating his reasons for that belief or even mentioning that plenitude plays no part in Kepler's thought.) But precisely how did Kepler arrive at his belief, and what kind of inhabitants (*incolae*) has he in mind?

Kepler's letter to Brengger was written in response to Brengger's

[53]Martin Hasdale, writing to Galileo on 15 April 1610, mentions a remark to this effect made by Kepler in connection with the *Sidereus Nuncius* (Galileo Galilei, *Opere*, Edizione nazionale 10:315). Kepler of course sees Galileo as standing in the same tradition (ibid.). The page numbers given below for references to Kepler and his correspondents refer to the *Gesammelte Werke*.

[54]Kepler refers to Copernicus as a disciple of Aristarchus (2:224). In one passage in his manuscript of *De Revolutionibus*, Copernicus himself confirms this, but he deleted the passage; see Copernicus, *Complete Works*, vol. 2 (Warsaw and Cracow, 1978), p. 25. A belief in the plurality of worlds was sometimes attributed to Aristarchus.

criticism of *De Stella Nova in Pede Serpentarii* (1606). It was in this work, about the supernova that had appeared in 1604, that Kepler had first set out his reasons for believing in a plurality of worlds. These reasons were not metaphysical but purely scientific. In the letter, his view that the stars are "similar" to the Earth is merely put in the form of an opinion (*mihi videntur*); but the treatise itself shows how this view follows as a consequence from Copernicus's cosmology, which "admitted the Earth to citizenship of the heavens" (*jus civitatis in coelo dedit*, 1:246). Conversely, and leaving all metaphor aside, the stars, including the new star in the constellation of the Serpent Bearer, consist of the same material, the same elements or substances, as the planet Earth. They are therefore essentially similar to Earth, although Kepler, like Bruno, makes a distinction between fiery and watery worlds, according to the element that predominates (1:247). And as this universal material is seen on Earth to possess a "vital faculty," a power to create and form (*architectonica naturalis facultas*), which has caused living plants and animals to come into being, one can see no reason why the same should not be true of other heavenly bodies: "Therefore, whatever faculty this Earth, being one of those bodies [*globorum unus*], possesses, it is reasonable to suppose that the others, too, are provided with such faculties" (1:267–68).

This idea of life having originated in some spontaneous, mechanical fashion is a bold and theologically highly contestable one. On this occasion Kepler speaks only of frogs, fishes, flies, and beetles, and not of the origin of man. But having rejected the biblical account of the creation of the animals, it is hard to see any logical grounds for calling a halt there: the next step must be for science to displace mythology in relation to the creation of man himself. Kepler had already ventured a little closer to this danger zone in his *Astronomiae Pars Optica* of 1604. There, starting out from Copernicus's refutation of Aristotle's qualitative division of the universe into the Earth on the one hand and all other celestial bodies on the other, he discussed its implications, not for the stars this time, but for the moon. (The implications were of course essentially the same for both.) The moon, he said, was constituted similarly to our Earth; it had mountains, oceans, continents. Plutarch, he recalls, had already held this view, and Maestlin, Kepler's teacher at Tübingen, had proved it by his observations and thus experimentally confirmed Copernicus's theory (2:218, 224). But in between these two passages about the similarity between moon and Earth, Kepler lets slip a hint that there might be human beings, as well as animals, on the moon. Once again it was Plutarch's *De facie in orbe lunae* that gave the cue. Physical calculations have shown, says Kepler, that the mountains on the moon are incomparably higher than those on Earth, and one must assume, with Plutarch, that the proportions of "men [*homines*] and animals" bear some relation to the geophysical proportions of their planet. Therefore the creatures living on the moon must be considerably larger than those on Earth, which would incidentally also make them better able to tolerate the greater fluctuations in temperature—*if*, he adds, there are any such beings there (*siquidem aliqui ibi sunt*, 2:220).

He shrugs this idea off as a joke (*iocemur*) and briskly passes on: "But to return to our topic." But his imagination, fired by Copernicus, cannot quite relinquish the thought of human beings on other heavenly bodies. In studying the universe, Kepler, unlike Bruno, is always seeking new knowledge of man. Thus when, two years later, in *De Stella Nova* he speaks of how greatly Copernicus's theory has increased the assumed size of the universe, he moves straight on to consider man's position, although the treatise is supposed to be on a purely astronomical subject. Copernicus's measurements, he says, have placed the fixed stars "incredibly" far out in space, making man and *his* star far smaller than they already were. For Robert Burton, author of the *Anatomy of Melancholy* (1621), this larger universe was to be one more reason for him to accept, with some misgivings, the notion of a "plurality of worlds" in distant solar systems. His misgivings contradict the widely held view that the enlargement of the universe caused no anxiety to sixteenth- and seventeenth-century thinkers;[55] and that view is belied by Kepler's thinking too. Kepler *does* express anxiety, both here and elsewhere, particularly in the *Dissertatio cum Nuncio Sidereo* (1610). But in *De Stella Nova* the hint of anxiety is quickly followed by a somewhat labored argument for the Earth's preeminence: its smallness makes it all the more noble, all the more divine. The largest objects, the fixed stars, do not move, and this is an imperfection in them. The tiny Earth does move; and that is not its only distinction. Here too Kepler refers to that formative power which, of itself (*de se ipsa*), has brought forth and is still bringing forth all the abundance of plants and animals. But on this occasion the creative power is presented as belonging exclusively to the Earth. For we are told that it confers on the Earth such nobility (*nobilitas*) that the Earth may hold all other solid bodies in the universe in contempt. And finally, the human beings on Earth—"those particles of dust which

[55]For instance Korninger, *Naturauffassung*, chap. 1, esp. p. 22; R. G. Collingwood, *The Idea of Nature* (London, 1945), pp. 96–97; Nicolson, "Discovery of Space." Burton, *Anatomy of Melancholy* 2:54–55 (this passage first appeared in the 5th ed., 1638).

are called men"—"are lords of all that solid mass" (*domini quodammodo sunt totius molis*). But as for their origin, Kepler somewhat hastily attributes it not to a natural "faculty" but to a "Creator," who made mankind in his own image to be the highest and most noble part of all Creation (1:237).

Just a few pages farther on in the same treatise, Kepler plainly sets forth the logical deduction that what is true of the Earth must apply equally to other bodies, at least as far as the origin of plants and animals brought into being by the Earth's "vitalis facultas" is concerned. Yet here, astonishingly, he appears determined to overlook that point. It is equally surprising that whereas the other passage speaks only of animals, this one includes man. But for man, Kepler assumes a Creator, not spontaneous generation, and thus he preserves man's uniqueness in the universe, which is, of course, the point of his "ennoblement" of the Earth.

At the end of *De Stella Nova*, however, Kepler abandons his flight into the safe theology of one, unique mankind: his anxiety about a possible plurality of mankinds rises to the surface again. In the closing chapter he tackles the question of what the appearance of the new star signifies—"for what purpose" (*quem ad finem*) God has sent it (1:339). In considering this, he assumes as a matter of course that other heavenly bodies have intelligent inhabitants; for the question is, in the first instance, whether this new star is intended as a message from God to us or to those other beings. Changing his tune, Kepler now considers the Earth to be far too small and paltry (*nimis exilem*) for the significance of this extraordinary astronomical phenomenon to be considered solely in relation to it. The universe is vast, and both the philosophers of antiquity and Tycho Brahe held the view, by no means absurd, that the other heavenly bodies had "inhabitants—not men, but other creatures" (*incolas, non equidem homines, at creaturas alias*). But we must assume these creatures at least resemble humans; for Kepler insists they too are subject to the "providence of the supreme Guardian," and he implies that they have something like human reason when he affirms the possibility that this astronomical phenomenon, so difficult to account for, may be meant "equally" (*aeque*) for them and for us. Indeed, he goes so far as to conjecture that its significance "may be more amenable to their understanding than to ours" (*signa . . . magis forsan ipsorum appropriata captui, quam nostro*, 1:339). In other words, the inhabitants of other planets may not merely approach the level of human beings; they may possibly be superior. Does this mean that we are not, after all, lords of the universe? This thought was to give Kepler no peace once he learned that Galileo's telescopic observations had confirmed conclusively that the Earth and other planets were physically alike. That was still some years ahead; but already in *De Stella Nova* he was unmistakably asking that question.

Kepler's answer is fascinating because of its desperate attempt to reconcile new science and traditional theology. For in the final chapter, he does not simply fall back on the theological argument that man was created in God's image, which in the sixteenth chapter already referred to, he had considered adequate to explain the unique preeminence of man and the Earth in Creation as a whole. Here—in the philosophical section of his treatise, which he says cost him more effort, trouble, and "anxiety" than all the rest (1:151)—it is purely scientific and not theological arguments that lead him, after much deliberation, to the conclusion that after all, the nova can only be a sign intended for our Earth. His reasons for this view are based wholly on physics. The appearance of the nova in October 1604 coincided, in time and space, with the unusual conjunction of Saturn, Jupiter, and Mars. This grouping, however, could be seen as such and recognized as a remarkable event only by observers on Earth. Therefore the appearance of the nova, too, could have significance only for us on Earth (1:339). (It could not be intended for *other* humankinds, as Dick incorrectly states in his brief account—the only one—of Kepler's thought in the context of early ideas on plurality.)[56]

So the anthropocentric view of the universe is salvaged once again, but we should note the manner of the salvage. At the end of the paragraph, Kepler reminds the reader, almost with a note of triumph, of another defense of this view made earlier on in the treatise: we may be a mere speck of dust in the vastness of the universe, he had said, but we are created in God's image. There he invoked a fundamental tenet of religious belief. This time he argues not from a principle but from specific astronomical facts, which is a very different matter. These arguments may suggest that the astronomical event in question is a sign intended for us on Earth, but they by no means invalidate the idea of rational or even intellectually superior beings on other planets: it is simply that this particular nova is not intended for them. On the subject of what these beings may be like, and whether or not they too are created in God's image, Kepler maintains a total silence, just as he did in the letter to Brengger about the inhabitants (*incolae*) of other planets, which was the starting point of this discussion.

Such scattered observations, with their muddled combination of theology and science, show the uncertain groping of Kepler's thoughts

[56]Dick, *Plurality of Worlds*, p. 73.

in these early years; they create more problems than they solve. Perhaps what is most noteworthy about them is Kepler's boldness in pursuing such ideas at all. We have to remember that all these texts predate Galileo's observations, which were to give Copernicus's cosmology so much more credibility and show once and for all the error of the Aristotelian system, which denied the possibility of plurality.

The same applies to Kepler's *Somnium*, a fictional account of a journey to the moon. This is probably his most important contribution to the subject of the plurality of worlds, and it is more truly a *literary* contribution than Bruno's didactic poem *De Immenso*. Marjorie Nicolson, who deserves our gratitude for rescuing the *Somnium* from oblivion, places the work in a false historical context when she calls it a representation of the moon-world "as the telescope had shown it."[57] This is precisely what it is not: it predates the telescope, which makes it all the more remarkable.

Kepler's *Dream* was not published until 1634, but it was written in 1609 and circulated in manuscript form. It is the first literary work that, inspired by Copernican science, treats of aliens in outer space. This is another difference between Kepler's slim volume and Bruno's bulky didactic poem *De Immenso*. Bruno had studiously avoided any discussion of the nature of the inhabitants of the heavenly bodies and merely told of hymnic praise issuing forth from infinitely many "worlds," which themselves remained an abstract and vague philosophical construct. The *Somnium*, in contrast, directly addresses the nature of the aliens, and consequently of man too. In this it also differs totally from mere social satire or social allegory in the manner of Lucian's *Vera historia* (True History), which simply transposes all-too-human, all-too-topical issues to the moon. This is not Kepler's way. Certainly he had thought of writing a sociopolitical allegory or satire (note 82),[58] and in the (lost) first version he had partly carried out this plan (note 213). But he abandoned it in the final text, and also in the notes he added later and in the geographic appendix (see notes 82, 83). His reasons are explained in a letter to

[57]*Voyages to the Moon* (New York, 1960), p. 46; idem, *Science and Imagination*, p. 27.

[58]Strangely enough Winter, in *Compendium Utopiarum*, p. 66, considers that the *Somnium* does in fact incorporate social criticism. The *Somnium* consists of the text of 1609 and Kepler's notes, which were added c. 1622–30. The geographic appendix, which I discuss later, has its own set of notes. This work has not yet appeared in the *Gesammelte Werke* but can be found in Kepler, *Opera Omnia*, ed. C. Frisch, vol. 8(1) (Frankfurt, 1870). An English translation by Edward Rosen, *Kepler's "Somnium,"* with outstandingly good critical apparatus, was published by University of Wisconsin Press (Madison, 1967). For the quotations or references that follow, where two page references are given, the first is to *Opera Omnia*, the second to Rosen.

his friend Matthias Bernegger written on 4 December 1623. Campanella wrote a work describing a "Sun-state," he says, so why should he himself not write about a "Moon-state"? He could describe the "cyclopic" manners of the age but transpose the scene to the moon as a precautionary measure. But, he continues, such subterfuges did not help either Thomas More or Erasmus, and so he would rather remain wholly in the "delightful fields of philosophy" (18:143). His "philosophy," however, the product of both literary imagination and astronomical calculation, is not primarily cosmologic, like Bruno's, but anthropological. It revolves around the question What is man, viewed in the light of the plurality of worlds? And what is perhaps most surprising is that Kepler, who at the time of his studies in Tübingen had still intended to become a Lutheran clergyman, passes over in silence the specifically dogmatic issues raised by the subject, such as those pointed out by Melanchthon. This was bound to heighten the tension between Christian dogma and the scientific doctrine of plurality, and it may help to explain why there has been such a resurgence of interest in Kepler's *Dream* in recent years. Today it is universally recognized as a major pioneering work of science fiction and as a highly idiosyncratic classic of world literature. It has even been suggested that it should be required reading in schools.[59]

On the title page of the first edition, however, the typography chosen ensures that literary imagination and even philosophy take second place to the scientific basis of the work. Strikingly large lettering pronounces the book to be an "astronomical work" by the Imperial Mathematician, while small print modestly characterizes it as a mere dream—"Somnium." And indeed this is in quantitative terms a fair representation of the content; for strictly speaking, only the opening section and the close, which contains an account of the flora, fauna, and quasi-human beings on the moon, can be described as literature. The main part of the book is an astronomical treatise, admittedly most

[59]This suggestion was made by John Lear, "The Forgotten Moon Voyage of 1609," *Saturday Review* 46:18 (1963), 44. See also Nicolson, *Voyages to the Moon*, p. 41: "fons et origo of the new genre"; Arthur Koestler, *The Watershed: A Biography of Johannes Kepler* (Garden City, N.Y., 1960), p. 247; Siegfried Mandel, "From the Mummelsee to the Moon: Refractions of Science in Seventeenth-Century Literature," *Comparative Literature Studies* 9 (1972), 410; Gale E. Christianson, "Kepler's *Somnium*: Science Fiction and the Renaissance Scientist," *Science Fiction Studies* 3 (1976), 81, 89; Lewis Mumford, *The Myth of the Machine*, vol. 2 (New York, 1970), p. 45; Hélène Tuzet, *Le cosmos et l'imagination* (Paris, 1965), p. 225: "Cette oeuvre est . . . si moderne, que le langage des années 1960 vient sans cesse au bout de la plume pour traduire son latin savoureux." The most significant analysis of the *Somnium*, one that does not, however, make any reference at all to the question of the plurality of worlds, is that by John Lear, *Kepler's "Dream"* (Berkeley and Los Angeles, 1965).

imaginatively presented—a "lunar geography," as Kepler himself sometimes calls it. In fact the *Somnium* grew out of an astronomical dissertation, now lost, that Kepler had written as a student in 1593 but, because of its Copernican basis, was not allowed to defend in public in staunchly Lutheran Tübingen. Sixteen years later he incorporated it, largely unaltered, in the "novel." It is no coincidence that Kepler might have found the central idea for the book—an ingeniously simple idea—in Giordano Bruno, whose writings he knew well. At one point in the dialogues *De l'infinito universo e mondi*, Bruno almost roguishly challenges the Aristotelian and Christian doctrine of the centrality of the Earth with the objection: "Without doubt those who inhabit the moon believe themselves to be at the centre ⟨of a great horizon⟩ that encircleth this earth, the sun and the other stars."[60] Kepler elaborates this notion quite seriously, as a means of advancing the cause of Copernicanism. Many of the more than two hundred notes, which he added between about 1622 and 1630 to strengthen the scientific basis of the *Dream*, make it perfectly clear that this was his aim. Just as the inhabitants of Earth, deceived by appearances, believe that the Earth stands still and is circled by the sun, moon, and stars, so the moon dwellers believe that the whole universe revolves around a stationary moon. Both these beliefs are shown to be invalidated by the Copernican theory of the Earth's movement on its own axis and through the zodiac, and Kepler's triumphant demonstration of this is raised to a high degree of elegance by his vivid depiction of the *reality* of these motions, which was denied by orthodox thinkers such as Osiander.[61]

Despite this scientific aim, however, Kepler is ultimately concerned with man—man on Earth and man on the moon. As a student at Tübingen he had learned from the Copernican Michael Maestlin that the Earth and the moon were not made up of essentially different substances, and in his own earlier Copernican works, starting with the first one, *Mysterium Cosmographicum* (1596), he had never tired of reiterating this point. This being so, one had logically to admit the possibility that the moon was inhabited by beings similar to man, that terrestrial man was not "alone," and that it might not be true that he was the apple of the eye of a Creator who had made the whole universe

[60]*On the Infinite Universe*, in Singer, *Giordano Bruno*, p. 280.
[61]See for instance Kepler's notes 4, 104, 110, 128, 146. On Osiander see this chap., sec. 1. In 1609, the same year in which he wrote the *Somnium*, Kepler published his *Astronomia Nova*, in which he called Osiander to account for presenting Copernicus's theory about the real nature of the universe as a mere mathematical hypothesis with no claim to represent reality (*Gesammelte Werke* 3:6).

solely for his use and pleasure. This recognition, of fundamental significance, is the message that finally emerges from the *Somnium*, and it is only in some of the later notes that it is cautiously hedged about with an "if" or a "perhaps" (notes 115, 146). At Kepler's time—difficult as it may be for us fully to appreciate today—this was an astonishingly bold position to adopt. Not only did it contravene orthodox theology; it was also more daring than that other strain of anthropocentrism which was already foreshadowing the Enlightenment: the assertion that the proper study of mankind was man, an assertion that failed to consider that man might best be understood, either fearfully or joyously, in relation to a nonterrestrial counterpart, whose existence the sixteenth-century Scientific Revolution had shown to be at least conceivable. If one did take seriously the possibility of an extraterrestrial counterpart, it might well be that questions of Earth-bound anthropology and also of biblical theology would be overshadowed by more urgent and more interesting ones. This thought may have had some part in Kepler's decision to leave aside specifically theological questions as well as social comment. Instead, scientifically based speculations about extraterrestrial life form the climax of the *Somnium*. Such speculation was clearly a matter of deep concern for him; otherwise he would hardly have devoted so much intensive study to the question throughout his life. (He finished correcting the proofs of the *Somnium* only shortly before his death.)

The introduction, admittedly, reads like the conventional opening to a work of fiction. One night, we are told, the narrator, after observing the moon and stars, retires to bed and dreams that he is reading a book he has brought home from the bookfair. The contents of this book form the main part of the *Somnium*. The first few pages tell us, with various more or less transparent autobiographical allusions and allegorical motifs, of a young man from Iceland named Duracotus. A later note (the fourth) reveals that he is the personification of Science. A storm at sea drives him to the shore of Denmark, and here he is introduced by none other than Kepler's teacher, Tycho Brahe, to the "most divine science," astronomy, which smooths his path to "higher things." These evidently include the insights he gains with the help of his mother after his return to Iceland. His mother, Fiolxhilde, is what is popularly known as a witch. She has wise spirits (*sapientissimi spiritus*) at her command, and she conjures up one of them by pronouncing a magic formula consisting of twenty-one letters. This formula is later revealed in Kepler's thirty-eighth note: it is "Astronomia Copernicana." The spirit summoned by this formula comes from Levania, the moon, which Plutarch (whom Kepler mentions as a precursor) had made the home of the souls of the dead and of those as yet unborn. Kepler calls this spirit a "Daemon," a word he derives from *daiein*, "to know" (note 51), thus implying this spirit has special knowledge of the moon. And indeed this expert proves most reliable—more so than Kepler's knowledge of Greek. The spirit's description of the physical and climatic conditions of the moon, which makes up most of the rest of this short novel, is a detailed and accurate astrophysical account that tallies point for point with the then current state of knowledge in Copernican astronomy.

Thus the "story within the story" is in effect a scientific lecture given by the Daemon. It serves to remind us that Kepler was writing his novel primarily for scientists, to convince them of the truth of the Copernican theory. As for the more imaginative passages toward the end of this inner story, they benefit by being set in this rigorously factual context. Extrapolating from the Copernican theory, these passages speak of the flora, fauna, and human inhabitants of the moon, and although in themselves they must be classed as imaginative writing, the fact that they follow upon the strictly scientific information given by the Daemon enhances their credibility.

Before the Daemon turns to the subject of the moon's inhabitants, he discusses at some length the problems of space travel, the geography of the moon, and the celestial objects that can be seen from the moon. The moon itself is divided into two hemispheres, one of which is always turned toward the Earth (Volva) and one away from it; they are called Subvolva and Privolva. For Subvolva, the Earth is the moon. A day and night on the moon last as long as a month on Earth. The sun appears to revolve around the moon twelve times in a year; the sphere with the fixed stars, thirteen times a year; or to be more precise, they seem to revolve 99 and 107 times respectively in an eight-year cycle, or 235 and 254 times respectively in a nineteen-year cycle. This latter period contains twenty summers and winters in the moon's polar regions and forty summers and winters at its equator, with an intermediate number in the regions in between. This sober yet continually startling account of the moon's geography is greatly detailed; there is also a mathematically precise description of the fluctuations in the motions of celestial bodies as perceived from the moon. But we need not consider these passages in detail here.

Of greater relevance to the subject of the plurality of humankinds is the fact that anyone reading this description of a highly outlandish world suddenly realizes that what is being presented with such breathtaking matter-of-factness is a world inhabited by intelligent beings for whom it is their familiar environment. Subvolva has its population of

Subvolvans, Privolva its Privolvans, and between the two are the border dwellers, "the frontiersmen who inhabit the divisor circle" (pp. 35/21).[62] It is these inhabitants who bear the fierce heat of the day and the frosty nights and who perceive Mars and Venus as being now larger, now smaller, the sun as circling sometimes faster, sometimes more slowly. The Subvolvans observe the great Copernican spectacle of the Earth's turning; they use the Earth's position to calculate the longitude of their own position on the moon, and the waxing and waning of Volva tells them the time of day. Kepler mentions this almost in passing, thereby tacitly conveying the information that these moon-creatures possess a kind of intelligence remarkably similar to that of humans. The cosmic clock is read differently by each group of moon dwellers (according to the latitude from which they observe the Earth), but in each case correctly: "For in general, for those who live between Volva and the poles on the mid-Volvan circle, new Volva is the sign of noon; the first quarter, of evening; full Volva, of midnight; the last quarter, of returning sunlight. For those who have Volva as well as the poles located on the horizon and who live at the intersection of the equator with the divisor, morning or evening occurs at new Volva and full Volva, noon or midnight at the quarters" (pp. 36/23). The moon dwellers can make exact subdivisions of the time in between these fixed points by noting the regular reappearance of the spots that are the Earth's continents; a complete revolution is counted as an hour and is their one uniform measure of time (pp. 36/23). This method of measuring time is fairly elementary, even though it does show a specifically human sort of intelligence at work; but there are apparently certain observers who are more intelligent and "more careful," who notice that the Earth, Volva, appears sometimes larger and sometimes smaller and who understand the reasons for this (pp. 37/25). So the moon dwellers possess not merely marginal intelligence but a knowledge of the "most divine science," astronomy.

In passages like these, Kepler is, however, more concerned with the astronomical phenomena than with those who observe them, and so the lunar beings are present only as more or less disembodied minds, able to perceive these phenomena but lacking any physical attributes that would enable the reader to visualize them. Only toward the close of the novel does Kepler turn his attention to the practical living conditions and habits of the moon dwellers, and now at last we move from scientific instruction to pure science fiction. Every detail of the moon dwellers' way of life is determined by the difference

[62]See footnote 58 for method of documentation.

between the two hemispheres. The Subvolvans enjoy a relatively temperate climate because of the Earth's influence and the slightly greater distance from the sun during their day, whereas the Privolvans are exposed to untempered extremes of heat and cold. A further difference is caused by the variable distribution of water on the moon. Like many of his contemporaries, including Galileo, Kepler believes there are oceans on Levania. When the Earth and the sun are both on the same side of the moon, their combined gravitational force causes all the water on the moon to flow to the hemisphere facing the Earth, so there is a flood that leaves only the mountain peaks uncovered. But when the sun and the Earth are on opposite sides of the moon (for about the other half of the month, in *our* time) some of the water flows back to the Privolvans' hemisphere. A special feature of these regular inundations is that Kepler believes the moon's surface is porous, riddled with caves and grottoes into which the water can flow. These, incidentally, also provide the inhabitants with shelter from the heat and cold (pp. 38/27).

What are the inhabitants like? Immediately after describing these dramatic physical conditions, Kepler reveals that whatever is born (*nascitur*) on the moon is enormously big (*monstrosae magnitudinis*).[63] Growth is rapid; life, correspondingly short. The reasons for this rather bald assertion (not expressed as a mere supposition) Kepler adds in the 1620s, in notes 212 and 213. The main substance of these notes is, once again, the strange theory that the sizes of all things are in proportion to each other: the moon is topographically perfectly similar to the Earth, and because of its higher mountains and deeper valleys, we expect to find larger creatures, with correspondingly large functions such as hunger, thirst, waking and sleep, work and rest. For their short life-span, in contrast, Kepler offers the rather original explanation that a day on the moon is about thirty times as long as a day on Earth. This reasoning set out in the notes is indeed curious. But in the novel itself (and the Daemon's exposition now becomes progressively more imaginative and even eventful, reading more like a novel), the narrator does not linger over such environmental causes but moves straight on to the moon dwellers' way of life.

We need, by the way, have no qualms about referring to them as "men" on the moon. The view, which we find even in the standard work by Marjorie Nicolson, that Kepler's lunar world contains no human beings, no civilization, no reason, but only snakelike creatures

---

[63]*Opera Omnia* 8(1):38; Rosen, *Kepler's "Somnium,"* p. 27. All further references to the 1609 text of the *Somnium* are to *Opera Omnia* 8(1):38–39 and Rosen, pp. 27–28.

of grotesque appearance, is an error that has seriously misled students of the history of thought, and frequent repetition of the idea in critical literature has not made it any more plausible.[64] It appears improbable in the light of those scattered observations (already quoted) that Kepler made before he wrote the *Somnium*, even more so in the light of his later remarks in the geographic appendix and in the notes to the *Somnium*, where (as we have seen) he refers to "work" (*labor*) done by these beings. In the *Somnium* itself, too, we have seen unmistakable references, in the section dealing with matters of physics, to a form of Levanian life that is intelligent, aware of astronomy, and thus similar to man. In the biologic section, these references become noticeably more specific.

These people on the moon, then, have habits of life adjusted to their environment. The Privolvans are nomads; the extreme climatic conditions rule out a fixed abode (*habitatio stata*). During the daytime (equivalent to fourteen days and nights on Earth), they wander in hordes over the whole of their hemisphere in search of cooler regions. Some travel on foot (their legs, Kepler states, are far longer than those of our camels); some, with wings (artificial ones?—*pennis*); others, again, let the current of the departing waters carry them along in boats (*navibus*), which they would hardly do if they were mere animals; or, "if a delay of several days is necessary," they crawl into the caves of the moon's surface to escape from the sun's rays. Kepler, the first writer of science fiction to identify himself with the Scientific Revolution of the Renaissance, lets his imagination work by logical extrapolation while at the same time giving due weight to practical considerations. Startling new notions are presented in a matter-of-fact style. "Most of them are divers," we are told with disarming bluntness. Their breathing is by nature (*naturaliter*) slow, so they can stay for a long time under water. Where some depth of water has collected, they cool themselves down; on the surface, the water evaporates, except where the moon people have made channels to direct it into the caves. In the evenings they come up to the surface to seek food.

We would expect the life of the Subvolvans to be more interesting and more highly developed than this very limited existence led by the nomadic Privolvans. Kepler does hint that this is so, even implying that there is culture and civilization: just as Privolva is comparable to our deserts, woods, or open country, he says, so Subvolva compares to our cantons, towns, and gardens. In the "Geographical . . . Appen-

[64]Nicolson, *Voyages to the Moon*, pp. 46–47; *ELH: A Journal of English Literary History* 7 (1940), 99; the same view as recently as 1976 in Christianson, "Kepler's *Somnium*," p. 85.

dix," written more than a decade after 1609, he even speaks of the fortifications of the moon's cities and of the engineering techniques the inhabitants used. But in 1609, in the novel itself, Kepler unfortunately makes only that one reference (cantons, towns, gardens) to the human life style of the Subvolvans. The dreamer has something to report about the animals on the moon: they have a spongelike, porous skin, are mostly snakelike, shed their skins in the evening, and so on. But hardly has he turned his attention to the Subvolvans themselves—or at least to their climate (constant cloud cover, frequent showers)—when he is awakened from his sleep by the patter of rain. This deprives him, and us, of any further instruction by the Daemon, who up to now has been such a mine of information.

This abrupt conclusion was no doubt prudent. Up to this point in the exposition of his ideas, Kepler might have countered the likely theological objections (Are all these beings descended from Adam and Eve? Is Christ their Savior too?) with the excuse that the Privolvans do not display fully human characteristics. But the human traits of the Subvolvans—their towns and gardens, the administrative division of their territory into cantons, their knowledge of astronomy—could not be similarly explained away. On the contrary, between the lines preceding the interruption—and also, surely, by the very fact the account is left open ended—Kepler raises the question of man's bio-logic and theological uniqueness in the cosmos: Are the Subvolvans part of God's Creation? Or did they originate spontaneously out of primeval matter as, according to Kepler's early writings, other crea-tures on other stars did? Are they gifted with an intelligence equal to ours? Are they perhaps even more intelligent than we are (a possibility briefly hinted at in relation to extraterrestrials in *De Stella Nova*)? Are they in any case subject to Divine Providence (as we have seen that Kepler insisted at the end of *De Stella Nova*)? In one of the later notes (on the primitive Privolvans, at that), Kepler almost falls into the trap of expressing an opinion on the Privolvans' origins. The Privolvans, we are told, are able to survive the long drought because they are equipped with the ability to travel sufficiently fast to keep up with the receding water. Note 214 explains this by saying that, after all, God (*Deus*) has given the animals on our planet the qualities necessary for them to withstand hostile environments. So are we to assume that it was also God who gave the Privolvans their qualities—and what about the more highly developed Subvolvans? With studied nonchalance Kepler confines himself to suggesting these questions without answer-ing them.

He has, however, far more to say about the Subvolvans—at any rate, about their conditions and way of life—in the "Geographical . . . Appendix" to the *Somnium*. This he wrote in the 1620s, and he probably combined it with the text of the *Somnium* only in 1630,[65] when he already had the proofs of the novel before him and perhaps himself felt it left too much unsaid. This would be two decades after Galileo's observations had shown in 1609/10 that the moon, with its mountains, valleys, and seas, could be *seen* beyond any doubt to be similar to the Earth. After this, literary writers treating the scientifically inspired theme of the plurality of worlds were able to give far freer rein to their imagination. Kepler himself, in the preface to his *Dioptrice* (1611), is inspired to metaphor: it is, he says, as though Galileo had opened a new gate to the heavens, through which one could now see with one's own eyes what was previously hidden (4:342). Characteristically, though, he adds in the same breath that the new knowledge challenges reason (*ratio*) to answer such questions as For what purpose or for whose benefit (*cui bono*) there are mountains, valleys, and seas on the moon? and "may not some creature less noble than man be imagined such as might inhabit those tracts?"[66] The rational thinker, then, would suppose creatures of some sort in outer space. But would not such a conclusion immediately call forth Christian counterarguments, perhaps disguised as purely scientific studies? In fact, Galileo's telescopic observation of the moon had so increased the plausibility of the idea of plurality that such anxiety would in any case be rife in the years after 1609/10. It is therefore worth looking briefly at Galileo's significance in this context, before going on to discuss Kepler's post-Galilean reflections on the plurality theme.

6. *The Telescope Frees the Imagination: Galileo*

In the winter of 1609/10, Galileo Galilei, professor of mathematics at the University of Padua, was one of the first to train the newly invented telescope on the night sky, and what he saw produced a change in perception that was truly epoch making. For his findings, unlike those of Copernicus, were not only of interest to other astronomers but—as he emphasized on the title page of his account of his observations, *Sidereus Nuncius* (1610)—were of concern to "philoso-

---

[65]Rosen, *Kepler's "Somnium,"* pp. xxi, 149, 233.

[66]*The Sidereal Messenger of Galileo Galilei, and a Part of the Preface to Kepler's Dioptrics*, ed. and trans. Edward Stafford Carlos (London, 1880), p. 81. *Gesammelte Werke* 4:342: "Quaerimus . . . an non ignobilior aliqua Creatura, quam homo, statui possit, quae tractus illos inhabitet."

phers" and indeed, he claimed, to "everyone." Through his observations made by means of the *perspicillum*, which now rapidly became a highly fashionable article, Galileo provided empirical confirmation of Copernicus's refutation of Christian-Aristotelian geocentrism, and his observations thus suggested a question that really did concern everyone, or at least the more educated: the question of whether our Earth alone supported life of any kind and, more particularly, whether man himself was alone and unique in Creation. A scholar studying Copernicus's theory might recognize that this question was implicit in it; but as a result of the observations made by Galileo, an avowed follower of Copernicus, the question acquired vivid reality, was within the understanding of everyone, and became intensely topical. For now anyone who had eyes with which to look through the telescope could *see* that the Earth was not a unique celestial body. Copernicus's theory had made the ancient speculation about a plurality of worlds and mankinds seem more plausible, but Galileo's empirical research made this idea, the "new heresy," far more credible still. The plurality of worlds was now a more realistic proposition than it had ever been, more realistic than the appearance of *De Revolutionibus* in 1543 had made it. It is at this point that the old idea becomes a significant phenomenon in the history of thought, though not yet the mass phenomenon familiar to us in the latter half of the twentieth century.

In addition, the telescope enlarged and enriched the skies far beyond the firmament visible to the naked eye. The centuries-old belief in a closed system of spheres, and even Copernicus's finite universe bounded by the sphere of the fixed stars, were replaced by an unimaginably vast expanse of space that contained myriad suns but had no midpoint or center of gravity and was perhaps of infinite extent. (Galileo avoided committing himself as to whether the universe was finite or infinite.)[67] But whether the universe was in fact infinite or only of uncertain size, the Earth seemed like a mere grain of sand tossed about somewhere within it; thus, to the layman, Bruno's views too seemed confirmed.

This enlarged view of the universe intensified doubts about the centrality and uniqueness of man. It could hardly be maintained any

[67]*Opere*, Edizione nazionale 6:518, 529; 18:293–94 (letter to Fortunio Liceti, January 1641). On the astronomical developments see Koyré, *From the Closed World*, pp. 95–99; on Copernicus, pp. 28–34. Pietro Redondi's sensational discovery that the church brought Galileo to trial for astronomical heresy in order to avoid charging him with the more serious heresy of his atomistic view of matter does not, of course, invalidate the point made here—which was, naturally, not lost on Galileo's contemporaries. See Redondi, *Galileo Heretic* (Oxford, 1987).

longer that God had created those stars, hitherto invisible, for the "benefit" or "use" of man, who would one day invent the telescope![68] So were they created for the delight of a plurality, or even an infinity, of other inhabited worlds? If so, the ordinary Christian would have difficulty in believing himself to be the pinnacle and the final cause of Creation. With his direct and indirect attacks on anthropocentrism, Galileo ranks with Darwin and Freud as a destroyer of man's self-confidence; Copernicus himself made a less dramatic impact, and it is significant that his book was not banned by the Catholic church until 1616, when the Inquisition was already paying close attention to Galileo. As Bertolt Brecht says, with some exaggeration, in his play about Galileo, astronomy in his day "reached the market-places" (sc. 14); and as laymen became aware of the new astronomical insights, they must also have thought of that fascinating but also disquieting subject, the plurality of worlds.

Scarcely any other idea has so fired the imagination of writers, not only astronomers like Kepler and scientifically oriented philosophers like Bruno, but more purely literary writers too. We see this happening in England and France in the seventeenth century and in the German-speaking countries in the eighteenth. Galileo himself, soon widely hailed as the "new Columbus,"[69] was sufficiently worldly wise to make a point of evading the delicate question of whether the planets were inhabited. Literary authors, however, drawn to the theme by philosophical or scientific interests, saw it as their business to explore imaginatively the features of these new planet-worlds and above all the nature of their human or quasi-human inhabitants. The subject offered undreamed-of scope for new adventures of the imagination and at the same time a sort of legitimization of that imagination. No less a "philosophical poet" than Tommaso Campanella, writing to Galileo soon after the publication of the *Sidereus Nuncius*, pointed out the attractions of this exploration of hitherto untrodden territory.[70] In the wake of the "new Columbus," writers of fiction girded themselves for the encounter with rational beings on other planets—in other words, extraterrestrial "humans." But *would* they be humans?

Analogical thinking, the scientifically respectable approach we have seen Bruno and Kepler use, pointed the way for these explorations. Galileo's discoveries, resulting from his use of the telescope, made the drawing of inferences by analogy an even more irresistible

---

[68] *Opere* 7:394–95. See also Christian Huýgens, *Kosmotheoros* (The Hague, 1698), p. 8.
[69] Marjorie Nicolson, *Science and Imagination*, pp. 18–20.
[70] Galileo, *Opere* 11:21–26 (letter to Galileo dated 13 January 1611). See Nicolson, *Science and Imagination*, pp. 24–25.

temptation than had his predecessors' calculations—which is why Galileo's colleague, the Aristotelian Cesare Cremonini da Cento, refused even to look through the new instrument. Galileo had pointed it toward the moon, Jupiter, and the Milky Way; what it showed him lent itself only too readily to analogical inferences that, without exception, favored the assumption of a plurality of worlds.

Thus the telescope revealed that the surface of the moon was not perfectly smooth and rounded, as Aristotle's doctrine, charged with philosophical and symbolic significance, had claimed; it was similar to the Earth; in fact it was "as it were a second Earth" with its mountains and valleys, seas and ravines. Indeed, as Galileo rather rashly remarked in the *Sidereus Nuncius*, the main crater of the moon bore a striking resemblance to the geographic character of Bohemia.[71] Why, then, should one not assume that this new world on the moon was also inhabited, by individual beings similar to humans? (Galileo's friend Giovanni Ciampoli had very quickly pointed out to him that this analogical inference might be drawn—and that it would be theologically dangerous: what about descent from Adam?)[72] Moreover, as Galileo was the first to see, at least one of the other planets, Jupiter, was itself circled by "planets" (i.e., moons), so that another aspect of the Earth's supposed uniqueness was illusory. Why, then, should not Jupiter too be "like another Earth,"[73] for whose inhabitants those moons shed light—after all, who else could benefit from it? Why should not the other planets also be inhabited? And if, finally, the "innumerable fixed stars" were to be regarded as suns,[74] why should not they have their own planetary systems containing inhabited worlds?

How did Galileo himself answer these questions? In the *Sidereus Nuncius* he avoided not only giving answers, as I have said, but even posing the questions. Only three years later, however, under the pressure of his readers' irresistible and embarrassing urge to make analogical inferences, he was forced to take a position. In his *Istoria e dimostrazioni intorno alle macchie solari* (1613) he takes up the subject very abruptly and drops it again with equal abruptness. The view that the moon and planets are inhabited by "animals like ours, and particularly men," is, he says, false and to be condemned. If, however, one were to picture these creatures as wholly different from those on

---

[71] *Opere* 3(1):65, 68. Kant was still to compare the "crater-like spot Tycho" with the "Kingdom of Bohemia": *Werke*, ed. Ernst Cassirer, vol. 4 (Berlin, 1913), p. 205.

[72] *Opere* 12:145–47 (letter to Galileo dated 28 February 1615).

[73] *Dialogo dei due massimi sistemi del mondo*, in *Opere* 7:368.

[74] *Opere* 3(1):59–60; 7:354. Page references in the text are to *Opere*.

Earth and from any that our imagination can conceive, then—he would neither agree nor disagree but would leave the decision to "wiser men" (5:220; see also 11:467; 12:240–41). In the *Dialogo dei due massimi sistemi del mondo* of 1632 his position is still basically unchanged, except that here he is more obviously on the defensive. His answer follows a statement by Simplicio, the traditional believer, that to imagine human beings on the moon and planets is sacrilegious (7:85). In their replies, Sagredo and Salviati, who speak for the author, tread with the utmost wariness. No, Sagredo does not assume that the moon harbors creatures similar to those on Earth. But it does not follow that there may not be things (*cose*) that are not merely different from those we know but quite beyond our imagination. *If* that is so, agrees Salviati, but without asserting that it *is* so (cf. 7:125), then such creatures (*creature*) would admire the grandeur and beauty of the universe and, as the Scripture ordains, praise the Creator for ever (7:86–87). This is, as it were, a second line of defense, and later a third is added: only by assuming the nature of such creatures to be quite beyond our imagining do we do justice to the omnipotence of the Creator (7:126).

These are answers, certainly, but the sort that take back with one hand what they give with the other. Or does Galileo, to whom his contemporaries sometimes attributed a belief in extraterrestrials (of some sort), in fact reveal his true opinion between the lines, in a sort of code, when he has Sagredo speak of the possible extraterrestrials whom we could not even imagine: in the same way, Sagredo says, a man who had been born and raised deep in the forest could not imagine fish, oceans, and ships, or men, towns, and palaces beyond the confines of the forest (7:86). Is Galileo implying that disbelief in the plurality of worlds is the mark of the narrow-minded and ignorant—of the backwoodsman, as it were?

Be that as it may, the question had now been asked, and in the light of empirical science it was now, as never before, an incontrovertibly legitimate one. As with Faustian boldness it swept aside the warning given by Milton in his anxiety to preserve the faith, "Dream not of other worlds, what creatures there / Live, in what state, condition or degree,"[75] it had an electrifying effect on the minds of philosophers and writers. Now the old topic of plurality was discussed, often with specific reference to Galileo, more frequently and more urgently than ever before, and with a new insistence on its legitimacy. Writings by philosophical libertines and Anglican bishops, abbés and creators of

[75]*Paradise Lost*, bk. 8, lines 175–76.

utopian fantasies revolve around a central, if unspoken, theme, an unprecedented event: an imagined meeting between *Homo sapiens* and his counterpart in outer space. This, as always, also involves a reappraisal of man's own nature. The maxim characteristic of that time, that the proper study of mankind is man (coined by Paul Scarron in 1601), might appear to have been displaced by what Thomas Mann called "Milchstraßenspekulation" (Milky Way speculation), but this is not so. Alexander Pope, who took up that maxim in his didactic poem, the *Essay on Man* (1733–34), makes this point right at the beginning of the poem (and Kant was to take these lines as a motto for the part of his *Allgemeine Naturgeschichte und Theorie des Himmels* [Universal natural history and theory of the heavens, 1755], which deals with the plurality of worlds):

> He, who through vast immensity can pierce,
> See worlds on worlds compose one universe,
> Observe how system into system runs,
> What other planets circle other suns,
> What varied Being peoples every star,
> May tell why Heaven has made us as we are.
> (1, lines 23–28)

In the light of the awareness created by the new science, any attempt to define man has to include comparison with his extraterrestrial counterpart, once the object of pure speculation but now, after Copernicus and Galileo, a scientifically plausible possibility. This is above all Galileo's legacy, to philosophy and literature in particular, whose inexhaustible subject is human nature. His *Sidereus Nuncius* (together with the *Dialogue on the Two Chief Systems of the World*, which made the same discoveries accessible to a wider public), is "the most important single publication . . . of the seventeenth century, so far as its effect on imagination is concerned."[76]

7. Literary "Mysteries of the Heavens": Kepler at the Telescope

The liberating effect of the telescope on imaginative writing, even where fantasy is kept firmly tied to scientific probabilities, is strikingly apparent when we compare Kepler's thoughts on plurality before

[76]Nicolson, *Science and Imagination*, p. 4.

Galileo's discoveries and those he was prepared to risk after 1610. In the *Somnium* he had condemned himself to wake up just when the reader was expecting the promised information about the Subvolvan culture. By contrast, such information is freely divulged in the "Geographical, or rather Selenographical, Appendix" he attached, again with numerous notes, to the *Somnium* after Galileo had revealed to him the "mysteries of the heavens" (*Ges. W.* 16:389). This appendix already reflects Kepler's own telescopic observations, although of course Galileo's pioneering work is acknowledged. Characteristically, however, Kepler presents the findings of his experimental scientific work in the form of a discourse that aims to offer "literary enjoyment."[77] In this literary-cum-scientific fantasy he begins with the towns on the moon (*oppida lunaria*). In the *Somnium* he had aroused the reader's curiosity by mentioning them, only to disappoint it. Now he makes ample amends, giving a wealth of astonishing detail. The towns are the craters the telescope has revealed. The Subvolvans laid them out as protection against both the marshy dampness of the terrain and the heat of the sun, and against enemies as well. In a matter-of-fact tone, Kepler even tells the reader *how* these fortifications were built: by the citizens working communally, as befits these "Swiss" among the moon dwellers:

> The design of the fortification is as follows. They drive a stake down in the center of the space to be fortified. To this stake they tie ropes which are either long or short depending on the size of the future town. The longest I have detected is five German miles. With this rope fastened in this way, they move out to the future rampart's circumference, as defined by the ends of the ropes. Then the entire population assembles to do the digging for the rampart. The width of the ditch is not less than one German mile. They take all the excavated material inside some towns. In others, they have built partly outside and partly inside; thus the rampart is constructed in two sections, with a very deep ditch in between. The individual ramparts return upon themselves, as though a pair of compasses had made them perfectly round. They achieve this circularity by equalizing the ropes stretched from the stake in the center. Thus the result is that not only is the ditch pushed down quite deep, but also the center of the town looks sunken like a chasm, as though it were

[77] *Opera Omnia* 8(1):67: "fructum aliquem voluptatis literariae." All further comments on the appendix refer to pp. 67–68 and to Rosen, *Kepler's "Somnium,"* pp. 151–52. In the following discussion of the notes to the appendix, page references are given first to *Opera Omnia*, then to Rosen.

the navel of a puffed up belly, while the entire periphery is raised up high by the pile of material removed from the ditch. For from the ditch all the way to the center would be too long a distance from which to clear away all the material. Hence the moisture of the damp ground collects in this ditch. Whatever space is enclosed by the ditch is drained by it. When it overflows with water, it becomes navigable. When it has dried out, it can be crossed as a land route. Therefore, wherever the sun's strength assails them, those who are in the center of the space betake themselves toward that part of the circular ditch in the shadow of the outside rampart; and those who are beyond the center take refuge in the part of the ditch turned away from the sun, in the shadow of the inside rampart. And thus for the fifteen days during which the place is uninterruptedly scorched by the sun, they follow the shadows as peripatetics in the true sense of the word, enduring the heat. (Rosen, *Kepler's "Somnium,"* pp. 151–52)

While giving these vivid sketches of communal life on the moon, Kepler seems wholly oblivious of the fact that he is writing not an observational report but pure science fiction. All at once, as though awaking from a trance, he puts in a sobering reminder that these statements must be regarded as hypothetical ("haec problematis in morem tibi proposita"). Each one still needs to be confirmed by the phenomena the telescope has revealed; that is, those phenomena have to be shown, on the basis of optical, physical, and metaphysical axioms, to accord with these conclusions. Then, with equal abruptness, the appendix breaks off with the words "But these are playful remarks, etc." (*Sed haec ludicra sunt etc.*).

But this is not a genuine retraction. The notes on this text, innocent of any flights of fancy, take the idea of manlike beings on the moon perfectly seriously; they even demonstrate its validity by means of arguments that can only be described as philosophical deductions from both empirical observations and metaphysical axioms. There is nothing lighthearted or playful here. Kepler states as a matter of principle that where there is order in nature, it can be attributed only to a rational mind (*mens*) if it cannot be satisfactorily explained by reference to natural causes, such as the spontaneous movement of matter. The perfectly circular craters and the regularity of the distances between them cannot, unlike the other features of the moon's surface, be explained by natural causes. They must therefore be of artificial origin, arranged by intelligent beings for a particular purpose: they are the work of an "architectural mind" (*ab aliqua mente architectrice*, pp. 71/160). So the philosophical conclusion is that there are creatures capable of reason (*rationis capaces*) living on the moon (pp. 71/162). This in effect means human beings, or at any rate beings similar to humans; for reason was traditionally seen as the faculty that distinguishes man from the animals, and Kepler gives not the slightest hint of wishing to depart from this conception. Other information he gives about the inhabitants of Subvolva confirms this impression. As the moon has many such artificial craters, one must assume a large population. The fact that their crater towns are placed at such regular distances from one another, according to some "definite law" (*certa lege*), shows that there is mutual agreement (*mutuus consensus*) between the various groups, which does suggest that the basis for a quasi-human culture is there (pp. 72/162). This is a conclusion already suggested by the references to the collective town-planning and land drainage schemes, with masses of individuals being organized to carry out the work.

Kepler gives further such indications in the notes, and immediately afterward he makes the inevitable comparison between dwellers on the moon and on Earth: "Once a comparison is instituted between the populations of the moon and of the earth, the judgment about similar things is the same" (pp. 73/169).[78] For instance, as the craters are "civilized places" (*partes cultae*) with "civilized people" (*cultiores*), pastures, fields, and farmsteads, Kepler assumes that the rugged mountain areas are inhabited by savage bands of robbers. No mountain region is complete without bandits! Keeping out these robbers, then, is one of the functions of those circular fortifications, and Kepler is surprisingly well-informed about the surveying techniques of the skilled engineers who build them (pp. 73/169–70). The main point, however, is that Kepler believes the moon to be inhabited by "creatures [that] are to a certain extent endowed with reason" (*creatura rationis participe quadamtenus*, pp. 74/172). And he paints a vivid picture of these humanlike beings, spending their days fleeing from the sun by moving on foot or by boat round the great ditch, following the shadow cast by the outer or inner ramparts, until the fourteen-day-long lunar night falls—really, Kepler implies, the only rational way to cope with such an environment. And at midday, when there is no shadow, it is logical to assume (*consentaneum est*) that the moon people take refuge in caves that, sensible as they are, they have specially hollowed out of the steep ramparts . . .

In the paragraph concluding the explanatory notes on the geo-

---

[78]"Dato semel initio comparationis inter gentes, lunarem et terrestrem, jam de similibus idem est judicium."

graphic appendix, Kepler says he believes these notes have supplied all the proofs of which he spoke at the end of the appendix itself. He believes, then, that his statements there about life on the moon— statements that were dismissed there as mere jokes—have now been backed up by sufficiently strong arguments, based on telescopic observations as well as on principles of optics, physics, and metaphysics, for the reader to have been convinced. Finally, he refers his reader to Plutarch's *De facie in orbe lunae,* which forms the conclusion of the *Somnium* and itself contains some realistic details about everyday life on the moon.

The first point that strikes us about these thoughts of Kepler's is that he is able to imagine a living being in space only as a *creatura rationalis* on the human model. For him there is only *one* kind of reason; he is not able to conceive of anything "wholly other." He even has difficulty in imagining different *degrees* of rationality (of that quality, that is, which forms part of the definition of man). True, he does speak on one occasion of the moon dwellers' possessing reason only "to a certain extent"; but his own arguments seem to equate earthly and lunar reason. All the lunar phenomena revealed by the telescope— the fortifications, the hill-like mound at the center of the craters, the circular ditches—must, he thinks, have a rational cause; and he finds this cause by consulting his own reason. The moon dwellers he describes occupy themselves in a way that exactly accords with *our* reason; they behave precisely as *we* would behave in similar terrain. They differ from us in their larger proportions, which Kepler again refers to here: there must, he says—perhaps not wholly seriously— be a correlation between the proportions of the geographic features of a "world" and the stature of its inhabitants (pp. 74/171). But his idea of the physical nature of the moon dwellers, as well as of their intelligence, is clearly formed in the image of earthly man: they are travelers by boat, builders of fortifications, robbers, town planners, farmers, and indefatigable walkers.

What is the explanation for this similarity in both intelligence and physical form? Can this parallel life-form have developed on the moon as a result of the kind of spontaneous generation Kepler referred to in *De Stella Nova*? Or did God create more than one rational humankind? Are being a child of God and possessing reason the same thing, so that man must let the moon dwellers share equally in the glory of being the pinnacle of Creation? Or does Kepler perhaps regard terrestrial man as superior? After all, the moon dwellers' dreary way of life, with its endless movement in search of shade, can hardly allow them to achieve any perceptible *higher* culture (though on the positive side, they are perhaps mercifully free of the darker aspects of earthly civilization, such as the persecution of witches in the name of religion, which cast its shadow over the Kepler family).

In the geographic appendix, written after Galileo's discoveries, Kepler continues, just as in the *Somnium*, to hint at such questions rather than to ask them directly, let alone answer them. But in the notes to this appendix he refers the reader more than once to his *Dissertatio cum Nuncio Sidereo* (1610), in which, just a few weeks after the publication of Galileo's revolutionary account of his observations, he had expressed his thoughts on its philosophical and theological implications. Here he does indeed formulate the delicate philosophical questions and suggest some answers to them. (He was writing, incidentally, in the capital of Bohemia, that country which, in Galileo's eyes, so strikingly resembled a particular region of the moon.) Although the answers Kepler offers have a markedly pious ring to them, the reasoning by which he arrives at them is more scientific than theological, as it was in *De Stella Nova*. Here, however, he addresses himself more to fundamental issues.

The encounter between the two great figures of modern science in this "conversation" is a meeting of opposites: on the one hand the professor of mathematics at the University of Padua, soberly observing, cautiously shunning all philosophical speculation; on the other, the "Mathematician to His Holy Imperial Majesty," who is, for all that, strongly inclined toward philosophy. While lacking Bruno's religious fervor, Kepler boldly extrapolates from the facts established by Galileo the same conclusions he had previously drawn from Copernicanism on a purely theoretical basis, right up to the text of the *Somnium* the year before. It is this interest in the wider philosophical implications, rather than the detailed discussion of mathematical and physical matters, that is the dominant characteristic of this booklet. Just as he had indicated in the letter to Bernegger that his aim in writing the *Somnium* was philosophical, so here Kepler announces at the very outset that he intends not only to go through Galileo's report point by point, but also, in marked contrast to Galileo himself, to comment on those areas of *"philosophy"* that the *Sidereus Nuncius* either threatens to invalidate, or confirms or elucidates. He immediately adds, as though to reassure his readership, that his aim in doing this is to preserve the reader from (clearly not groundless) "uncertainty"—from doubt about Galileo's credibility, but also from any inclination to "spurn the philosophy which has hitherto prevailed"[79]—as if that were possible! Here Kepler

[79]Kepler, *Conversation with Galileo's Sidereal Messenger*, trans. with an introduction

has put his finger on the crucial issue on which thinkers had to take one side or the other, the crossroads where the choice was between medieval and modern modes of thought.

Kepler's "philosophy" of modern science has two aspects, one negative, the other positive. The negative one is the refutation of the "dreadful" philosophy of Giordano Bruno (pp. 37/304), that is, of his assumption of an infinite cosmos containing innumerable worlds. Before he read Galileo's work, Kepler says—adding drama to his encounter with the *Sidereus Nuncius*—he had feared, and his colleague Johannes Matthäus Wackher von Wackenfels had hoped, that the newly discovered "planets" might be satellites of a remote fixed star and there might be no end to the wandering stars still to be discovered in an infinite universe. But to Kepler's great relief, Galileo irrefutably showed his "planets" to be moons of Jupiter, thus preserving the finite Copernican universe and proving to Kepler's satisfaction that Bruno's dizzying philosophy could for the time being be dismissed as mere fantasy (pp. 36–38/304–5). Needless to say, while Kepler rejects Bruno's belief in an infinite universe, and with it the belief in an infinity of inhabited worlds, he shares his conviction that other celestial bodies, if only those within our solar system, are inhabited. How firm this belief of Kepler's was is shown with almost comic effect in the *Dissertatio cum Nuncio Sidereo* when, with disarming confidence, he argues from that assumption as if it were an established fact. Galileo, he says, has empirically confirmed that the moon has its own atmosphere, which he himself had deduced by theoretical reasoning. And how could it be otherwise? How could the moon's inhabitants endure the heat of the sun if its intensity were not often diminished by thick mist, "as happens among the Peruvians" (pp. 29/300)? Peru and the moon are not so very different from each other, even if they are more exotic than Bohemia.

If, however, the moon is inhabited, then—now that Copernicus has done away with the division of the cosmos into Earth and heaven—there is no reason why other heavenly bodies should not also have inhabitants of their own. Kepler confines his comments to those of our solar system. Since the planets discovered by Galileo revolve around Jupiter just as the moon revolves around the Earth, it is "not improbable" that Jupiter too is inhabited (pp. 39/305). By what sort of creatures? Kepler gives no direct answer, but he does give the

---

and notes by Edward Rosen (New York and London, 1965), p. 15. Original: *Gesammelte Werke* 4:291. In the following discussion, page references are given first to Rosen, then to *Gesammelte Werke* 4.

impression that he has some variety of human species in mind: he speaks of space travel, which might enable "settlers from our species of man" to go there, and he also compares such journeys (as Spenser did *before* the work of the "new Columbus" was known) with the discovery of America, where Europeans were met by beings whom from 1537 onward they had papal permission to regard as humans. This seems to have struck Kepler too: surely it is significant that he immediately and abruptly puts in a disclaimer. That passage, he says, was only a lighthearted incidental comment on the bold spirit of his adventurous contemporaries: "For [!] the revered mysteries of sacred history are not a laughing matter for me" (*non sunt enim mihi deridiculo veneranda sacrae historiae mysteria*, pp. 40/305). He must be referring to the doctrine of the unique status of the Earth and of man. He protests too much, betraying that he *was* thinking of men living in outer space and not, say, of reptiles, which are one form of life mentioned in the *Somnium*.

This suspicion is confirmed by the teleological nature of the arguments Kepler uses to support his analogical conclusions. (For a long time, teleological arguments were to be the main pillar of the theory of plurality.) God, "the almighty and provident Guardian of the human race" (pp. 40/305), does not cause anything to happen without purpose (*frustra*). "There are in fact four planets revolving around Jupiter at different distances with unequal periods. For whose sake [*cui bono*], the question arises, if there are no people on Jupiter to behold this wonderfully varied display with their own eyes? For, as far as we on the earth are concerned, I do not know by what arguments [*rationibus*] I may be persuaded to believe that these planets minister chiefly to us, who never see them" (pp. 40/306). The moons of Jupiter are therefore "with the highest degree of probability" or indeed "undoubtedly" ordained (*comparatos*) "for the Jovian beings [*creaturis*] who dwell around Jupiter" (pp. 42/307; 41/306). In an analogous way, the moon gives us light at night, and other planets, of which Kepler specifically mentions Saturn, are similarly provided for. The basic idea underlying this chapter is made satisfyingly plain when he adds that this analogical thinking is even more convincing to those who "accept Copernicus's system of the universe" (pp. 41/306).

Once the analogical conclusion has thus been given such a firm foundation (a Creator still seems to be presupposed), Kepler can no longer evade the question of how we compare with those planetary beings who, if we accept the analogical principle, are at least similar to ourselves. Once again a cosmologic train of thought leads back to man. Does this new perception mean that man forfeits the theological comfort of knowing himself to hold the first place in God's love? This question leads to the "positive" element in Kepler's "philosophy" in the *Dissertatio*. And this is the area on which he concentrates. The nerve Galileo's treatise touched in Kepler was evidently his imaginative philosophy of the plurality of worlds, together with the anthropological problem implicit in it. Kepler formulates this problem in a passage from which I have already quoted and which, partly because Burton reproduced it in his *Anatomy of Melancholy* (1621), was echoed by his contemporaries and by later writers right up to H. G. Wells: "Well, then, someone may say, if there are globes in the heaven similar to our earth, do we vie with them over who occupies the better portion of the universe? For if their globes are nobler [*nobiliores*], we are not the noblest of rational creatures. Then how can all things be for man's sake? How can we be the masters of God's handiwork?" (pp. 43/307).

This is a difficult knot to unravel, says Kepler, as we have "not yet" all the necessary knowledge at our disposal. Nevertheless, he proposes to set out "those philosophical [!] arguments which, it seems to me, can be brought to bear. They will establish, not merely in general, as was done above, that this system of planets, on one of which we humans dwell, is located in the very bosom of the world [*praecipuo mundi sinu*], around the heart of the universe, that is, the sun. These arguments will also establish in particular that we humans live on the globe which by right belongs to the primary rational creature, the noblest of the (corporeal) creatures" (pp. 43/307–8). This makes it quite clear that Kepler not only clings to the notion that the sun occupies a central position in the whole universe and that therefore man too is close to the physical center; he also postulates that we, by virtue of the special position of the Earth in the solar system, are the noblest beings, at least among the corporeal rational beings (angels are incorporeal). The Christian view of man has duly been salvaged—but how?

Kepler adduces three arguments for the sun's being at the center of the cosmos. All three are scientific ones, and yet at the same time they invest the physical facts and spatial relationships with a symbolic dimension that ultimately points back to theology, or at least to the geometric sense Kepler confidently attributes to God. First, he speaks of the numerous fixed stars whose existence Galileo has revealed and claims that, far from confirming Bruno's beliefs, they confirm that the sun, the center of our planetary system, occupies the central position in the universe as a whole. Kepler refers the reader to his treatise *De Stella Nova*, in which he had maintained that the sun is incomparably more distant from the fixed stars than they are from one another. It follows from this that the place where we mortals dwell is at least close to the heart of the universe; for it is only from the center that the firmament can be seen as a spherical wall of stars (pp. 43, 34/308, 302–3). In thus making use of the symbolic significance of man's being at the center, or almost at the center, of the universe, Kepler is clearly echoing the kinds of arguments, attaching metaphorical value to spatial relationships, that played a part in orthodox opposition to Copernicus (see sec. 1 above). Here, however, Kepler argues against the strictly geocentric (or rather infernocentric!) convictions of those thinkers, turning, as it were, their own weapons against them.

Second, Kepler believes he can prove that our sun, as befits its central position, also shines incomparably more brightly than all the fixed stars put together (pp. 43/308), "and therefore this world of ours does not belong to an undifferentiated swarm of countless others" (pp. 36/303). The third argument for believing that our sun occupies the place of honor in the cosmos rests on geometry, which is "in the mind of God" and which man, made in God's image, shares with him (pp. 43/308). This argument centers on the geometry of Euclid's five regular solids: cube, tetrahedron, dodecahedron, icosahedron, and octahedron. For Kepler, these are the fundamental structural elements of our solar system. The sixth regular solid, the sphere, is present as the sphere of the fixed stars; the other five, in a certain order, fit into the spaces between the orbits of the six planets. This arrangement constitutes perfection. If there were other inhabited solar systems, we should be forced to assume either that they and their inhabitants are all alike, which would be pointless (why, Kepler asks, implicitly arguing with Saint Thomas Aquinas against Bruno, should one envisage a multiplication of perfection?), or alternatively, that they and the creatures on them are less perfect than our solar system and ourselves (pp. 44/308).

Having thus shown that our solar system is the central one and therefore the best and most noble of all systems that might possibly exist, Kepler goes on to prove that the Earth occupies a special position of preeminence—as the symbolic "abode of the dominant creature" (pp. 45/308)—vis-à-vis the other planets of our own solar system. For, since man was created for contemplation (*contemplationis causa, ad quam homo factus*), God has not placed him on the sun, which is God's own abode—or at any rate, appropriately in terms of Copernicanism, his symbol—God has placed man on the Earth, that ship which journeys around the sun. But why is the Earth the best of the six planet-ships? Again the answer is to be found in geometry. For one thing, the Earth occupies the middle position between Mars, Jupiter, and Saturn on the one hand and Venus, Mercury, and the sun on the other. It is curious to see how, after Copernicus had made it impossible to believe that the Earth was at the center of things, the idea surfaces again in a new guise, set in a context of symbolic theological thinking based on geometry! Moreover, Kepler continues, the Euclidean solids can be divided into two groups, primary and secondary, and the orbit of the Earth passes between them. "Merely by its position amidst the solids, the sphere of the earth is more distinguished than the other spheres" (pp. 46/309): yet another variant of the argument that links superiority with some sort of spatial centrality. Finally, the Earth is the only planet from which the whole solar system can be observed without one planet being made invisible by the overpowering brilliance of the sun. The inhabitants of Jupiter, for instance, have a far more limited view; to compensate for this, God gave them four moons instead of one. The upshot is that we, beneath our single moon, are the favorites of God: "We humans who inhabit the earth can with good reason (in my view) feel proud of the preeminent lodging place of our bodies, and we should be grateful to God, the creator" (pp. 46/309).

An unequivocal answer, certainly: not only are we uniquely privi-leged in being the inhabitants of the most perfect planet in the most perfect solar system, but we ourselves are, in symbolic accord with our place in the universe, the noblest (nobilissimae) and the "primary" (primariae) creatures (pp. 43/308). Yet, coming after all the groping, uncertain questionings of Kepler's earlier works, this sudden confi-dence and forcefulness make one suspect a defensive tactic by one who fears that his religious orthodoxy may be seriously questioned. As a supporter of the new Copernican science, Kepler had been held in suspicion by the church authorities ever since his student days. His declared aim, in the Dissertatio, of reconciling the view of the universe confirmed by Galileo with the "philosophy which has prevailed hith-erto" must, in those days of inflexible dogmatism, have amounted to an attempt at squaring the circle. The stridently triumphant tone of his assertions (which did not prevent his works from being placed, as Copernican heresies, on the index of banned books) tries in vain to drown out all doubts. One purely scientific doubt arises from his use of the regular Euclidean solids as the framework of the solar system. He had first put forward this idea in his Mysterium Cosmographicum in 1596, but how could he retain it after he had, in 1609, rejected circular planetary orbits in favor of elliptic ones (in the Astronomia Nova and the Somnium)? On the philosophical level, too, substantial doubts arise. No sooner has Kepler mentioned that Copernicus's theory allows the Earth to take up a modest place among the planets, than suddenly the Earth (and with it mankind) is promoted to the highest rank of all. At first sight this looks like the attempt of a Christian believer to save man from the danger of losing himself in the Copernican universe. But Kepler's representation of the perfect human being is too strongly reminiscent of the humanist conception of man presented by Pico della Mirandola, who was persecuted for it by the "only saving church." Would a Lutheran Christian, conscious of his sinfulness and not superlatively endowed with intelligence, recognize himself in Kepler's Renaissance evocation of "the noblest of rational creatures"? Hardly.

A further point is that if we are the "primary rational creature, the noblest of the (corporeal) creatures," (pp. 43/308), that indicates the inhabitants of other planets are also endowed with reason. Does Kepler envisage differing degrees of reason? Are the dwellers on other planets more or less intelligent than we are, or are they equal to us in intelligence but not so highly favored by God and, according to the symbolic interpretation of their location, not quite so perfect? The dreary civilization of the town builders on the moon comes to mind. But in that case, how can God be considered just? Is he in fact the creator of these "others"? If he is, then Kepler is implicitly undermining the "revered mysteries of the sacred history" (descent from Adam and Eve, redemption by Christ), a topic he carefully avoids broaching directly. If he is not, then Kepler must fall back on spontaneous generation, which indeed he does allude to at one point in this work, when he assumes that similar worlds would contain similar creatures (pp. 44/308). But the church whose displeasure Kepler is trying to avoid would regard the idea of spontaneous generation as blasphemous.

Why, in fact, does Kepler assume the existence of extraterrestrials? One wonders what emotional motivation or compulsion underlies his belief. Given that the extraterrestrials are deemed to be less highly favored, is this an instance of the conquest mentality so prevalent in this age, extended into space? If so, is Kepler driven by theological "jealousy" to assert man's superiority, prompted by the unacknowledged sense that the post-Copernican universe, transformed and alive with other beings, constitutes a threat? Earlier in his life, Kepler, like Bruno, had in passing implied that there might be creatures superior to us, *higher* beings (an idea that was to become familiar to later generations), but significantly, he does not return to it. As for the idea, current in the Enlightenment, that the greatness of the Lord is enhanced by his having created life everywhere, and perhaps even higher forms of life than on Earth—it was voiced by Bruno but it could not be further from Kepler's thoughts. It reflects a liberalization of

Christian attitudes that for thinkers of Kepler's day (Bruno is a special case) was not yet a real possibility. The forced way in which Kepler applies his symbolic reasoning shows clearly that, as he mounts his scientific defense of the Christian theological position threatened by the consequences of Copernicanism as revealed by Galileo, his paramount aim is the appeasement of the "prevailing philosophy." In the *Dissertatio*, the universe is still seen as having been created essentially for us, who are in some undefined sense the highest form of life; it is well within the grasp of the human mind, a finite universe constructed on symbolic geometric principles. Its sphere of fixed stars gives the lie to Bruno's dizzying, "dreadful" philosophy of an infinite universe, an idea to which Kepler reacts with a horror as intense as Pascal's—more intense than his adopted role of Christian apologist would seem to require (see also 1:253). What is also revealed by this rather contrived piece of reasoning is just how unsettling the implications of the Copernican theory, now empirically confirmed, must have been, and how necessary it was to fend off those extrapolations which reason inevitably made and which must have haunted Kepler like bad dreams.

These, then, are the questions and answers, expressed or implied, that spring from this first encounter between a thinker inclined toward theology and philosophy, but also toward literature, and the new science, which had become a matter of urgent concern. Echoes of Kepler's *Dissertatio* and of his "dream" of the moon occur remarkably often, with frequent references to Kepler by name, in the writers who succeed him, including literary authors in the broadest sense: Wilkins and Donne, Campanella and Burton, Henry More, Borel and Huygens.[80] After Kepler's *Dissertatio* and Galileo's *Sidereus Nuncius*, human and humanlike beings on the moon and planets are often depicted very vividly and in great detail, and these representations most likely owe something to the *Somnium* of 1609 (or 1634). During the same period, however, as the results of Galileo's observations became more and more widely known, the theological objections, biblical and christological, that Kepler (and Bruno, Hill, Benedetti, and others) had gone to such lengths to allay, now increasingly made themselves heard. Only now did biblical truth and scientific truth confront one another head on, though even Kepler had on one occasion held up one against the other and made pointed references to Saint Augustine's

[80]On the far-reaching influence of the *Dissertatio* see the afterword in *Gesammelte Werke* 4:459; on that of the *Somnium* see for instance Nicolson, *Voyages to the Moon*, p. 47: "familiar to all later writers of cosmic voyages during the seventeenth and eighteenth centuries." See also Nicolson's index under "Kepler, influence," and idem, *Science and Imagination*, pp. 77–79.

condemnation of the belief in antipodeans, which was often seen as a parallel to belief in the plurality of worlds (3:33–34). Now believers in plurality seeking to reconcile their views with Christianity subverted the traditional dogmatic positions by their claims that belief in a plurality of worlds is not incompatible with Scripture, liberally interpreted. But this was also a time of increased vigilance, particularly on the part of the Catholic church, which had in the meantime become fully aware of the threat posed by the philosophy of plurality in its post-Copernican and post-Galilean form and which rejected the concept on principle as irreconcilable with the Bible and Aristotle. The Jesuit Julius Caesar La Galla, mentioned earlier in this chapter, wrote in this vein in his *De Phaenomenis in Orbe Lunae* only two years after the *Sidereus Nuncius* had appeared. In 1616 the pope officially denounced as heresy the Copernican doctrine of the planetary movement of the Earth, which was the basis for any concept of plurality; in 1633, Galileo was forced by the Inquisition publicly to recant his teaching. The consequent notoriety of the plurality of worlds as a belief held by heretics gave it an irresistible attraction for all independent thinkers, who eventually made it a tenet of *their* religion. In an atmosphere of such tension, the idea of plurality flourished with remarkable vigor.

CHAPTER 3

The Baroque Period:
Between Heresy and Piety

1. Bafflement, Orthodoxy, Half-heartedness, and Fervor: Donne, Pascal, Kircher, Gryphius, Milton, and Others

In Galileo's wake, many amateurs began to study the skies, and the "new astronomy" became a fashionable science. This meant that the "new philosophy" implicit in it also reached the marketplaces, as Brecht said—or at any rate, the educated public; and the upshot of the "new philosophy" was the "realistic" suspicion that "we are not alone." When the *Sidereus Nuncius* had challenged orthodox belief by its suggestion of a plurality of worlds, Kepler had contrived, with difficulty, to salvage man's central and preeminent position and its theological significance; but its preservation now became increasingly difficult, if not impossible. Kepler had, by a feat of intellectual contortionism, reconciled the divine truth of the Bible (*one* world) and the scientific truth of nature as observed through the telescope (many worlds); but now the gulf between the two kinds of truth gaped wider and wider. The patent contradiction between them imposed a choice: either Scripture, the revealed Word of God, was the source also of cosmologic information that must be accepted as literally correct; or else each kind of truth could be regarded as possessing validity in its own separate domain. A third option was that the two kinds of truth (the Word of God and the works of God) could be brought into harmony by a nonliteral interpretation of Scripture, or a broadening of the basis of Christian doctrine. One could argue that the more planetary humankinds God had created, the greater his stature, and even suggest that Christ had redeemed humankinds on other planets as well as ours (unless these had their own redeemer or had no need of redemption). Such matters of theology and dogma, indeed the very foundations of Christian teaching, were never far from the minds of those who debated the plurality of worlds in the seventeenth century. Gradually it became the case that to express belief in the plurality of worlds was not necessarily to break with the church. John Wilkins became an Anglican bishop; Gassendi was and remained an abbé of the Catholic church: this clearly reflects the changed intellectual climate.

In the years between Galileo's "sidereal message" (1610) and the conclusive proof of Copernicanism provided by Newton's *Principia* (1687), there was of course a wide variety of reactions to the idea of "new worlds"—an outstanding instance, in fact, of the ability of old and new ideas to coexist simultaneously. Many of the most creative thinkers chose, even after Galileo had provided such sensational and direct evidence in favor of heliocentrism, to remain pre-Copernicans; the temptation to "dream of other worlds," against which Milton's Archangel had warned, held no attraction for them. Shakespeare and the philosopher and theologian Pierre Charron are among this number;[1] they turn a blind eye to the new ideas because their own interests lie elsewhere. "Other worlds" are of no significance to them. As late as 1625 there appears a Spanish account of a space journey which does not acknowledge the impact of Galileo by so much as a word. This is Juan Enríquez de Zúñiga's *Amor con vista*. Not only is the universe still the system of spheres with its various "heavens," but these heavens are used allegorically as abodes of the gods of classical mythology.[2] In Germany the wars of religion delayed the assimilation of the new science, and the tendency to ignore the new worlds remained the norm in German literature and philosophy right up to the early eighteenth century.

At the opposite end of the spectrum we find other educated people reacting to the telescope's revelations with extreme anxiety and bewilderment, not least because of the idea of the "new worlds" these discoveries presented to both the intellect and the imagination. Histories of thought, especially those written in German, often insist that it was only with the publication of Fontenelle's *Entretiens sur la pluralité*

---

[1]On Charron see Henri Busson, *La pensée religieuse française de Charron à Pascal* (Paris, 1933), p. 286; on Shakespeare, W. G. Guthrie, "The Astronomy of Shakespeare," *Irish Astronomical Journal* 6 (1964), 201–11; on Milton see the discussion later in this section.

[2]See Monroe Z. Hafter, "Towards a History of Spanish Imaginary Voyages," *Eighteenth-Century Studies* 8 (1975), 267–69.

des mondes (1686) that the new science first began to affect people with feelings of fear and vertigo induced by the sense of being cast adrift in boundless space. But this is not strictly true.[3] I have already pointed to Kepler's deep sense of disorientation; whether expressed or not, it underlies every one of his philosophical utterances from the date of the *Sidereus Nuncius* onward, and perhaps even before. In English literature it is John Donne who presents the most compelling instance of the overwhelming shock caused by Galileo's discoveries; in French writing it is Pascal.

Donne knew Kepler's works, and he is the first writer after him who is troubled by the philosophical implications of the new science and does not simply reach for a ready-made solution to the problem. His anxiety too can be traced directly to the impact of Galileo's 1610 report of his observations. Donne, who was an unusually keen follower of new developments in astronomy, must have read the *Sidereus Nuncius* in the very year of its publication because he mentions Galileo in his satire on the Jesuits, *Ignatius His Conclave*, published in 1611. For Donne, too, Galileo is above all the discoverer of new worlds. He describes Galileo's achievement as that of having "summoned the other worlds, the Stars, to come neerer to him, and give him an account of themselves," and he refers to them as "the other starrs, which are also thought to be worlds."[4] This work contains no hint that the implications of these new worlds affect the author personally; but soon afterward he does face up to the questions they force upon his attention. The reasons for this change are to be found in Donne's inner

---

[3]The idea that cosmic vertigo and the cosmic humiliation of man were unknown before Fontenelle is stressed particularly by Hans Blumenberg, in *Die kopernikanische Wende* (Frankfurt, 1965) and also in his edition of Galileo's *Sidereus Nuncius* (and other writings, Frankfurt, 1965), p. 26. Siegfried Korninger says that there is "absolutely no question" of despair at the findings of the new science occurring in seventeenth-century England (*Die Naturauffassung in der englischen Dichtung des 17. Jahrhunderts* [Vienna, 1956], pp. 11, 30). This opinion is shared by Douglas Bush, *Science and English Poetry: A Historical Sketch, 1590–1950* (New York, 1950), p. 29. Wolfgang Philipp, in *Das Werden der Aufklärung in theologiegeschichtlicher Sicht* (Göttingen, 1957), esp. pp. 78–82, argues that a "cosmic shock" (*kosmisches Erschrecken*) in the face of the Copernican world view is observable at least in the theology of the baroque period; see p. 154 for his comment on Blumenberg's argument. The view, rejected by Korninger, Bush, and others, that many laymen were affected by this fear may be found in, for instance, Marjorie Nicolson, *Science and Imagination* (Hamden, Conn., 1976), p. 30, and Charles Monroe Coffin, *John Donne and the New Philosophy* (New York, 1937), p. 133. Grant McColley's article, "The Seventeenth-Century Doctrine of a Plurality of Worlds," *Annals of Science* 1 (1936), 385–430, is essentially a collection of relevant passages; in fact only about the last fifteen pages deal with the seventeenth century.

[4]*Complete Poetry and Selected Prose*, ed. John Hayward (London, 1939), pp. 359, 399. Only texts not in this edition are quoted from *Works, 1621–1631*, ed. Henry Alford (London, 1839).

development, in the very personal crisis that led him from Catholicism to Anglicanism. The new scientific view of the universe is not the cause of his melancholy at this period, but it intensifies it. The new cosmology, far from being a mere metaphor, makes him recognize that he lacks a firm sense of belonging in the world of traditional religion—but that he feels equally insecure in that of the new science.[5] This crisis is reflected in "The First Anniversary: An Anatomie of the World" (1611). On the surface, this famous poem is a lament for the daughter of his patron, Sir Robert Drury, who had died a year earlier; but its true subject is, rather, a variation on the Christian theme of the decline of the world as it moves closer to the end God has predestined for it. Abruptly, however, this orthodox pessimism is reinforced by a reference to the unorthodox "new astronomy," or "new philosophy" (which is the same),[6] that now threatens to undermine everything: "And new Philosophy calls all in doubt" (line 205). Donne is quite clearly referring to Copernicanism ("the Sun is lost, and th' earth") and more specifically to Copernicanism as revealed, concretely and vividly, by Galileo's telescope, which suggested to Donne's contemporaries the existence of a plurality of worlds:

And freely men confesse that this world's spent,
When in the Planets, and the Firmament
They seeke so many new . . .
    (Lines 209–211)

The cause of Donne's inner turmoil is evidently not so much Copernicus's calculation of the noncentral position of the Earth as it is Galileo's recognition of the similarity of the moon and planets to the Earth. The new science acquires significance for Donne only in the light of Galileo's work; and because Galileo's observations make a plurality of worlds seem probable, this significance is of a threatening

[5]See Marjorie Nicolson, *The Breaking of the Circle* (New York, 1960), pp. 83, 155; Coffin, *John Donne*, p. 137: " 'The Sun is lost, and th'earth, and no mans wit / Can well direct him where to looke for it.' In writing thus Donne is not simply versifying a popular pessimistic doctrine, but rather exploiting that doctrine in his original way to make clear the fit of disillusionment which has seized him. Nor is this to deny that his gloom was deepened by the influence of the scientific discoveries alluded to. Always there is a reciprocal action to be taken into account: though the materials from which the imagery is drawn are properly subordinated to the major purpose of analyzing a condition of mind, they nevertheless are made to share the responsibility for the existence of that condition."
[6]Compare Donne's poem addressed to the countess of Bedford: "As new Philosophy arrests the Sunne, / And bids the passive earth about it runne" (*Complete Poetry and Selected Prose*, p. 165).

nature. In the light of this view of the universe, what is our place in God's plan of salvation? Is a retreat to a literal understanding of the Bible still possible? If God's work of salvation is so immeasurably extended, what is our destined role in it? At this crisis point, Donne can give no answer. He does not adopt the orthodox Christian course of insisting on the literal truth of the Bible and rejecting the plurality idea out of hand. Nor does he accept the equally convenient solution offered by those pious agnostics for whom God's unfathomable greatness can encompass even a plurality of worlds. He is simply at a loss. With the exception of Pascal, this is probably the clearest example of the bewildering effect (still often disputed) that the new astronomy could have on educated thinkers in countries receptive to new scientific ideas. The harmonious cosmos was in ruins, and there was no prospect of a new order to replace it.

In 1615, three years after "The First Anniversary," Donne took holy orders as an Anglican priest. His personal crisis was over; he had conquered that philosophical disorientation which had been at least exacerbated by the distressing possibility of a plurality of worlds. On occasions when Donne returns to the topic in the years that follow, it is only in a figure of speech used for hyperbolic effect, as when he says in a sermon of 1617 that the passion of Christ is so immeasurable that it could save "millions of worlds";[7] or else he is simply reporting the opinion of others, without endorsing it: "Men that inhere upon *Nature* only, are so far from thinking, that there is anything *singular* in this world, as that they will scarce thinke, that this world it selfe is *singular*, but that every *Planet*, and every *Starre*, is another *world* like this; They finde reason to conceive, not onely a *pluralitie* in every *Species* in the world, but a *pluralitie of worlds*."[8] In a sermon of 1629 on the subject of the Creation he remarks, carefully distancing himself from the view expressed, that "subtle men" have "with some appearance of probability imagined, that . . . there are many earths, many worlds, as big as this, which we inhabit";[9] but even this is only a passing remark and merely shows how little interest he now has in an idea that had, while the impact of Galileo's discoveries was still fresh, dealt him a devastating blow.

In Donne we see disorientation superseded by faith, a crisis followed by its solution; in Pascal's *Pensées*, in contrast, there is continuous and unresolved tension between his conflicting reactions to the

[7] *Works, 1621–1631* 6:6–7. See Nicolson, *Science and Imagination*, pp. 54–57, on cosmologic themes in Donne's late work.
[8] *Complete Poetry and Selected Prose*, p. 514.
[9] *Works, 1621–1631* 4:491.

new cosmology. There is a similarity, however, in that for Pascal, as for Donne, the idea of a plurality of worlds adds to his feeling of being threatened by the cosmology of the new science, and in his case too this effect is subtle rather than obvious. "This is as far as our instinctive knowledge leads us," so begins (in Pascal's original version) the famous pensée about the two infinities, the infinitely great and the infinitely small, between which man is placed.[10] Looking through the telescope one is overwhelmed; the Earth shrinks to a mere dot in immeasurable space. "What is a man in the infinite?" No other idea could lower man's status as effectively as this awareness of infinity: Pascal sees the discoveries of science as "a great cause of humiliation."

But he reacts in more than one way to this basic situation of the human being confronted with the "abysses of infinity and nothingness." The attitude that predominates in the pensée about the two infinities is one of silent awe in the face of the wonders of nature and of its omnipotent Creator. In the equally famous pensée about man as the thinking reed, on the other hand, man triumphs over the universe that obliterates him by his ability to think and to comprehend what is happening to him: "But if the Universe were to crush him, man would still be nobler than his killer. For he knows that he is dying, and that the Universe has the advantage over him; the Universe knows nothing of this" (no. 264/198). And another pensée (which in part echoes the wording of the one just quoted) draws the conclusion: "It is not in space that I must seek my dignity, but by . . . my thought. . . . In space the Universe encompasses me and swallows me like a dot; in thought I encompass the Universe" (no. 265/111). But this triumphant recognition of the sublimity of the human mind is not always attainable. All too often there is only the sense of "being engulfed" (*engloutissement*), the terror of being overwhelmed by the universe. This reaction, too, is captured in pensées that have rightly become part of our common heritage. Their theme is a major aspect of "man's wretchedness without God," which according to Pascal's plan was to be the subject of the first part of the work. (The second part was to deal with "man's happiness with God" [no. 73/2].) "I see those fearful expanses of the Universe which hedge me in, and I find myself fixed in one corner of this vast space," says Pascal, adopting the role of an unbeliever, a man far from God (no. 335/429). But this

[10]*The Pensées*, trans. J. M. Cohen (Harmondsworth: Penguin, 1961), no. 84; *Pensées*, ed. Zacharie Tourneur and Didier Anzieu (Paris: Bibliothèque de Cluny, 1960), no. 197. References in the text are to these two editions, in this order. On Pascal's attitude to science in general see Harcourt Brown, *Science and the Human Comedy* (Toronto, 1976), pp. 45–74.

role comes naturally to him, as we infer from a pensée that has become a familiar quotation: "The eternal silence of those infinite spaces strikes me with terror" (Le silence éternel de ces espaces infinis m'effraie, no. 91/199). This is not the evocation of a passing mood, but the classic expression of the insecurity brought about by the cosmology of the new science. It is the new science that Pascal is warning against when he recommends submission to the miracles of God (no. 84/197); "Nature offers me nothing that is not a matter of doubt and disquiet" (no. 414/430).

One aspect of this anxiety in the face of immensity is the thought that it may contain other worlds. There is a hint of this in Pensée No. 335/429, where the passage previously quoted continues, "without knowing why I am placed here rather than elsewhere." Does this imply that other regions in the universe are also inhabited? Neither here nor elsewhere does Pascal enter into direct and detailed discussion of the plurality of worlds. Pensée No. 90/41, a single exclamation, is, however, highly suggestive in this context: "How many realms know nothing of us!" It makes sense to read this fragmentary thought in conjunction with Pensée No. 88/68, which contains a verbal echo of it and also echoes Pensées Nos. 335/429 and 91/119, thereby establishing a link between the thought of "realms" that do not know of our existence and the terror inspired by the immense spaces Galileo's telescope had brought to the layperson's awareness: "When I consider . . . the small space that I fill or even see, engulfed in the infinite immensity of spaces unknown to me and which know me not, I am terrified and astounded to find myself here and not there." Evidently the infinite spaces we do not know *and that do not know us* are not necessarily to be thought of as mere emptiness but rather as "realms," as "worlds," to use the term current among scholars of the day— inhabited worlds. Bafflement in the face of the universe of the new science is also bafflement as to the existence of a plurality of worlds. This pensée ends characteristically with a question that throws clearly into relief Pascal's existential position between "doubt and disquiet" on the one hand and belief in God the Creator on the other: "Who put me here? By whose order and design have this place and time been allotted to me?" Behind this question lurks an unspoken question about the place of the "others" in that design, linked with anxiety about the uniqueness of man.

As the idea of plurality gained ground, this anxiety must have become more widespread than can be shown by documentary evidence. Orthodox thinkers could of course dispel it simply by pointing authoritatively to Holy Writ, which spoke of only one Creation, one human race, and one Redeemer, thereby making the mere suggestion of a plurality of worlds patently absurd. This orthodox reaction had not changed basically since Melanchthon's day: in the age of Galileo the same biblical and christological arguments were rehearsed again and again by both Jesuits and Protestants: de Cazre, J. B. Morin, Juan Caramuel, Thomas Milles, Caspar Bartholin, Athanasius Kircher, Samuel Sorbière, and others.[11] And just as one may suspect that Melanchthon's rejection of the plurality of worlds was inspired partly by the fear that it implied a plurality of religions—a natural fear in the age of conquest and discovery—so in the seventeenth century the same association of ideas still usually lurked at least at the back of people's minds. Robert Burton, for instance, expresses it openly: "If there be infinite planetary and firmamental worlds, as some will, there be . . . infinite religions."[12] In the age of the "new Columbus," this fear may have been a spur to greater subtlety in the way orthodox writers handled the apologetic arguments for one world, which they had inherited from the sixteenth century in a rather unsophisticated form. The best example of an attempt to influence the public consciousness by theological argument at a higher level can be found in the monumental work *Quaestiones Celeberrimae in Genesim* (1623) by Marin Mersenne, a Parisian father of the order of Minims and a student of many branches of science, who was in close contact with Descartes, Galileo, Gassendi, Huygens, Hobbes, and other leading thinkers of his day. Mersenne thus brings a knowledge of both theology and astronomy to his careful, not to say hair-splitting, discussion of the plurality of worlds, which finally rejects the idea as both theologically and scientifically unacceptable. What is remarkable is not the fact but the manner of this rejection.

Mersenne was skeptical of the very scientific basis of the theory of plurality, the doctrine that the Earth moves annually through the zodiac. He doubted it not primarily for theological reasons but because he could counter Copernicus's arguments with other arguments drawn from physics: as long as the heliocentric view of the universe could not be proved beyond all contradiction, he considered that it should be rejected—though not as a heresy; for the church's anathema of 1616 had not made the contrary view of the universe a matter of faith. Mersenne did not on principle bar the way to acceptance of Copernicanism: he did not, with all due respect to the decrees of the

[11]On de Cazre see chap. 2, sec. 1 above. On the others see Busson, *Pensée religieuse française*, pp. 291–94; McColley, "Seventeenth-Century Doctrine," p. 426.
[12]*The Anatomy of Melancholy*, ed. Holbrook Jackson (London: Everyman, 1968), 3:377.

One True Church, rule out the possibility of a scientific proof (such as Newton was to furnish two generations later). At the current stage of the scientific debate, however, he considered the arguments for heliocentrism, like those against it, to be inconclusive; and for this reason alone he judged it best to align himself with the church's judgment against Copernicus and his followers. Yet for all the casuistry of his thinking, he was not merely doing formal justice to the new developments but was genuinely receptive to them. This is evident from the fact that he read Galileo's writings with interest and admiration and indeed did much to publicize his ideas even after the Inquisition had already pronounced its verdict.[13]

On the truth of Copernicanism, then, Mersenne advocates a policy of "wait and see"; but to assume that he therefore automatically advises against speculation about a plurality of humankinds is to underestimate the scrupulous, almost pettifogging fairness with which he weighs the evidence. On the contrary, he devotes a whole chapter of the *Quaestiones* (cols. 1081–96) to this topic. The first section of this chapter, the "article" on the question of "whether besides this world, which we see to have been completed on the seventh day, there is any other unknown to us" (cols. 1081–86), shows him to be both familiar with the history of the theory of plurality and the arguments for it and aware in particular of the increased impetus given to the idea by Galileo's telescopic discoveries and the conclusions drawn from them by Kepler in his "Conversation" with Galileo about the *Sidereus Nuncius*. (That Mersenne should not have recognized or should have "circumvented" the relevance of Copernicanism to the theory of the plurality of worlds is out of the question.)[14] Still, Mersenne is not convinced of the logic of the plurality idea, but his reasons are not, like Melanchthon's, purely dogmatic. On the contrary, in the second part of the chapter, "in which it is proved that the world is unique, and objections are answered" (cols. 1086–96), he plays the role of devil's advocate and begins by pointing out the weakness of the official orthodox position. On the question of whether we are "alone," he does not regard the Bible as a source of reliable and unequivocal information, since this matter, like many others, is referred to only ambiguously, if at all. Moreover, he argues, this question has never been the subject of a judgment by an official council of

[13]William L. Hine, "Mersenne and Copernicanism," *Isis* 64 (1973), 29–31.
[14]Steven J. Dick, *Plurality of Worlds: Origins of the Extraterrestrial Life Debate from Democritus to Kant* (Cambridge, 1982), p. 95; cf. idem, "Plurality of Worlds and Natural Philosophy" (Ph.D. diss., Indiana University, 1977), p. 162: "Mersenne did not yet see the relevance of his discussion . . . to the Copernican theory."

the church, nor can it be decided by arguments based on officially established articles of faith. The rejection of the idea of plurality by the church fathers cannot be regarded as binding, since they have been known to be mistaken, for instance in the matter of the antipodeans. (This point had already been made by Kepler and Campanella, and it was to recur in the writings of Wilkins, Godwin, and others in the course of the seventeenth century—as a leitmotif often appearing in association with the name of Columbus.) From this reasoning, Mersenne concludes that the uniqueness of the world cannot be regarded as a doctrine of the Christian faith: "For these reasons it appears to me that one must conclude that this truth—that there are not numerous worlds [*plures mundi*], or, which is the same thing, that this our world, whose parts we see, is unique—is not a matter of faith" (col. 1088). For a prominent cleric, described as "the oracle of the learned world,"[15] to voice such an opinion only a generation after Bruno had died at the stake as a partisan of plurality, seven years after the church's condemnation of the heliocentric theory, and ten years before Galileo's enforced recantation, shows astonishing boldness.

In a second step, however, Mersenne does in fact "prove" the uniqueness of our world, but he does this by arguments of a general theological or philosophical nature, not by *dogmatic*—or, for that matter, scientific—ones. Contrary to Bruno's opinion, God's infinite *power* need not, he says, be active in the creation of an infinity of worlds. (Mersenne takes up the Thomist objection that many identical worlds would be pointless, while many different ones would be imperfect and would raise the question of God's motivation—of "why he had not created better worlds.") The *unity* of God finds its most appropriate expression in a single world. Finally, Mersenne invokes God's *free will* when, remembering the official Catholic argument of 1277 (that if God's free will is not to be impugned, the plurality of worlds must be admitted as a possibility), he argues that this very freedom means that God does not need to create everything that it is possible for him to create and that he exercised this august power of choice in creating only one world.[16]

---

[15]Busson, *Pensée religieuse française*, p. 287.

[16]*Quaestiones Celeberrimae in Genesim* (1623), cols. 1090–91: "Ad 3 & 4 dico mundum vnicum à Deo creatum esse, vt in ei vnitate diuina vnitas aliquo modo repræsentaretur, adde quòd non fuit Deo melius, vt plures mundos, quàm vnicum faceret, aut etiam nullum produceret, cùm enim omnia, quæ sunt in Deo, sint ex omni parte infinita, quomodo vel vnus, vel plures mundi, qui semper finiti essent, & velut nihilum ante Deum, illi meliores essent? Deinde non essent isti mundi meliores, quoad rationem productionis suæ; quicquid enim product Deus, infinitâ suâ potentiâ, & sine dependentia ab vllo producit: quòd si quid melius in eis esset, quoad res productas, homines

Orthodox scientists of the early seventeenth century took up a stance similar to that of the orthodox theologians: either starkly rejecting or more flexible, but ultimately negative. The plurality concept, which now posed such a threat as a result of Galileo's work with the telescope, was just as unacceptable as the Copernican theory that provided the basis for it. Even a thinker as receptive to new ideas as Francis Bacon was among the scientists who took this attitude.[17]

Against this background, literary accounts of space journeys were bound to adopt a viewpoint opposed to all speculation about the plurality of worlds. Such a rejection of plurality takes the form of out-and-out satire in Charles Sorel's *Vraie histoire comique de Francion* (1623–26), a picaresque novel whose episodes include two imaginary journeys into space (bks. 3 and 11). It is particularly interesting that in the first of these, Sorel takes the traditional view of the universe—the Ptolemaic system with its transparent spheres and the fixed stars attached to the firmament—and pushes it ad absurdum, whereas in the second space journey he is equally ready to make fun of the conclusion so frequently drawn from the destruction of the old cosmos, that the "Earth-like" planets of the new cosmology, and also the moon, are inhabited. The height of absurdity is reached with a book that Francion's friend Hortensius plans to write about the manlike inhabitants of the moon. And, curiously, Sorel introduces, as a ridiculous notion, an idea that was to become a serious theme in science fiction, treated by H. G. Wells, Kurd Lasswitz, and many others: the invasion of the Earth from outer space. (Incidentally, a catastrophic outcome is averted here, just as in *The War of the Worlds*, by the intervention of a feverish infection.) This episode in Sorel's book is no more than an amusing satire on the unorthodox idea that there may be intelligent beings in outer space: an encounter with aliens simply represents the peak of absurdity. We know from Sorel's scientific book, *La science universelle*, that his attitude toward both heliocentrism

---

vlteriùs inquirerent, cur non meliores mundos creasset, vsque ad infinitum; quod cùm in rebus creatis admitti non debeat, vel possit, vt passim D. Thomas cum Patribus edisserit, stultè quæritur, quamobrē vnicum crearit. Quamuis autem infinitos creasset, nondum fortè quiesceret humana curiositas, sed cuperet inuestigare, cur mundus non esset æqualis Deo, & cætera, quæ indigna sunt, quæ vel efferantur, vel cogitentur. Denique vbi causa simul infinita est, & liberrima, frustra quæritur, qua de causa hoc, aut illud fecerit: cùm enim nulla re indigeat, certissimum est eorum, quæ facit, solam causam esse voluntatem ipsius, qua nihil prius, nobilius, maius, vel melius excogitari potest."
[17]*Descriptio Globi Intellectualis et Thema Coeli*, chaps. 5, 6, in *Scripta in Naturali et Universali Philosophia* (Amsterdam, 1653). Other opponents are named by Otto Zöckler, *Geschichte der Beziehungen zwischen Theologie und Naturwissenschaft*, vol. 1 (Gütersloh, 1877), p. 539.

and the plurality of worlds was one of skepticism and rejection (he chose to accept Tycho Brahe's geocentric compromise between Ptolemy and Copernicus) and that his reasons for this rejection were derived not from theology but from physics.[18]

The most interesting example of primarily *theological* polemics against plurality in a story of a space voyage is to be found in a book that remained popular well into the eighteenth century, *Itinerarium Exstaticum* (1656, titled *Iter Exstaticum Coeleste* in later editions), by Athanasius Kircher, a Jesuit father who was extremely prominent in the intellectual life of his day. It is interesting because the author was not only an orthodox dogmatician in the Scholastic tradition but also a noted if somewhat amateurish scientist who kept abreast of developments particularly in post-Galilean astronomy and, incidentally, also made telescopic observations himself.[19] Of course Kircher's concern with developments in cosmology was largely a reconnoitering of enemy territory; both publicly and privately he anathematized Galileo's Copernican conception of the motions of the Earth.[20] Like Sorel, Kircher too favored Tycho Brahe's system (1660, pp. 38–39), which had the sun circling the stationary Earth, while the planets circled the sun; this was at least compatible with the church's cosmologic symbolism and to that extent not unacceptable to orthodox churchmen. For Kircher too the Earth's central position in the universe was linked with man's theological centrality—and uniqueness. Indeed, in terms of the ideological content of this "journey into the heavens," this is the book's overriding theme. In all cosmologic matters, as Kircher programatically emphasizes in his "admonitory prelude," he follows the doctrines of Holy Writ and of the church fathers. (In fact he does depart from the Aristotelian-Ptolemaic conception on some points, not only where Tycho's beliefs diverge from it, and he was criticized for this by upholders of orthodoxy.)[21]

Kircher demonstrates his own orthodoxy equally clearly, however, in his approach to modern scientific theories and hypotheses, particularly in his treatment of the plurality of worlds. The subject is raised by Theodidactus, the disembodied soul who undertakes the

[18]Beverly S. Ridgely, "The Cosmic Voyage in Charles Sorel's *Francion*," *Modern Philology* 65 (1967), esp. p. 7.

[19]On Kircher's high reputation among his contemporaries see John E. Fletcher, "Astronomy in the Life and Correspondence of Athanasius Kircher," *Isis* 61 (1970), 52–53. *Iter Exstaticum Coeleste [Itinerarium Exstaticum]*'s subsequent editions are dated 1660, 1671, 1729, and 1753; references in the text are to specific editions.

[20]Unpublished letters, quoted and paraphrased by Fletcher, "Astronomy," p. 56.

[21]See Conor Reilly, *Athanasius Kircher, S. J., Master of a Hundred Arts, 1602–1680* (Wiesbaden, 1974), pp. 164–68.

"ecstatic journey" to the moon, sun, planets, and fixed stars, guided by his guardian spirit, Cosmiel. At every place they visit, he asks his mentor whether there is life on these bodies, and he receives the same reply each time: God in his inscrutable wisdom destined only the Earth to be the abode of life and of man (e.g., 1656, pp. 352–53, 390–91)—*even though*, Kircher says, agreeing with Copernicus in rejecting Aristotle's qualitative division of the universe, the matter found in the rest of the universe is the same as that "below the moon" (1656, pp. 101, 302, 398). This is why some of the heavenly bodies they visit do appear to the Catholic soul to be strongly reminiscent of Earth—first the moon with its mountains and seas, rivers and islands, and then also Venus, Mercury, and Jupiter with the exotic appeal of their land-scapes. But these fairy-tale worlds lack even the simplest forms of life, whether created or arising spontaneously. Even vegetation is absent: as in the world of Algabal created by Stefan George, the colors of the landscape are only those of precious stones. In the case of the repulsively inhospitable worlds of Mars and Saturn, however, it is immediately obvious that they are dead and hostile to life, with flam-ing volcanoes and sulphurous clouds on the one and glaciers on the other. Kircher shows some courage in assuming planetary systems belonging to the fixed stars, but his verdict on life on these planets is similarly negative.

What is the purpose of these innumerable uninhabited worlds? God created them "for the sake of man," Theodidactus is told at the end of his grand tour (1656, p. 432). Christian dogma speaks only of the heavenly bodies providing us with light and the means of measur-ing time; if Kircher implies something more, it can only be the astrolog-ical influence exerted by these bodies on terrestrial man through the spirits or angels whom Kircher does place there. It was, of course, not unorthodox to assume inhabitants of this kind, superior to man, in outer space. Though these theologically sanctioned planetary beings are handled allegorically, Kircher's portrayal of them sometimes lends them a touch of humanity (though never of human frailty); but on each occasion the spirit-guide quickly reiterates the orthodox assurance that there is only one human race, the one that inhabits our Earth—which in this pious cosmology is decidedly *not* a planet. Neither the theology of plenitude nor the new science can persuade Kircher of the existence of a plurality of worlds. He himself states this explicitly (as does his pupil Melchior Cornäus) in the "Apologeticon" that forms the conclusion to the second and third editions ("That God could not have made more than one world [*plures Mundos*]," 1660, p. 493). This assertion was intended to guard against the imputation, which did indeed arise, that he might in some way be aiding and abetting the "new heresy" (1660, pp. 493–99, 509–12).

That Kircher should adopt this position must have irked scientists not so strongly committed to a particular philosophical viewpoint, and before the century was out Huygens's *Kosmotheoros* (1698), a classic work in the history of the idea of plurality, would criticize Kircher's work in a tone of amused condescension. Kircher's overanxious attention to dogmatic rather than scientific considerations resulted in absurdities, among which Huygens singles out (as Johann Christoph Gottsched and Otto Zöckler were also to do[22]) Kircher's reflections on whether baptism in the crystal-clear waters of the rivers on Venus could be valid. Interestingly, however, Huygens adds that Kircher would have spoken very differently had he been in a position to speak freely, without theological constraints and simply on the basis of his scientific knowledge and his own reason.

Such free speech was, in the early decades after Galileo's revelations, exercised by numerous amateur scientists less strongly bound by dogma than Kircher, who was a professor at the Collegium Romanum. It is noteworthy, however, that they did not, as Huygens would have expected, automatically and triumphantly conclude that there *is* a plurality of worlds. It is far more common to find various rather lukewarm responses; and perhaps one should regard these as the typical expression of the mood of the time, rather than either orthodox opposition to or enthusiastic assimilation of the plurality idea. Andreas Gryphius, for instance, was definitely receptive to the new insights into cosmology; as a young man he wrote a thesis "de igne non elemento" (that fire is not an element), which caused offense in orthodox circles, evidently because it cast doubt on the traditional view that the Earth beneath the moon's trajectory was surrounded by a sphere of fire. This means in fact that he, like Donne ("the Element of fire is quite put out"), doubted the old cosmology and aligned himself with Copernicus; indeed, in a well-known poem he paid Copernicus enthusiastic tribute: "Du dreymall weiser geist / Du mehr den grosser Mann. . . . / Der du der alten träum und dünckell widerlegt" (Thou thrice-wise intellect, thou more than great man . . . who hast refuted the dreams and vain conceit of the ancients). And yet Gryphius, at least in his published writings, did *not* draw the conclusion so often drawn from Copernicanism, that the Earth was not the

---

[22]*Kosmotheoros* (The Hague, 1698), pp. 90–91; Gottsched's translation of Fontenelle, *Gespräche von mehr als einer Welt*, 3d ed. (Leipzig, 1738), p. 107; Otto Zöckler, *Geschichte*, vol. 2 (Gütersloh, 1879), p. 56.

only inhabited planet. As Hugh Powell says, "He accepted the 'New Philosophy' and yet remained a sincere believer in Christian doctrine. Andreas Gryphius was indeed a child of his age."[23]

With Robert Burton, whom I have already mentioned several times, the picture is more complex. In each of the six editions of his *Anatomy of Melancholy* for which he himself was responsible (which appeared between 1621 and 1651 [eleven years after his death]), he deals, in the "Digression of Air," with post-Copernican, post-Galilean cosmology, revising his text each time to reflect the latest developments and showing a particular interest in the plurality of worlds as an implication of that cosmology. Conscientiously he reviews what has been written on the subject and enunciates some of the philosophical questions that arise from it, recalling, among other things, Kepler's unease in his *Dissertatio cum Nuncio Sidereo*. But Burton's own view on the matter is not so easy to discover. Over a period of two decades, he expands his commentary on post-Galilean developments without systematically thinking it through again, and this results in some confusion, which may have been not only that of his educated contemporaries but probably his own as well. The cosmologic excursus is a maze in which different critics searching for an underlying design reach totally different conclusions: that Burton's view of the universe is essentially still the medieval one; that it vacillates inconclusively between old and new; or that, at any rate by the time the last edition is reached, it accepts the plurality of worlds in the light of Copernicanism as confirmed by Galileo.[24]

This last conclusion is the newest interpretation, reached by comparing the variants in the six editions. It is correct in claiming that Burton unequivocally links the plurality idea with Copernicanism, as an implication inherent in it. But the most important passages adduced as evidence of Burton's gradual conversion to both theories were already present in the first edition, in which, according to this view, Burton is still supposed to have rejected the concept of plurality associated with the new science.[25] What is more, the relevant statements

---

[23]Hugh Powell, "Andreas Gryphius and the 'New Philosophy,' " *German Life and Letters*, n.s., 5 (1952), 277. Donne, "The First Anniversary," line 206.

[24]Charles Trawick Harrison, "The Ancient Atomists and English Literature of the Seventeenth Century," *Harvard Studies in Classical Philology* 45 (1934), 11; Robert M. Browne, "Robert Burton and the New Cosmology," *Modern Language Quarterly* 13 (1952), esp. p. 145; Richard G. Barlow, "Infinite Worlds: Robert Burton's Cosmic Voyage," *Journal of the History of Ideas* 34 (1973), esp. p. 301.

[25]See the quotation "But *hoc posito* . . ." in Barlow, "Infinite Worlds," pp. 297–98, and "We may likewise . . . ," ibid., p. 299; see also Browne, "Robert Burton," pp. 140–41. Barlow's account of Burton's development leading to his acceptance of the new

serve in all editions only as suppositions for the sake of argument; and these earnest intellectual exercises are both preceded and followed, even in the last edition, by references to the "prodigious paradoxes" and "absurdities" inherent in Copernican astronomy, with its elevation of the Earth to the status of a planet (2:53, 56). These considerations lead Burton, an Anglican parson, to conclude that it is better to leave all "mysteries" of this kind to the discretion of God, who has purposely kept them hidden but may one day reveal them to us (2:60). So states an addition to the fifth edition, the last to appear in the author's lifetime (1638). The cosmologic excursus in the *Anatomy of Melancholy* (which was probably used for decades by educated people as a kind of reference work) thus gives us a useful insight into the period, revealing not bafflement and fear like that experienced by Donne or Pascal, but a kind of confusion and indecision that is intriguing rather than existentially threatening, a vacillation between distrust and acceptance, which must have been the reaction of many of those who were receptive to the modern theories.[26]

Milton is another writer whose piety makes him respond less than enthusiastically to the new ideas, and in the manner of his response he too is representative of his time. In 1638/39 Milton had visited Galileo, the prisoner of the Inquisition, at his house at Arcetri, near Florence. One can picture him reacting to Galileo's beliefs about cosmology, the basis of the modern concept of plurality, with a polite shrug of the shoulders. Whereas Burton, at least up to the fifth edition, which added a passage expressing pious trust in God, was seriously seeking a definitive conclusion to end his uncertainty, Milton is sufficiently calm and detached to be able to find amusement in the conflict of hypotheses. He is moderately interested, but basically it does not matter to him whether or not God created living beings in other worlds. His convictions as a Protestant, like Kircher's as a Catholic, cause him to focus his attention solely on the centrality of man in God's view—mankind as descended from its first ancestors in the Garden of Eden and redeemed by Christ. He shares this attitude, if nothing else, with Satan in his *Paradise Lost* (1667). Satan flies through the infinite expanse of space:

---

astronomy and its implications is not clear; see pp. 292, 293, 301. Page references in the text are to the edition cited in n. 12 above.

[26]Douglas Bush, *English Literature in the Earlier Seventeenth Century*, 2d ed. (Oxford, 1962), p. 295: "Probably no passage in the literature of the time gives us a more vivid picture of the intellectual confusion caused by the new astronomy—for relatively few persons of course—than Burton's galloping survey of the multitudinous theories that were, in more than one sense, in the air"; one of these is the theory of the plurality of worlds, "the idea that especially attracted and repelled him."

Amongst innumerable stars, that shone
Stars distant, but nigh hand seemed other worlds,
Or other worlds they seemed, or happy isles,
Like those Hesperian gardens famed of old,
Fortunate fields, and groves and flow'ry vales,
Thrice happy isles, but who dwelt happy there
He stayed not to inquire. . . .

(3, lines 565–71)

It is interesting, however, that Milton, too, specifically links the "other worlds" with Galileo. In book 1, he speaks of Galileo on the hill of Fiesole in Tuscany observing the moon through his "optic glass," "to descry new lands, / Rivers or mountains in her spotty globe" (lines 291–92). "The glass / Of Galileo, less assured, observes / Imagined lands and regions in the moon" (5, lines 261–63). Whether these lands are inhabited or not Milton does not specify; but in another passage, no less a personage than the Archangel Raphael says that perhaps every star has its inhabitants: "and every star perhaps a world / Of destined habitation" (7, lines 621–22).

But these are merely a handful of fleeting references to a topic that was already exerting its fascination on an ever-growing number of people. Only once does Milton discuss it more fully, and then only to make the point that the idea is of little importance; he gives no verdict as to its plausibility. The occasion is the astronomical conversation between Adam and the Archangel in book 8 (lines 1–178). It is somewhat reminiscent of the dialogues between Theodidactus and his guide Cosmiel in Kircher's *Iter Exstaticum*. The "novice" in each case voices heretical doubts about the Christian-Aristotelian view of the universe; but in Milton, the answers do not seem, at least at first sight, to follow orthodox dogma quite so closely.

At the beginning of the conversation, Adam's ingenuous logic leads him to question the geocentrism and anthropocentrism vouched for by traditional theology and science. "Something yet of doubt remains": it seems to Adam disproportionate that all the shining stars of the firmament with their impressive circular orbits should be subservient to the immobile and insignificant Earth—a mere dot or "atom" in the universe—and hence subservient to man. The Earth, though inferior to the stars, is "served by more noble than herself." Adam's still-innocent, naive imagination does not, of course, go so far as to envisage "other worlds" lighted by these distant stars, perhaps in a whole series of Copernican universes. Raphael, conversant with modern scientific theories, does acknowledge this possibility, but only as a possibility: he does not even speculate on whether it might be a fact. He introduces Copernicanism and the idea of the plurality of worlds only hypothetically, in order to dispel Adam's doubts. What if the Earth were not the center of the universe? This "what if" opens the door, if only by a crack, to the contemporary debate about plurality. (It has been shown that the actual wording of certain passages of books by John Wilkins and Alexander Ross, arguing for and against the planetary nature of the Earth, must have been in Milton's mind when he wrote this conversation.)[27] What, then, if the Earth were a planet? In that case one must accept the analogical conclusion, familiar from the earliest discussions of Copernicus's theory, that the planets and even the moon are earths, possibly inhabited earths, and that there may be similar ones in the planetary systems of other "suns":

> What if that light
> Sent from her [the Earth] through the wide transpicuous air,
> To the terrestrial moon be as a star
> Enlight'ning her by day, as she by night
> This earth? Reciprocal, if land be there,
> Fields and inhabitants: her spots thou seest
> As clouds, and clouds may rain, and rain produce
> Fruits in her softened soil, for some to eat
> Allotted there; and other suns perhaps
> With their attendant moons thou wilt descry
> Communicating male and female light,
> Which two great sexes animate the world,
> Stored in each orb perhaps with some that live.
> For such vast room in nature unpossessed
> By living soul, desert and desolate,
> Only to shine, yet scarce to contribute
> Each orb a glimpse of light, conveyed so far
> Down to this habitable, which returns
> Light back to them, is obvious to dispute.
>
> (8, lines 140–58)

What if . . . ? The idea, based on Copernicanism, that we are not the only children of God is introduced only as a passing conceit. Why does Milton introduce it at all, if he means only to touch on the controversy without pursuing it in any depth? Adam is, after all, only

[27]Grant McColley, "Milton's Dialogue on Astronomy: The Principal Immediate Sources," *PMLA* 52 (1937), 728–62. On the controversy between Wilkins and Ross see below, at the beginning of sec. 3.

too eager to learn. The hypothesis is mentioned purely so as to put all such cosmic speculation, whether for or against the plurality of worlds and of humankinds, firmly in its place:

> But whether thus these things, or whether not,
> Whether the sun predominant in heav'n
> Rise on the earth, or earth rise on the sun,
> . . . . . .
> Solicit not thy thoughts with matters hid,
> Leave them to God above, him serve and fear;
> Of other creatures, as him pleases best,
> Wherever placed, let him dispose: joy thou
> In what he gives to thee, this Paradise
> And thy fair Eve; heav'n is for thee too high
> To know what passes there; be lowly wise:
> Think only what concerns thee and thy being;
> Dream not of other worlds, what creatures there
> Live, in what state, condition, or degree,
> Contented that thus far hath been revealed
> Not of earth only but of highest heav'n.
>   (lines 159–61, 167–78)

We are back, once again, with divine inscrutability. Like the Anatomist of Melancholy (in the late revision of his cosmologic excursus), Milton's angel and Milton resort to belief in a Protestant "hidden God." Indeed, as Raphael remarks at the beginning of his discourse, God deliberately made the Book of Nature illegible to us, for two reasons. First, this illegibility is a cipher for the inscrutability of the superhuman; it may be that God is amused by man's efforts to discover a meaning ("perhaps to move / His laughter at their quaint opinions wide"). The very fact that different cosmologic theories (not least those about the plurality of worlds) contradict each other may serve to remind man that, faced with something far beyond the scope of his reason, it behooves him simply to "admire" the edifice of the heavens created by the Divine Architect—and also to marvel at his own supremely privileged theological significance in God's plan of salvation. "Not to earth are those bright luminaries / Officious, but to thee earth's habitant." This seems to indicate that Raphael can interpret the structure of the heavenly edifice at least to the extent of recognizing that it is geocentric and thus in accordance with anthropocentric theology (in spite of his own declaration, "Whether heav'n move or earth, / Imports not"). Thus geocentrism, with its one world, is implicitly established, and this accords well with the second reason for the indecipherability of the Book of Nature: that its illegibility is a reminder of the other Book of God, namely the revelation contained in the Bible. The Archangel's speech culminates in the admonition to man to be content with what "thus far hath been revealed," which can only refer to God's revelation in Holy Writ.[28] And as far as Milton is concerned, the Bible contains nothing about "more than one world." It speaks of God and of man in *this* world, and the relationship between the two is a major one of those "things more worthy of knowledge" to which this very Protestant angel wishes to draw Adam's attention ("Argument," bk. 8).

However much Milton stresses the inscrutability of God and the unimportance of knowing how many worlds or humankinds there are, this cryptic reference to the revealed Word of God shows that he is in fact predisposed to reject the idea of plurality, even though here in book 8 he seems half-heartedly prepared to consider it. In fact the cards have been heavily stacked against the idea from the start. For although this Christian epic does not set out to depict a world exactly corresponding to one specific scientific cosmology, the action is in fact set in a universe unmistakably conceived as geocentric, even if it is larger than was usually supposed in pre-Copernican times.[29] Milton allows the controversy between Wilkins and Ross to surface briefly in the conversation between Adam and the angel and appears to leave the question open, being in any case more preoccupied with quite different matters; but his inclination is to resist Wilkins's enthusiastic campaign for plurality. (He also strongly objected to Wilkins's leading role in encouraging more intensive scientific research by members of the Royal Society; most particularly to his assertion that where there was a conflict between astronomy and the literal sense of the Word of God, he would opt for astronomy.[30] Wilkins's adoption of this principle had of course influenced his attitude to the plurality of worlds.) On the face of it, then, both Milton and Burton fail to give an answer to the problematic question of plurality and fall back on the inscrutability of God; but the similarity between them is, as we now see, only apparent. Of the two, Milton appears more conservative, less re-

[28]Ibid., pp. 756–57.

[29]See for instance William Fairfield Warren, *The Universe as Pictured in Milton's "Paradise Lost"* (New York, 1915), p. 17; Grant McColley, "The Astronomy of *Paradise Lost,*" *Studies in Philology* 34 (1937), 209–47, esp. pp. 244–45; Kester Svendsen, *Milton and Science* (Cambridge, Mass., 1956), pp. 84–85; Nicolson, "Milton and the Telescope," *Science and Imagination,* pp. 80–109.

[30]McColley, "Milton's Dialogue," pp. 759–61.

ceptive to the no-longer-quite-new ideas. But they are alike in that their thinking is still visibly dominated by the conflict between orthodoxy and the new heresy, which for both is represented by Wilkins's *Discovery of a World in the Moone* (1638). Each is preoccupied with the question of how far one can remain faithful to the official teachings of the church and yet venture on to the new territory opened up by science. The idea that this heresy could be a form of piety did not occur to either of them.

Thirty years after his philosophical exploration of the moon, John Wilkins, the vicar, became a bishop of the Church of England—despite his advocacy of the plurality of worlds and without being forced to retract it. That this was possible reflects the fact that in the midseventeenth century, at least in the Protestant countries, the "new heresy" was beginning to transform itself into a "new gospel." This gospel—obviously influenced by Galileo's "sidereal message," which was gradually permeating the general consciousness—proclaimed that the heavens now declared the glory of God in a new way: his praises were now being sung by the humankinds of all the worlds to whose possible existence Galileo's work had alerted "everyone." Mersenne, no doubt mindful of Bruno's death at the stake—which must have been a traumatic memory for many others too—had not, in 1623, accepted this as a logical inference: he saw no need for God's omnipotence to manifest itself in an infinity of worlds (cols. 1083, 1090–91). In the mid- to late-seventeenth century, in contrast, particularly after Newton had confirmed Galileo's findings once and for all, it dawned on more and more people that this heresy could be interpreted as a gesture of homage. Ralph Cudworth and Huygens, Borel and Otto von Guericke, and Wilkins before them, were the apostles of this new religion, which was still suspect in some quarters of course but all the more exciting for that. Later sections of this chapter, and the next chapter, contain a fuller account of these writers and the development of their ideas. To round off this preliminary survey of the various philosophical reactions to the idea of plurality in the period between the *Sidereus Nuncius* and the *Principia Mathematica*, here are just two less well known examples. The fact that both exhibit a certain banality shows the extent to which the transformation of Bruno's heresy into an accepted idea in liberalized theology had already become a commonplace.

The first is from Eusèbe and Théophraste Renaudot's *Recueil general des questions traittées és conferences du Bureau d'Adresse* (vol. 2, 1655). In one of the lectures given at this semicommercial, semiphilosophical institute to a general audience, the unidentified author remarks quite in passing:

> For this opinion about the plurality of worlds—which cannot in any way be dangerous in itself, but only through the conclusions that the human mind in its frailty will want to draw from it—far from being contrary to faith, as people imagine, is, rather, an assertion of the divine omnipotence and excellence of God, which is all the greater if it shows itself in the production of more creatures: so much so that it seems that to have created only one world and only one space would have been a restriction of that immeasurable goodness. (2:699)

Such expressions of vague religious enthusiasm, not inhibited by orthodox dogma, are no longer rare in the second half of the seventeenth century either in France or in England.[31] (Incidentally, the Renaudots' text was also published in English in 1664, as *A General Collection of Discourses of the Virtuosi of France . . .* "rendered into English by G[eorge] Havers," p. 537.) It is worth noting, however, that at this time it is still often considered necessary to show that the new "Galilean" gospel (a recurrent pun) does not conflict with the revealed Word of God. This is done, for instance, by one Robert Wittie, "Dr in Physick in both Universities, and Fellow of the Colledge of Physicians in London," in his handy little octavo volume *Ouranoskopia, or a Survey of the Heavens: A Plain Description of the Admirable Fabrick and Motions of the Heavenly Bodies, as they are Discovered to the Eye by the Telescope, and Several Eminent Consequences Illustrated Thereby* (London 1681). One of these "eminent consequences" is the plausibility of there being many worlds in the heavens, an idea the continuation of the text on his title page explicitly links with the Copernican hypothesis. If, he says, it is probable that the moon is a world, "why I pray *pari ratione* may not the other Planets be Worlds too, and have Inhabitants to exalt the great Name of their and our Creator?" (p. 26). Not only does the Bible not contradict this, as has been claimed for so long; on the contrary, Wittie goes on to show how some obscure biblical passages become clear if one interprets them literally in the light of that assumption (pp. 26–28, 31–34). Those orthodox polemicists who insist on a literal interpretation of the Bible now find their own method being used against them: faith, claims Wittie (in contrast to, say, Athanasius Kircher), enjoins us to believe in the plurality of worlds. But—and this

[31]For further examples see McColley, "Seventeenth-Century Doctrine," pp. 427–28.

addition is characteristic of the more liberal tendencies in the theology of the day—reason too finds the idea of plurality acceptable. Reason sees the analogy between Planet Earth and some of the other celestial bodies and recognizes that these would have no purpose if they were not inhabited by "Rational Beings" like those on Earth.

The exercise of reason—as practiced most notably by the members of the Royal Society (p. 36) in their study of nature—itself leads back to the God of Holy Scripture; for it is this God who, in his wise providence, ordained the functions of the heavenly bodies. "Nor do I see what absurdity in Reason or Religion can arise from such reasoning, to think that the All-wise God doth use the same Methods of Providence throughout the whole Universe, for like ends" (p. 30). This supposed heresy has no room for atheism (p. 41). On the contrary, the greatness and wisdom of the revealed Divinity are only enhanced by the plurality of humankinds (pp. 38, 44). And once God is credited with such inconceivable greatness, even the doctrine of the redemption by the Son of God, hitherto often used as an argument against the plurality of worlds, can be reconciled with it (p. 39), especially as numerous biblical passages taken in their literal sense (e.g., Eph. 1:9–10 and Col. 1:16) confirm that "they [the aliens] and we have no other way to come to God but by *Jesus Christ*" (p. 33). As early as 1622, Campanella's thoughts had, very hesitantly, moved in this direction, but he had taken care not to call the beings in outer space "men," as Wittie confidently does (see sec. 2).

If, then, the doctrine that the way to God is through Christ applies to extraterrestrials too, does this mean that other worlds also suffered a fall and that Christ was crucified in other worlds as well? Such questions about specific points of dogma—at one time sufficient reason for orthodox theology to dismiss the plurality of worlds and the antipodeans as unthinkable—have continued to exercise the minds of theologians right up to C. S. Lewis. Wittie avoids discussing these matters. His acceptance of the plurality of worlds urgently raises again, however, the troubling question of man's status, which had already caused Kepler such embarrassment, and on this, he does comment. It is significant that unlike Kepler he no longer attempts to prove terrestrial man's preeminence in God's favor; nor, on the contrary, is he convinced, as later writers are, that we are inferior to the "others." But he does strike a blow at man's theological vanity that would have been unthinkable for Kepler: "Does it not savour of too much Haughtiness, and too high an Opinion of our Selves, and our Services to God, to suppose that the great God made all those immense Bodies, that yield so fair a Luster, and that immeasurable space possessed by them, only for the use of Us on this pitiful invisible point; and that the Infinite Deity of Heaven should have no active Service or Adoration in all those Bodies, save only from us, poor Worms!" (pp. 30–31).

The baroque topos that in relation to God, man is a mere "worm" here acquires a surprising new dimension in the context of the new science. It was left to nineteenth- and twentieth-century science fiction to trace the humiliating and alarming implications of this thought to their logical conclusion, an attitude of cosmic defeatism. In the seventeenth century this aspect of the matter was accorded less attention than the glory of God "revealing" itself in a multiplicity, perhaps an infinity, of worlds inhabited by human beings. Two or three generations after such flights of speculation led to Bruno's death at the stake as a heretic, the same ideas are presented as nothing less than *The Reasons of the Christian Religion*—the title of an apologetic work, published in 1667, by the dominant figure of English Protestantism in the seventeenth century, the Puritan preacher Richard Baxter. While not citing biblical passages as evidence for the plurality of worlds, he is nevertheless as firmly convinced as Wittie that an interplanetary chorus sings the praises of God.[32] In the eighteenth century this idea enters devotional literature.[33] What was once a mortally dangerous heresy has become a fervent expression of piety.

The first step in the conquest of inhabited space by scientifically inspired philosophy was the placing of intelligent beings on the moon. Kepler's pioneering creation, the *Dream*, had set an example; Galileo's likening of the moon to the Earth had carried especially strong conviction through the direct impact of visual evidence. The *Selenographia* (1647) by the Danzig astronomer (and correspondent of Kircher's) Johannes Hevelius, a reference work much used by both professional and amateur scientists, confirmed the widespread assumption of water on the moon (p. 298) and the possibility also of intelligent creatures there ("If there should be living creatures [*res creatae viventes*] on the moon . . . ," p. 294). To be sure, Galileo had already (in 1616, the year when Copernicus's theory was banned) denied this possibility; he no longer believed, as he originally had, that the conditions necessary for the existence of water on the moon were in fact present.[34] He was

---

[32]In *The Reasons of the Christian Religion* (London, 1667), Baxter makes frequent mention of this topic, linking it with recent scientific findings; see for example pp. 97–98, 100–101, 180, 388–390.

[33]Zöckler, *Geschichte* 2:425.

[34]*Opere*, Edizione nazionale 12:240–41.

followed in this revised judgment by the Jesuit and anti-Copernican Giovanni Battista Riccioli in his standard work *Almagestum Novum Astronomiam Veterem Novamque Complectens Observationibus Aliorum* (1651) and later by the Utrecht philosopher Gerhard de Vries in his polemical *Dissertatio Academica de Lunicolis* (1678)[35] and by Christian Huygens in his *Kosmotheoros* (1698, p. 115). Even so, the idea of inhabitants of the moon had its supporters right into the eighteenth century, especially among nonspecialists. It developed into quite a fashion, so much so that it became the subject of satires and burlesques, particularly toward the end of the seventeenth century. Ben Jonson's masque *News from the New World Discovered in the Moon* (1621) discusses, among other things, the question of whether there are inns and alehouses on the moon; other examples include a posthumously published satire in verse by Samuel Butler (1612–80) entitled "The Elephant in the Moon"; Nolant de Fatouville's comedy *Arlequin empereur dans la lune* (1684) and Aphra Behn's farce based on it, *The Emperor of the Moon* (1687); as well as Elkanah Settle's opera *The World in the Moon* (1697). Serious scientific interest in the plurality of worlds had shifted its focus from the moon to the planets well before the end of the century, and so it could not be accused of persisting in speculative fantasies that were scientifically obsolete. For the analogy between the Earth and the planets, based on Copernicus's removal of the division of the cosmos into two parts, remained valid regardless of whether there was water on the moon or not.

2. A Voice from the Dungeons of the Inquisition: Campanella

One of the first to recognize that Galileo's telescope was capable of giving an unprecedented impetus to the idea of the plurality of worlds was Tommaso Campanella. He was both a scientist following developments in post-Copernican cosmology with the close attention of a declared anti-Aristotelian, and a Dominican monk whose orthodoxy was called into question during his lifetime and is still debated even today.[36] He was thus very familiar with the sort of questions that occupied the middle ground where new astronomy and old theology met, notable among them the idea of the plurality of worlds. While

---

[35]Dick, *Plurality of Worlds*, pp. 120–23.

[36]Gisela Bock, *Thomas Campanella: Politisches Interesse und philosophische Spekulation* (Tübingen, 1974), p. 21, and see pp. 1–26.

not a "zealous prophet" of the idea,[37] he was its first classic exponent in the post-Galilean era, and he is constantly mentioned in connection with it in the literature of the day, either (for instance by Wilkins) as an authority, or (for instance by Mersenne) as a heretic. Burton, the skeptic, also refers to him in this context.[38]

In his literary works Campanella makes only passing use of the idea that Adam is not God's only child; for his novel of a utopian state, *Civitas Solis* (1623), is not a story of an encounter between humans and creatures from outer space. It is set somewhere in eastern Asia, and the only extraterrestrial element in it is the belief held by the Solarians, the citizens of the ideal state, that they are not "alone" in the universe. But even this belief is only briefly touched upon. The *way* in which it is introduced is, however, significant. Of the Solarians' cosmologic beliefs we are told that "they are uncertain whether there are other worlds besides ours, but they think it is madness to claim that there is nothing [in space]; for they say that neither in the world nor outside it is there nothing, and nothingness is incompatible with God, the infinite Being."[39] This formulation is clearly dictated by well-founded caution; what it gives with one hand it takes away with the other, thus avoiding a direct assertion. In addition it invokes the argument of God's infinite *potentia*, which for centuries had been the one possible Christian defense of the plurality idea—even though it had not saved Bruno. Campanella also avoids suggesting a causal connection between plurality and the new science: the citizens of his Sun-state are not certain whether or not the sun is the center of the universe, and they perform the feat of both praising Ptolemy and admiring Copernicus (*Civitas Solis*, pp. 457, 455). A biographical note is called for here: every word that Campanella utters about plurality was written in the prison of the Spanish Inquisition at Naples, where he spent twenty-seven years of his life (1599–1626). True, it was as a political agitator and not as a theological heretic that he had fallen into the hands of the Jesuits; but of course he had every reason to fear that his views on matters of doctrine, which had come under scrutiny at his trial, would be kept under close surveillance.

[37]Zöckler, *Geschichte* 1:533.
[38]On Campanella's influence see his *The Defense of Galileo*, trans. Grant McColley, Smith College Studies in History 22:3–4 (1937), pp. ix, xli–xliv.
[39]Translated from the first edition of *Civitas Solis*, printed in Campanella's *Realis Philosophiae Epilogisticae Partes Quatuor* (Frankfurt, 1623), p. 457: "In ancipiti versantur, num alii mundi extra nostrum sint; ac furoris esse arbitrantur asserere, nihil esse: quoniam inquiunt, nihil neque intus neque extra Mundum est. Deusque ens infinitum non compatitur secum nihilum."

138 THE LAST FRONTIER

Although his many writings do not include a space novel, Campanella, as a philosophical poet and novelist, saw with especial clarity that the philosophical implications of Galileo's discoveries were of a kind particularly apt to inspire the imagination of literary and philosophical writers to further exploration, above all to further inquiry into the nature of the inhabitants of those other heavenly bodies. Appropriately, it is in a letter to Galileo himself that he develops this idea, and here it is taken as a matter of course that the idea of extraterrestrial beings is an inference from Copernicanism as made visible by Galileo. This long letter, written on 13 January 1611, gives Campanella's reaction to the *Sidereus Nuncius*.[40] Campanella's thoughts in this letter are not logically arranged, but they may be outlined as follows. For him, as for Kepler, the *Sidereus Nuncius* bears a message about the "mysteries of God" (*arcana Dei*). Campanella, however, is one of the first to make the comparison, which was to become common, between Galileo and Columbus. Here, as elsewhere, the introduction of this comparison points to a fact rarely admitted openly: that the idea of plurality was the main reason for the philosophical excitement Galileo's work aroused. Just as Amerigo Vespucci gave his name to the new world on Earth, so, Campanella hopes, Galileo will give his name to the new world beyond the Earth (*novo caelesti* [*mundo*]). For he made visible what had been invisible: a new Earth, *at least* one, "a new earth on the moon." Campanella sees this as the start of nothing less than a new and different phase in history and (for of course he assumes that the new world is inhabited) an undreamed-of new era in human consciousness, which certainly would not be without its problems. He foresees that the theologians will "grumble." But he points out that theologians, even church fathers, are fallible, and he cites the example—which springs to mind as soon as Columbus is mentioned—of the antipodean question. They had said that antipodeans could not exist because it was impossible to see how people living so far across the sea could be descendants of

[40]Galileo, *Opere*, Edizione nazionale 11:21–24. On Campanella's modification of his earlier pluralism as a result of his acceptance of Galileo's Copernicanism, see Arno Seifert, "Andere Welten und Gegenwelten," *Historisches Jahrbuch im Auftrag der Görres-Gesellschaft* 106(1) (1986), 95–96. Seifert generally emphasizes, similarly to Lovejoy, the neo-Platonist as well as animist and cabalist roots of some of the plurality speculation around 1600, at the expense of the impetus this speculation received from Copernicus. Such analysis, however, tends to make altogether too much of the similarity believed to exist in the pre-Copernican assumption of "angels" or "intelligences" or "spirits" as the inhabitants of the celestial bodies, on the one hand, and the assumption of analogues of human beings in the wake of Copernicanism, on the other hand. My point is that the change brought about by Copernicus is decisive. See also ibid., pp. 87–88. A few of Seifert's suggestions have resulted in minor changes in wording.

Adam and Eve, or to accept that Christ was also crucified on the other side of the globe. In Campanella's day, however, the antipodean question is settled. So, to have doubts about the new worlds in the universe is, he implies, as backward as to doubt the existence of antipodeans in America.

What still requires investigation, though, is the nature and the characteristics of these inhabitants of the heavens (for instance whether they, like us, see themselves as being at the center of the world) and what the inhabitants of the various stars know of astronomy and astrology. Campanella evidently takes it for granted from the outset that these "inhabitants" (*incolae* or *habitatores*) are human, or at any rate similar to humans. It is only their nature that is in doubt. Here again it is characteristic that the terms of reference he applies are, first and foremost, of a kind that can only be called theological. For he considers it necessary to discuss not only in what kind of political state these inhabitants live, but also whether they are "*beati*" or like us. *Beati* (blessed) here is clearly a theological term indicating a state free from original sin. But no sooner has Campanella said this than he retreats from the tangled maze of theological anthropology and adds merely that if the moon is "inferior" (*vilior*) to the Earth, because it is smaller and, as a satellite, dependent on it, then the moon dwellers must accordingly be "less happy" (*infeliciores*) than ourselves.

But this theme, touched upon here in a private letter, obviously has an irresistible appeal for Campanella. He discusses it again, publicly and more fully, in his *Apologia pro Galileo*. He wrote this apologia in 1616 as a last-minute attempt to avert the church's condemnation of Copernicanism and to help Galileo in his struggle with the Inquisition, but it was not published until 1622 (in Frankfurt). It is above all a defense of scientific research as an activity sanctioned by God, and in particular a defense of Copernican and Galilean astronomy against the charge of being theologically subversive. But the strong impact of this short book,[41] which continued to be influential for half a century, was due not least to its cautious but spirited support for the plurality idea, which had been given a new scientific basis by the new astronomy. And rightly so. For the main thrust of the treatise is the demonstration, contrary to received opinion, that the findings of heliocentric astronomy and their philosophical implications ("that kind of philosophy which Galileo has made famous")[42] are compatible with the words

[41]See above, n. 38.
[42]Subtitle of the *Apologia*. In the following discussion, page references in the text are to the McColley translation (see n. 38 above), which indicates the pagination of the Latin original (Frankfurt, 1622). The translation is not always reliable. Page references

of the Bible, rightly interpreted. These words must be understood, depending on the context, either as a form of expression adapted to the level of understanding of the hearer—in other words, not literally—*or,* in certain other passages, literally and not merely figuratively. This procedure is particularly effective when used to argue for the belief that other heavenly bodies are inhabited. Whereas he had been more cautious in *Civitas Solis,* in this work Campanella makes it clear that for him, as for many others, this assumption follows conclusively from the Copernican analogies, made visibly manifest by Galileo, between the Earth and the moon, between the Earth and Jupiter (and other planets), and between the sun and the fixed stars with their (deduced) planetary systems (pp. 64–65). But by making this scientific assumption of the existence of "many worlds," the Dominican friar and prisoner of the church maneuvers himself into a dilemma: he must adhere either to the (suitably interpreted) Word of God or to the current doctrine of the One True Church, which, *primarily on scriptural grounds,* rejected the idea of a circling Earth and all that followed from it.

As far as Campanella is concerned, the findings of the new science and "new philosophy" (he uses this term in his preface) are unassailable. The problem is only how to reconcile the truth of the Bible and, he hopes, that of the church, with them. For Campanella, there is no question of science having fallen from grace, but rather of theology having in effect committed a sin of omission. To be sure, in a concluding formula he is at pains to stress that while defending Galileo he submits at all times to the censure and superior judgment of the Roman church. But to what extent this is mere lip service must remain open. He does make the point, in the *Apologia* as in the letter to Galileo, that the error of the church fathers in the matter of the antipodeans proves that clerical tradition is not infallible (p. 66).

Of the eleven accusations brought against Galileo as a supporter of heliocentrism, which Campanella deals with one by one, two have to do with the plurality of worlds. The eighth accusation was that in the *Sidereus Nuncius,* Galileo had assumed that there was water on the moon (though later, in a letter, he withdrew this view) and that he claimed to have seen continents and mountains there. In other words, Galileo had maintained that the moon was similar to the Earth, which contradicted the Aristotelian doctrine of the essential difference between the substance of heaven and that of Earth "below the moon" (p. 8). Biblical statements on this matter may be taken literally, Campa-

---

marked "1622" after quotations are to the first edition, a facsimile of which is in the appendix to the *Apologia di Galileo,* ed. Luigi Firpo (Turin, 1968).

nella decides, taking the opposite view from Calvin and Wilkins.[43] He
evidently decides purely on the basis of his *scientific* knowledge which
passages of the Word of God are to be taken literally and which, like
those that speak of a circling sun (p. 52), are not; at no point does he
comment on this crucial aspect of his method. It is therefore incorrect
to say either that Campanella holds that the Bible should be taken
literally[44] or that he was a leading figure in the rebellion against literal-
ism.[45] On the question of the presence of water in the heavens (specifically, on the moon), Campanella attacks the opposition with its own
weapons: it is the Bible itself, taken in its literal sense, which vouches
for these waters in the heavens, for instance in Genesis (1:6–7) and in
Psalm 148. Arguing further on the basis of this fact, Campanella
makes the point that where there is water, there must also be land,
surrounding it and holding it together (p. 56). And this deduction too
is confirmed by the Bible: Genesis 49 and Deuteronomy 33 speak of
fruit trees, mountains, and hills in the heavens. Thus the Old Testament, taken literally, agrees with Galileo's findings. Galileo therefore
deserves not criticism but gratitude, for saving the Word of God from
appearing ridiculous and for laying bare the folly of certain unnamed
biblical exegetes (p. 64). Thus the Bible, the Word of God, is at least
in parts also a scientific textbook. The plurality of worlds, regarded
with suspicion by official Catholic theology, is shown to be a revealed
truth.

So far there has been no mention of "men" on other planets. This
more dangerous assumption figures in Campanella's refutation of the
ninth accusation against Galileo, which was implicit in the eighth. If,
according to Copernican premises, the four elements that make up
our world are thought also to be present on the other heavenly bodies,
then it follows that "there are numerous [*plures*] worlds and earths
and seas, as Mahomet [!] asserts; and men living on them" (1622,
p. 8). As Campanella (like Melanchthon) points out, however, this
possibility is excluded by the Bible, which tells of only a single Creation
and a single crucifixion of Christ (p. 8). But in this instance, Campanella, the monk languishing in the dungeons of the Inquisition, does
not regard the adverse testimony of the Bible as conclusive. He begins
by clarifying his terms, making the distinction that Bruno had made.
What is suggested, following Galileo's observations, is not a plurality

[43]John Wilkins, *A Discourse Concerning a New Planet* (London, 1640), p. 54. He
invokes the authority of Calvin, whom he quotes.
[44]Dick, *Plurality of Worlds*, p. 92; cf. his earlier "Plurality of Worlds," p. 165: "should
always be interpreted literally."
[45]McColley in his translation of the *Apologia*, p. xxix.

of (invisible) world systems but a plurality of worlds within one immense heaven, one "world": not many Aristotelian *kosmoi* but many planetary worlds.[46] But this view, too, was heretical. So, in a second passage of arms, Campanella attempts to undermine the logical basis for condemning *this* conception of the plurality of worlds as a heresy. The fact that the Bible does not mention the creation of such other worlds is not, he says, logically speaking an argument against their existence. The Bible deals only with truths that concern our Earth. Moreover, he claims that dogmatic tradition, the other pillar of the Catholic church, also contains no actual prohibition of this conception of a plurality of worlds. This point is also stressed by Mersenne a year later (sec. 1 above), but unlike him, Campanella does not go on to say that the idea is to be repudiated all the same. On the contrary, Campanella claims that the Scholastics, notably Saint Thomas Aquinas, only condemned the plurality of worlds conceived of as a plurality of world systems in the Aristotelian mold, with its (Atomistic) suggestion of chance as the governing principle of the world. The concept of many small systems contained within one great all-embracing system created by God, with not one but many centers of gravity, is something quite different, Campanella goes on to say: it conflicts with Aristotle's doctrine (with its single point of gravity) but is not contrary to reason, the Bible, and dogmatic tradition.

But what about the next step, the assumption that there are *"human* species" (*hominum species*) in those worlds? It "does not follow" (*non sequitur*), Campanella says, from Galileo's discoveries. He stresses in the same breath, however, that this assumption of the existence of "men" does not mean, as has been objected, that Christ must also have been crucified on other celestial bodies, in order to redeem the human beings there. The argument, together with that of descent from Adam, was advanced by Saint Augustine and other church fathers against the idea of antipodeans—and was proved wrong, Campanella reminds his readers (see also pp. 15 and 35). Does this mean that Campanella, whether on the basis of Galileo or not, does in fact assume the existence of human beings in space? He is clearly inclined to do so. But he tries to play this down, saying that *if* they existed ("homines, si qui essent in alijs syderibus," 1622, p. 51), they would not be descended from Adam, would not have inherited Adam's sin, and therefore would not need to be redeemed, unless they had

---

[46]P. 64. The following account relates to pp. 64–67. In it, the order of Campanella's argument is not always precisely followed; sometimes passages that occur separately in Campanella's text are here placed together.

committed some other sin. This last possibility would accord with passages in the Epistles to the Ephesians (1:10) and to the Colossians (1:20) which state that Christ, by his blood, united Heaven and Earth with God the Father. "But we do not know these things; therefore we stand by the traditional teaching of the Fathers" (*stamus in antiqua expositione Patrum*, 1622, p. 51). This is a most remarkable sequitur when one recalls that in these very pages Campanella judges the church fathers' authority in such matters to be open to ridicule. Still, he is tactful: there is the other possibility, not quite so risky from a dogmatic point of view, that the creatures on other stars are not human. Galileo, says Campanella, expressly pointed out this possibility in his letters about the sunspots but at the same time expressed the opinion that the creatures of a "different nature" (*alterius naturae*), which might exist there, were "similar" (*analoga*) to us but not the "same" (*univoca*)—despite the playful creations of Kepler's imagination.

Strictly speaking, Campanella has still not quite managed to circumnavigate the rock of doctrine, for he does not explain what he means by "entia . . . nostris analoga entibus" (beings analogous to our beings, 1622, p. 52). This may be a way of evading the problems posed by original sin and redemption, but it raises and fails to answer the question of man's status in Creation, the question that had greatly troubled the conscience of Kepler, whose *Dissertatio* Campanella alludes to. Campanella passes over the question, trying to divert the reader's attention from it. Even if the doctrine of the plurality of worlds is untrue, he says, that does not lessen Galileo's achievement. This evasion, however, is an insult to the common sense of his contemporaries, for whom the plurality of analogous planets in the Copernican universe was a valid reason to assume a plurality of analogous species of men. The same is true of Campanella himself: before Galileo's observations, he says, he rejected the Copernican model of the universe (p. 70) and demonstrated the impossibility of assuming human beings to exist on other celestial bodies (p. 66). It is evident that *after* Galileo's observations he *does* consider it likely that the heavenly bodies are inhabited, but he does not say so explicitly. A man well able to think for himself, he is fully aware, as a prisoner of the Inquisition, of the danger inherent in this inference. This is why, for instance, we find no trace of the teleological argument that was soon to be an indispensable part of the case for the plurality of worlds: that stars invisible to our naked eye, as well as the planets, were not created by God without purpose, but "for" their inhabitants (not for us). Nor (unlike, for instance, Robert Wittie in 1681) does Campanella let himself be tempted to see the many inhabited worlds as a sign of the omnipotence of God, whose praise now resounds from a million stars. He confines himself to the narrower dogmatic matter of original sin and redemption by Christ. It would therefore be an exaggeration to say that Campanella deals "successfully" with the problems raised by the plurality of worlds.[47] Imprisoned by the Jesuits, he anxiously limits himself to the sort of reflections that the Catholic church would have tolerated at the time.

Protestant orthodoxy, which in Melanchthon's day had been wholly in accord with orthodox Catholicism, had by the early seventeenth century become considerably more permissive, and it is highly instructive to see how, through John Wilkins, Campanella's ideas were able to prosper and bear fruit in England.

3. Two Bishops on the Moon: Wilkins and Godwin

Like Campanella, John Wilkins (1614–72) was equally at home in theology and science. As a student at Oxford in the early 1630s, as a member of the Gresham College circle in London, as the leading spirit of a group with similar interests in Oxford, and as secretary in the 1660s of the Royal Society that grew out of this group, Wilkins's whole life was spent in the aura of the new philosophy Copernicus had inaugurated. Wilkins's studies and interests were also theological, however, and he made his career in the church, rising between 1637 and 1668 from the position of vicar of Fawsley to that of bishop of Chester. Shortly after embarking on this career, he published (anonymously, but with the imprimatur of the bishop of London) the book that contained his contribution to the debate about the plurality of worlds: The Discovery of a World in the Moone: or, A Discourse Tending to Prove, that 'tis Probable there May Be Another Habitable World in that Planet (London 1638).[48] To the third edition in 1640 he appended A Discourse Concerning a New Planet, Tending to Prove, that 'tis Probable our Earth is one of the Planets. Although this additional work does not speak of the plurality of worlds, it lays down the basis for the theory and the grounds for accepting it: Wilkins not only defends the Copernican

---

[47]Dick, "Plurality of Worlds," p. 177.

[48]Two editions of the Discovery were published in 1638, followed by further editions in 1640, 1684, 1707, and 1802; reprints of the first edition in 1972 and 1973. A French translation, Le monde dans la lune, was published in Rouen in 1655; a German one, Vertheidigter Copernicus, in Leipzig in 1713. Page references in the text are to the first edition unless another source is indicated.

world model but also shows that it is not a contradiction of Holy Writ to suggest that in doubtful cases one may regard the text not as literal truth but as a version adapted to the conceptions familiar to the reader. The very title suggests the nature of the link between this "permissible" Copernicanism and the complex of questions relating to the plurality of worlds: if "our Earth is one of the Planets," then other planets may well be Earths.[49] And the new heresy encapsulated in this much-used syllogism was indeed a major factor in the violent controversy unleashed by the publication of a work written to counter Wilkins's arguments. This book, by Alexander Ross, a "literalist" whose orthodoxy made no concessions at all to Copernicanism, was entitled *The New Planet No Planet: or, The Earth No Wandring Star; except in the Wandring Heads of Galileans* (1646).

But even apart from this sensational controversy, Wilkins had an uncommonly strong impact, both in England and on the Continent, with his "haute vulgarisation" of the new philosophy that only a short time earlier had had such a devastating effect on Donne: "Wilkins found himself at the center of the intellectual debate of that period, of which he quickly became a typical representative,"[50] indeed "the most dynamic force in seventeenth-century England."[51] He was certainly the most forceful advocate of Copernicanism and of plurality in England in the seventeenth century, and the most influential, particularly in the realms of popular science and literature; it has been possible to trace his extensive influence right up to Fontenelle's *Entretiens sur la pluralité des mondes* (1686).[52]

While Wilkins found in Copernicanism the physical basis for the theory of plurality, for him too it was, strictly speaking, the telescopic discoveries made by Galileo, the "eye-witness" of Copernicanism, that were decisive (p. 125). Wilkins speaks not only for himself but for his whole century when he says, "In my following discourse I shall most insist on the observation of *Galilaeus*, the inventour of that famous perspective, whereby we may discerne the heavens hard by us,

[49]This analogy is explicitly drawn in the *Discovery* (1st ed., 1638): "Now if our earth were one of the Planets (as it is according to them [believers in heliocentrism]) then why may not another of the Planets be an earth?" (p. 94).

[50]Waldemar Voisé, "John Wilkins, sur la science et le voyage dans la lune, à l'époque de la mécanique céleste postcopernicienne," *Revue de synthèse* 94(1973), 44. See also Grant McColley, "The Ross-Wilkins Controversy," *Annals of Science* 3 (1938), 153–88.

[51]McColley, "Ross-Wilkins Controversy," p. 155.

[52]See Barbara J. Shapiro, *John Wilkins, 1614–1672: An Intellectual Biography* (Berkeley and Los Angeles, 1969), pp. 33, 35, 38; Marjorie Nicolson, *Voyages to the Moon* (New York, 1960), pp. 93–98.

whereby those things which others have formerly guest at are manifested to the eye, and plainely discovered beyond exception or doubt. . . . So that what the ancient Poets were faine to put in a fable, our more happy age hath found out in a truth" (pp. 87–88, 91). "Fable" and "truth" obviously refer to the same thing: inhabited worlds on other heavenly bodies, whether on the moon or on the planets.

It was possibly Campanella who made Wilkins aware of this achievement of Galileo's, which was of such epoch-making significance for the idea of plurality. At any rate, it has been shown that Campanella's *Apologia pro Galileo* is the main source used for the *Discovery of a World in the Moone* (his "chiefe adversary" being Galileo's opponent Julius Caesar La Galla, who denied the analogy of Earth and moon in his *De Phaenomenis in Orbe Lunae*, 1612).[53] On the subject of plurality there are striking parallels, with Wilkins sometimes referring explicitly to Campanella and sometimes not. Among the elements common to both works are rejection of a plurality of *kosmoi* in favor of a plurality of planetary worlds; insistence on the compatibility of this second conception with the testimony of the Bible and the judgment of reason; consistent anti-Aristotelianism; the inclination to believe in creatures merely similar to man, not "human beings," in outer space; and last but not least, the caution with which both authors avoid openly asserting the existence of life in space although both clearly believe in it. (Wilkins's very title displays this caution: "*Tending* to prove, that 'tis *probable* there *may* be . . .".)

Against this background of similarities, however, the differences between the two stand out all the more clearly. First, the whole argument in the *Discovery* is of a different nature: Wilkins is attempting a scientific demonstration, while Campanella's work is a theological apologia. Second, where theological themes do impinge on Wilkins's argument, in connection with plurality, the accents are placed differently.

A good illustration of the first difference is provided by the contentious interpretation of the passage in Genesis about the waters—and therefore the "lands"—in the heavens. Campanella took the passage literally. Wilkins not only refuses to take it literally but also concludes the argument with the generalization that he does not believe "that any such thing can be proved out of Scripture" (pp. 110–12). What he means in this context is that the Bible is not a textbook on nature, that it cannot speak authoritatively on matters of natural philosophy. Theological arguments cannot establish the plausibility of the idea that

[53]Grant McColley, "The Debt of Bishop John Wilkins to the *Apologia pro Galileo* of Tommaso Campanella," *Annals of Science* 4 (1939), 150–68.

our world is not the only one, but scientific arguments—indeed, *only* scientific arguments—can. The one theological point Wilkins sees as relevant is the fact that the Bible does not categorically rule out the plurality of worlds.

On other matters too Wilkins uses exactly the same approach in the *Discovery*, one very different from Campanella's.[54] Despite the narrow focus of the title of his treatise he adduces reasons for assuming inhabited worlds not only on the moon but also elsewhere in our planetary system, though he does on the whole concentrate on the moon as one example that may stand for many (pp. 41, 178–79). Being a theologian, he naturally starts by stating the by-now-familiar theological and philosophical arguments for the doctrine of the uniqueness of our world upheld by the canonical tradition of the Roman church, Aristotelians, and orthodox interpreters of the Bible—but only in order to brush them aside. Wilkins too reminds the reader that the church fathers were mistaken about the antipodeans. Let another Columbus, then, discover a new world in the moon (pp. 6, 207). Aristotle's grounds for not believing in a plurality of *kosmoi*, especially his assumption of a single center of gravity in the universe, are, he says, contrary to reason, as Benedetti had thought (though Wilkins himself is arguing for a plurality of planetary worlds, not of *kosmoi*). Finally, Holy Writ, interpreted in a nonliteral way, by no means refutes the theory of the plurality of worlds (pp. 1–44). Wilkins disposes of these theological and philosophical considerations at the outset, so that he can proceed to the scientific arguments, which for him are the decisive ones.

The first interesting point to note is that he regards the idea of plurality as new, in fact as a "new *truth*" (p. 3). This means that, although he occasionally invokes supportive voices from antiquity, he does not see the plurality concept of his day as a simple revival of old speculations but as a new creation born of the spirit of the new science, in other words, of Copernicanism (see pp. 93–94). And indeed his proof is unmistakably based on Copernicus's arguments. Earth and heaven, or Earth, moon, and planets, are not essentially different in the sense of being composed of different elements, as Aristotle claimed; as the telescope shows, the moon has seas and continents, mountains, valleys, and plains; like the Earth it has an atmosphere, and even a moon of its own, namely the Earth. It also has seasons, and conditions of climate and light, which are at any rate comparable

---

[54]See for instance Campanella's announcement of his intentions at the beginning of the third chapter of the *Apologia*.

with those of Earth; and even meteors. But has it also inhabitants, and if so of what kind? All the other scientific questions lead up to this one; only the presence of inhabitants gives full meaning to the term *other worlds*. A discussion of the idea "that 'tis probable there may be inhabitants in this other World, but of what kinde they are is uncertaine" forms the whole of the last, and most interesting, chapter of the *Discovery*.

In this discussion Wilkins has to proceed with extreme caution: as a scientist he believes in the plurality of worlds, but as a clergyman he wishes to avoid offending the clerical authorities. He begins by saying that the subject could raise many difficult questions. These are the dogmatic questions characteristically associated with the topic in the seventeenth century: whether the moon's inhabitants are descended from Adam, whether they are to be thought of as being in a state of blessedness, or if not, how their salvation can be brought about, and "many other such uncertaine enquiries." Wilkins skirts these issues, leaving them to those more leisured or more learned than himself, and only cites the opinions of others (p. 186). But it is noteworthy that the writers whose views he chooses to quote are, without exception, convinced at least of the *probability* of life in outer space and on the moon in particular. To be on the safe side, he also precedes this review of the work of other scholars (a gallery of authorities almost in the Scholastic mold) with quotations from the *Docta Ignorantia* and the Bible that emphasize the limitations of man's knowledge, particularly with regard to whether or not "Heaven" is inhabited. This does not, however, prevent him from immediately adding that it is permissible to make the general assumption that such inhabitants *do* exist, nor from presenting, as a fact, another quotation from the *Docta Ignorantia*: that the moon is "a habitation of men, animals and plants" (p. 191). Campanella, he says, agrees with Nicholas of Cusa, and he gives an objective summary of Campanella's refutation of the ninth accusation against Galileo (sec. 2 above). Interestingly, however, Wilkins makes one critical comment: it was, he says, a mistake on Campanella's part to cite two passages from the Pauline Epistles in order to *prove* that Christ, on the cross on Golgotha, also redeemed other humankinds, essentially similar to ourselves, living elsewhere in space. This is wrong for two reasons. First, such use of biblical passages for apologetic purposes comes close to blasphemy ("I dare not jest with Divine truthes"). Second, to see Holy Writ as a revelation of worldly and natural things instead of divine things is to misunderstand it. Wilkins's view on this, as on the specific question of whether there is water in the heavens, is that the whole concept of the plurality of worlds is not contrary to the Bible, but neither is it capable of being proved by reference to biblical passages (p. 192). In other words, it is a purely scientific "truth" (p. 3). About how this truth can be reconciled with the truth of the revealed Word of God—beyond saying that it is not directly contradicted by it—Wilkins, keeping to his resolve to avoid theological controversies, says nothing. Thus the difference between the two classic advocates of plurality in Galileo's day can be summed up as follows. For Campanella, science and the Bible are compatible because each of the two "books" of God supports the other; for Wilkins, they are compatible because the Bible does not actually contradict the book of nature and has in any case a different subject. In his preface, Wilkins argues, like Campanella, for the freedom of scientific research, but he would accord it a far higher degree of autonomy vis-à-vis the other source of truth.

Nevertheless, for Wilkins too the scientific plausibility of "men" on other planets is a risky subject in the light of theological doctrine—reason enough for him to opt for Campanella's second possibility as being perhaps "more probable," namely, that the inhabitants of the moon are "not men as wee are, but some other kinde of creatures which beare some proportion and likenesse to our natures" (pp. 192–93). This, he surprisingly adds, had been more or less the opinion of Cusa (the cardinal!) too. In quoting Cusa's statement on this subject, however (that the inhabitants of the sun and, to a lesser degree, those of the moon are more spiritual, less material and gross than we), he brings up the far more delicate question, so troublesome to Kepler, of man's position in Creation. And this way of raising it is far more disturbing than Campanella's had been. If Nicholas of Cusa saw Earth's inhabitants as being inferior, less "intellectuales" than those of the "higher" stars, how can man still be the pinnacle of Creation? Wilkins, no doubt with relief in view of his position as a vicar, follows his stated intention of avoiding theological controversy and declines to discuss this dogmatic issue.

But when he was reworking the text for the third edition (1640), this evasive tactic seems to have arrested his attention, and he inserted a passage in which he takes the bull by the horns. He begins by considering a possibility suggested—and left open—by Galileo: "Or it may be, they are of a quite different nature from any thing here below, such as no imagination can describe; our understandings being capable only of such things as have entered by our senses." He is clearly more attracted, however, by the idea that other planets are inhabited by the kind of beings that could be assumed to rank between men and angels in the Great Chain of Being: "It may bee the inhabitants of the Planets are of a middle nature between both these" (1640, p. 190). This answer to the question of man's status would not necessarily be acceptable to Christian dogmatists. Wilkins forestalls their possible objections by referring to the greatness of the Creator.

Instead of theologizing, he now turns, in the 1640 edition as in the first of 1638, to Plutarch as another major witness on the question of the nature of the Selenites. He records Plutarch's view with the abstinence of a bibliographer, adding no comment of his own. With another abrupt change of direction, he cites various opinions on the location and capacity of hell and paradise (whether on the moon or elsewhere!). And he concludes this gallop through these abstruse theories with the declaration that he himself asserts nothing about "these Selenites," since no arguments seem to him strong enough. Yet in the same breath, he goes on to say that future generations "may invent some meanes for our better acquaintance with these inhabitants" (p. 207). He does not, then, doubt their existence; the problem is merely one of transport (p. 208), and in the third edition he devotes a special additional chapter to it. The chapter's every page breathes the confidence of the age of conquest. Already in the first edition of 1638 he mentions Columbus and says "I doubt not but that time who is still the father of new truths, and hath revealed unto us many things which our Ancestours were ignorant of, will also manifest to our posterity, that which wee now desire, but cannot know" (p. 209). There is no suggestion that the New World in the ether might in any way be threatening: we yearn to find it. Whether this is because of loneliness, or through an enthusiastic sense of cosmic brotherhood, is not made clear. Despite all Wilkins's rhetorical reservations, the fact that he firmly believes in the existence of an inhabited world somewhere in the universe emerges not only between the lines but on one occasion quite openly: he sums up at the end by saying that he has "in some measure proved what at the first I promised, a world in the Moone" (p. 210). But characteristically, the assertion is immediately retracted: that lunar world is only possible or probable, and "I shall willingly submit my selfe to the reason and censure of the more judicious." With this humility formula, the *Discovery* of 1638 ends. But, interestingly, in the 1640 edition Wilkins deletes the whole of that concluding passage and lets the chapter end with the prospect of "our better acquaintance with these inhabitants" in the future. The impression is inescapable that, perhaps as a result of the success of the first two editions, he has become bolder and more confident. The purely practical additional chapter about means of transport and the added passage about the possibly higher status of the extraterrestrials would both fit in with this assessment.

The submissive tone of the closing words of the first edition is reminiscent of the pious formula that concludes the *Apologia pro Galileo*, with the difference that Wilkins, though a clergyman, is submitting himself not to the judgment of a church but in a more general (and less orthodox) way to the reason of those capable of judging the matter. This difference is in line with the second general difference between Wilkins and Campanella noted above: a different placing of accents in their handling of the theological issues associated with the plurality of worlds. For in fact Wilkins does not wholly confine himself to pointing out that theological arguments based on the Bible cannot rule out the possibility of numerous worlds. For him, the plurality of worlds has theological aspects of which Campanella says nothing, and Wilkins seems totally indifferent as to whether these are sanctioned by the Bible or not. The assumption of a world on the moon conveys a heightened sense of the wisdom and greatness of God or Providence: "Neither can this opinion derogate from the divine Wisdome (as *Aquinas* thinkes) but rather advance it, shewing a *compendium* of providence, that could make the same body a world, and a Moone; a world for habitation, and a Moone for the use of others, and the ornament of the whole frame of Nature" (pp. 42–43). And in the passage added to the third edition, the idea of beings superior in status to ourselves is followed by the sentence "Tis not improbable that God might create some of all kindes, that so he might more compleatly glorifie himselfe in the works of his Power and Wisedome" (1640, p. 190). This is not plenitude theology,[55] but "physicotheology," in which the greatness of God is revealed in the greatness of his works. Later, particularly in the wake of the Newtonians, this is the theological motif typically associated with the plurality of worlds: as with Robert Wittie, God's praises are sung on innumerable planets.

At the root of this physicotheology is teleological thinking: God created nothing without purpose or "in vain." This argument is familiar from dogmatic theology; but now teleology no longer leads to the inference that everything was created for man on our Earth, but it may instead be used to argue, for instance, that the other planets were created for other beings similar to ourselves.[56] Like Kepler in his *Dissertatio* (see chap. 2, sec. 7 above), Wilkins shows that he is familiar

[55]As McColley, in "Seventeenth-Century Doctrine," claims, pp. 427–28.
[56]Richard S. Westfall, *Science and Religion in Seventeenth-Century England* (New Haven, Conn., 1958), p. 52.

with this idea; but it remains marginal. That there is a "world" on the moon is, he says, also made probable by the presence of mountains there; "for since providence hath some speciall end in all its workes, certainly then these mountains were not produced in vaine, and what more probable meaning can wee conceive there should be, than to make that place convenient for habitation" (p. 138). "For why else did Providence furnish that place with all such conveniences of habitation as have beene above declared?" (p. 190). But if the purpose of Creation is so readily accessible to our reason, then divine and human reason, which for Goethe are "two very different things,"[57] are being equated; and why not then equate them with "Nature" too? Later, Nature is indeed credited, for instance by Fontenelle, with thus acting rationally, with a particular end in view; and even in Wilkins (who was to be one of the cofounders of the Royal Society, which was suspected of atheism), we already find a reference to Nature instead of Providence in this context: "Nature frames every thing fully perfect for that office to which shee intends it. . . . 'Tis likely then that she had some other end which moved her to produce this variety, and this in all probability was her intent to make it a fit body for habitation with the same conveniences of sea and land, as this inferiour world doth partake of" (pp. 102–3). In this, Wilkins, the vicar, points far ahead to that secularization of the plurality idea which can be observed from the late seventeenth century onward.

If we look back once again to Wilkins's source, Campanella's *Apologia pro Galileo*, the qualitative difference is apparent. The Dominican painstakingly marshals biblical and christological arguments to make the assumption of a plurality of worlds acceptable to the upholders of dogma; the Anglican makes prudent reservations, aligning himself with current orthodoxy, but in fact boldly places his trust in scientific extrapolation regardless of whether it is confirmed by the Bible. Not only this, but in an equally bold departure, he introduces a new view of the wise and purposeful workings of Divine Providence—or even Nature.

Finally, Wilkins is also a pioneer in that he considers the future possibilities of space travel. Kepler did this, albeit only vaguely and in passing (chap. 2, sec. 7); but the days when a dream was an adequate vehicle are now past. In the chapter added to the third edition of the *Discovery* (1640), Wilkins discusses the physical possibility of various forms of interplanetary transport and thus the founda-

[57]Remark to Johann Peter Eckermann, 15 October 1825, in Johann Wolfgang von Goethe's *Conversations with Eckermann*.

tions of science fiction. By this time, however, a countryman of his, Francis Godwin (1562–1633)—also, curiously enough, an Anglican bishop—had preceded him in an imaginative depiction of space travel—though still of a fantastic rather than a realistic kind. This makes its appearance in Godwin's quasi-scientific novel of travel and adventure, *The Man in the Moone: or, A Discourse of a Voyage Thither, by Domingo Gonsales*. It was published in 1638, five months after Wilkins's *Discovery*, and also anonymously; but it had been written about 1627 or 1628.[58] It achieved a notable success, and its influence was perhaps even more pronounced than that of the *Discovery*. It went through numerous editions, right into the eighteenth century, and also appeared in several French, German, and Dutch translations and reprints of translations—more than two dozen editions in all.[59] The Domingo Gonsales who appears in Cyrano's lunar novel is of course none other than Godwin's hero, whose story left its mark in many other ways too on the *histoire comique*. Altogether, Godwin's novel seems to have been the classic example of the fictional space journey until way into the eighteenth century; even in 1739 the entry for "Mondreise" (voyage to the moon) in Johann Heinrich Zedler's encyclopedia consists wholly of a critical and detailed account of Godwin's novel.[60] Significantly, Wilkins, whose ideas about the world in the moon coincide in some respects with those of his clerical colleague, mentions him in the third edition of the *Discovery*. This reference occurs in the newly added fourteenth chapter, which has become fairly well known in science fiction circles. At the end of the chapter he writes, "Having thus finished this discourse, I chanced upon a late fancy to this purpose under the fained name of *Domingo Gonsales*, written by a late reverend and learned Bishop; In which (besides sundry particulars wherein this later chapter did unwittingly agree with it) there is delivered a very pleasant and well contrived fancy concerning a voyage to this other world" (1640, p. 240).

This comment suggests that Wilkins—like Zedler in 1739, and perhaps Godwin himself, judging by his subtitle—saw the detailed description of the actual voyage through space as the main achievement of Godwin's novel. There is some justification for this view. But

[58]Harold W. Lawton, "Bishop Godwin's *Man in the Moone*," *Review of English Studies* 7 (1931), 37; Grant McColley, "The Date of Godwin's *Domingo Gonsales*," *Modern Philology* 35 (1937), 47–60.

[59]See McColley, "Date of Godwin's *Domingo Gonsales*," p. 48; "*The Man in the Moone*" *and* "*Nuncius Inanimatus*," ed. Grant McColley, Smith College Studies in Modern Languages 19:1 (1937), pp. vii–viii.

[60]*Universal-Lexicon*, vol. 21 (1739), pp. 1100–1103. On the popularity of the novel see further Nicolson, *Voyages to the Moon*, pp. 85–93.

to pay attention only to this aspect of the book would be to ignore another vital aspect. For it is also the first book in world literature to depict (as Kepler's *Somnium* did not) an actual encounter between *Homo sapiens* and his extraterrestrial counterpart in the context of Copernican cosmology as revealed by Galileo. The depiction of the moon's surface undoubtedly shows specialized knowledge of Galileo's telescopic discoveries; this accords with the new dating of the novel (because of historical allusions in it) at 1627/28, whereas previously, various dates before 1603 had been put forward.[61] And it is appropriate that Galileo should be directly mentioned in the preface as the discoverer of the new world, which is compared to the world of the antipodeans discovered by Columbus. It has been pointed out that Bruno held his sensational disputations about Copernicanism at Oxford just at the time when Godwin was studying there; it is tempting to suppose that it was Bruno who sowed the seeds of Godwin's belief in an inhabited world on the moon,[62] but this cannot be proved. At all events, it was through Galileo that the idea of inhabitants on the planets and in particular on the moon really became widespread, so much so that it was retrospectively attributed to his teacher Copernicus. This attribution was made for instance by Grimmelshausen in his translation of Jean Baudoin's French version of *The Man in the Moone*, *Der fliegende Wandersmann nach dem Mond*.[63] Grimmelshausen says in his preface (1659) that the "most gracious" reader may find the story somewhat "fabulous" (*Fabelhaft*), but "the ancient philosophers of the Pythagorean sect were already of the opinion that there must be another Earth. Copernicus revived this opinion, and wished to place the Earth among the planets and claim that the Earth must shine and be luminous like the other planets, and that they in turn must, like the Earth, be real bodies (*corpora solida*) and be occupied by innumerable creatures." In other words, Grimmelshausen sees Godwin's novel as anticipating that encounter with "creatures" living on other planets which Wilkins too saw as a real possibility and as only a matter of time—an encounter which was now, as a result of the work of Copernicus (and Galileo), generally accepted as being at least conceiv-

[61]There are also references to other aspects of Galileo's physics; see McColley, "Date of Godwin's *Domingo Gonsales*," pp. 53–58. Nicolson's remark that the date of composition is immaterial fails to acknowledge the part Galileo played in the popularity of this theme as a literary subject (*Voyages to the Moon*, p. 71).

[62]Dorothea Waley Singer, *Giordano Bruno: His Life and Thought* (New York, 1950), pp. 183–84.

[63]Wolfenbüttel, 1659. On editions and another German version, probably by Balthasar Venator, see Michael Winter, *Compendium Utopiarum*, vol. 1 (Stuttgart, 1978), pp. 69–70.

able. This estimate, formulated from the viewpoint of the novel's contemporary reception, is more accurate than the view generally held today that the *Man in the Moone* is a utopian novel.[64] A closer look at this pioneering work of science fiction amply confirms Grimmelshausen's view.

Domingo Gonsales narrates his own story. It starts as a typical picaresque account of all manner of adventures set quite concretely in the final years of the sixteenth century. The young man, having run away from the University of Salamanca, enters, in Antwerp, the service of a French marshal with whom he takes part in the war between Spain and the Netherlands. He returns to Spain in the entourage of the duke of Alba, marries, flees to Lisbon after a duel, and takes ship for the East Indies, where he makes his fortune. On his return voyage to Europe, a storm drives his ship to Saint Helena. Here he discovers curious swanlike birds, "gansas," which he trains to pull a bizarre flying machine; after many exciting adventures at sea, he is carried by the machine, one day in 1599, to the Azores. After a narrow escape from the "savages" there, his birds then fly straight to the moon, where, we are told, they habitually spend the winter. "*Arrige aures*," Domingo tells the reader, "prepare thy selfe unto the hearing of the strangest Chance that ever happened to any mortall man" (p. 42).[65] But for his Spanish courage, he thinks, he would surely not have survived a flight so terrifying and so astonishing (p. 47). Its terrifying aspects include the devils and aerial spirits of superstition (who speak Spanish); the astonishing ones include the discovery, made with his own eyes, that the Earth is not the single Aristotelian center of gravity of the universe, is not enveloped in a circle of fire, and does rotate on its own axis, as Copernicus maintained (pp. 46, 53, 56, 65). As to whether Copernicus was correct in his belief that the Earth moves annually through the zodiac around a stationary sun, the Catholic, Domingo—or rather the Anglican, Godwin—deliberately leaves this question open (p. 60). In fact, however, his moon-world with its pronounced similarity to the Earth of course presupposes the Copernican doctrine that the Earth too is a "wandering star" and not a motionless and unique sublunary body.

After a journey lasting twelve days, Domingo lands on a hill on the moon, "that other world" (pp. 66–67). It is a world similar to our

[64]McColley's edition (see n. 59 above), p. viii; Nicolson, *Voyages to the Moon*, p. 71; Rita Falke, "Versuch einer Bibliographie der Utopien," *Romanistisches Jahrbuch* 6 (1953/54), 95.
[65]Page references are to the first edition (1638). The text is quoted from the McColley edition (see n. 59 above), which indicates the pagination of the first edition.

own, but on a far larger scale. The vegetation is gigantic, and so are the lunar animals—and people. No sooner has Domingo been wonderfully refreshed by the leaves of an unfamiliar bush than he finds himself surrounded by these people. Apart from their size and the indescribable material and color of their clothes, they are clearly exactly like us. In their homes and palaces, Domingo soon feels just as much at home as Gulliver among the giants of Brobdingnag. Their world is like Paradise, as we are told on two occasions (pp. 85, 104). Life is "wonderfull long," and learning and honesty are prized above everything (p. 77). The court of one of the subsidiary kings in this strictly hierarchical state is magnificent (p. 76). Gonsales finds favor with the highest monarch of the moon, whose power has almost a fairy-tale quality: "The gifts he bestowed on me were such as a Man would forsake mountaines of Gold for" (p. 97): they are jewels with magical properties. The most ordinary moon people are virtue personified; they lack nothing the heart could desire; their life is one of uninterrupted pleasure, in an eternal spring. Crime and illness are unknown. "But the chiefe cause, is that through an excellent disposition of that nature of people there, all, young and old, doe hate all manner of vice, and doe live in such love, peace, and amitie, as it seemeth to bee another Paradise" (p. 104).

Such a paradise raises theological problems: Why is there no original sin here? Is one to assume a second Creation? Are the moon dwellers the chosen people of God? The adventurer Domingo, whose studies at Salamanca were of brief duration, is not inclined to think about such matters. But at the beginning of the work, he does acknowledge them at least to the extent of expressing the hope that, in his true account of wonders no mortal has ever experienced, he will not be "over-liberall." In particular he hopes to cause no offense to the Catholic church, whose teachings he intends to promote to the best of his ability (p. 12). All he does in this regard, however, is to make a few passing references to the religion of the lunar people—in other words, Bishop Godwin thinks that these few references adequately fulfill his obligations to his church. The Lunars believe in God; their everyday greeting is "Glorie be to God alone" (p. 94); at the mention of the name of God ("*Martin* in their language"), they raise their hands and bow their heads (p. 83). What is more, they are Christians; at the name of Jesus Christ, they fall on their knees (pp. 73, 82). Melanchthon had feared speculation about other "religions" in outer space—which to him would be absurd; Burton had considered the existence of other religions possible. Godwin cannot face even the thought of them. But these brief allusions to religion are hardly more than a gesture of appeasement on the part of Domingo-Godwin, a gesture, however, that raises more questions than it answers. For instance, it is not at all apparent why the Lunars, who unlike us are evidently free from sin, should know a Redeemer.

Between us and the moon dwellers there is, however, not only a theological but also an anthropological difference that we have not yet examined. Their size, always greater than ours, is very variable and has symbolic significance: the greater their stature, the nobler they are. The smallest of the moon people, only a little taller than ourselves, rarely live longer than eighty years and are seen not as "true Lunars" but as "most base creatures, even but a degree before bruite beasts" (pp. 78–79). The "true Lunars," by contrast, are up to thirty times as tall, live up to thirty times as long, and lead a far more pleasant life, while the "dwarfe Lunars" (p. 84) are set to the most menial tasks. Despite the warm hospitality of the Lunars, Domingo cannot fail to recognize the inferiority of his status as an earthly human being, especially when he learns of the radical form of eugenics practiced by the moon people. If a Lunar lacks any of the perfections that are the norm among them, the defect is detectable at birth, and as they do not kill, the newborn is immediately dispatched by some means to the Earth, generally to North America. These rejected Lunars are in fact the changelings of superstition (p. 104), and *changeling* is the derogatory term (an exaggeration of their degree of inferiority) used for the meanest, smallest Lunars, whose physical size is nevertheless greater than ours (p. 79).

Written at a time when exploration of the subject of aliens was just beginning, this book, with its mixture of popular superstition and science fiction, is highly significant. For the first time, a literary work focuses on the possibility that beings living in outer space may be of a higher kind, may be superior to us and also happier than we are. Kepler had done his utmost to suggest that this was impossible. Wilkins had first, in the context of his quotation from Nicholas of Cusa, blatantly evaded the question but had later, perhaps partly in the hope of making the idea more or less acceptable, fallen back on the conception of the Great Chain of Being created by an unimaginably wise God. Unlike either Kepler or Wilkins, Godwin shows an actual encounter of man not only with a different kind of humanity, but specifically with a superior breed of human, or indeed a superior human species such as is regularly found in later literature of this kind. This is one major reason why *The Man in the Moone* is not a utopia but a work of science fiction, examining the nature of man in the light of the plurality of worlds.

Utopian literature seeks to show how human beings exactly like us could, under different social conditions, achieve a different, better, more perfect existence. There is no such suggestion in Godwin's novel; from the outset there is an unbridgeable qualitative gap between ourselves and the moon dwellers, who cannot be compared for instance with Campanella's utopian Solarians. It does not occur to Domingo to see the nature, way of life, and social conditions of the lunar people as a model for the societies with which he is familiar. Godwin's thought experiment, which is perhaps on the same level of "realism" as the creations of utopian writers, focuses on the possibility of a "different" kind of humanity in the context of the universe revealed by the new science; perhaps we can even detect in it the beginnings of evolutionary thinking, which was not to become common in science fiction until the nineteenth century. "What is presented here is [not fabulous creatures and figures of myth but] the possibility of a human-like race which has evolved to a higher level or in a better way and which, it is suggested, may actually exist somewhere. The ideal nature of its biological and psychic qualities [is] beyond anything that man could achieve through social change."[66] Thus the importance of *The Man in the Moone* in the history of science fiction can hardly be exaggerated, even if the author does prefer to point to problems rather than work out solutions to them.

4. Pious Atomists: More, Borel, Gassendi, and Guericke

While attempting, each in his own way, to reconcile the idea of a plurality of inhabited worlds with a more or less strict interpretation of Christian dogma, Campanella, Wilkins, and Godwin all repudiated, explicitly or implicitly, the possible suspicion that they were in some manner reviving the Atomism of Leucippus, Democritus, Epicurus, and Lucretius.[67] What they found threatening in Atomism was its materialistic atheism, its suggestion that the universe was subject to mechanical or accidental processes. The idea that the worlds—in other words, the heavenly bodies including the Earth—owed their origin to a chance arrangement of the smallest components of matter, the atoms, and that all processes in a world thus formed took place according to autonomous laws governing the motion of atoms in empty

[66]Winter, *Compendium Utopiarum* 1:71.
[67]E.g., Campanella, *Defense of Galileo*, p. 65.

space, was unacceptable not only to Catholics and Protestants alike but also to many secular philosophers such as Leibniz.

Yet it was inevitable that this Atomist way of thinking should be revived, by the early seventeenth century at the latest, by thinkers of the most diverse complexions. The destruction of the Aristotelian cosmology by the new science left a vacuum that Atomist ideas, hitherto suppressed by Aristotelianism, could all too readily fill. As early as 1611, Donne wrote in his "First Anniversary" (line 212) that because of the anti-Aristotelian "new Philosophy" the universe was "crumbled out again to his Atomies." And of course this revival of the Atomism of antiquity gave a new impetus to the idea of a plurality of (inhabited) worlds, though of course with the post-Copernican structural modification that is present from Bruno and Hill onward: For all the neo-Atomists discussed in this section (who are all convinced Copernicans), the "worlds" are no longer the coexistent or successive *kosmoi*, not seen but only assumed, of Atomism—each with its own Earth and sun, planets and fixed stars, replicating all that could be observed of our own cosmos. Instead they *are* the observable bodies of the heavens, or at any rate most of them, and also the invisible planets postulated ever more confidently, on the grounds of analogy, as satellites of the suns that are the fixed stars.

From this point of view, the corpuscular physics and philosophy of Descartes (which in the seventeenth century conquered the European universities and formed the basis of the most influential seventeenth-century theories of the plurality of worlds, those of Fontenelle and Huygens, but not only theirs) are nothing more than a variation of the neo-Atomism of the baroque period, fundamental differences notwithstanding. It is quite clear that the Atomism of antiquity was Descartes's theoretical starting point.[68] At various stages in the development of his thought, he tried to replace the Atomist assumption of atoms and empty space with the "plenum," the *full* space in which the smallest units of matter, the corpuscles or particles, are in close contact with one another while moving in a circular manner around a center; but to what extent he still retained the void of the Atomists as an auxiliary concept is at least an open question. In any case, the distinction between empty and full space was not of major significance for the concept of the plurality of worlds based on Atomist ideas: it was equally plausible either way. For in Cartesian physics each fixed

[68]See for instance R. H. Kargon, *Atomism in England from Hariot to Newton* (Oxford, 1966), pp. 94–96; Madeleine Alcover, *La pensée philosophique et scientifique de Cyrano de Bergerac* (Geneva, 1970), p. 137.

star was regarded, at least implicitly, as the center of a "vortex" (*tourbil-lon*, *vortex*), around which planetary satellites orbited, and Cartesian neo-Atomists saw the number of such systems as being theoretically infinite, just as the non-Cartesians did. Both groups visualized the universe as a plurality of world systems on the Copernican model.[69] It is therefore not surprising that in poetic and philosophical conceptions of the plurality of worlds, elements of Cartesian and non-Cartesian Atomist thought are often so intermingled as to be indistinguishable. This is the case with, for instance, Pierre Borel and Cyrano de Bergerac, and also with the "Cambridge Platonist" Henry More. More, in the preface to his main work on this subject, *Democritus Platonissans* (1646), invokes the authority not only of Democritus, Epicurus, and Lucretius but also of the "sublime and subtill Mechanick" Descartes and considers the philosophical differences between them to be "needlesse niceties."[70]

In this revitalization of ancient Atomism in the seventeenth century the real problem was of course its atheism. Most of the revivers of that suspect cosmology, therefore, even Descartes (and of course, considerably later, Newton, whose thought was firmly rooted in Atomism),[71] solved the problem by attributing to God the creation of the atoms and the origin of the laws of their motion. Sometimes an overinsistence on this point raised the suspicion that the author was paying mere lip service. But even where the sincerity of these pious assertions was not in doubt, there remained as a stumbling block the question of man's status. Without quite meaning to—indeed largely through the inner logic of the plurality idea itself—thinkers had populated the universe with many humankinds created by God, and thus they had called into question the preeminence of terrestrial man. Descartes himself, while invoking God as the Creator of the atoms and of the laws governing their motion, confessed that Atomism sounded the death knell of anthropocentrism.[72] And the idea, inconceivable to Kepler in 1610, that terrestrial man is a thoroughly insig-

[69]On the affinity between Atomism, Cartesianism, and Copernicanism see Thomas S. Kuhn, *The Copernican Revolution* (1957; Cambridge, Mass., 8th printing, 1976), pp. 235–42.

[70]The last quotation is from *Democritus Platonissans*, st. 39. Quotations are taken from *The Complete Poems of Dr. Henry More*, ed. Alexander B. Grosart (Edinburgh, 1878). On More's relationship to Descartes see Dick, *Plurality of Worlds*, p. 117; idem, "Plurality of Worlds," pp. 202–6. In his *Discours*, of which a discussion follows, Borel constantly quotes Lucretius, but he also published, in the preceding year, *Vitae R. Cartesii Summi Philosophi Compendium*. On Cyrano see sec. 5 below.

[71]See Kargon, *Atomism in England*, chap. 11.

[72]See sec. 5 below.

nificant creature whose high opinion of himself only makes him appear foolish, was a certainty to Cyrano writing in midcentury. Of course Cyrano is an extreme case in that he makes a point of *not* assuming a divine creator for his neo-Atomist universe. But the writings of even those neo-Atomists who do posit a creator of the atoms of matter, and thus of the "worlds," reveal the precarious status not only of the concept of a creator but also of anthropocentrism. Is the Creator still identical with the God of dogma, the Yahweh of Genesis—whose supreme purpose in Creation was to create the one human race, and who created everything else for its benefit—and with God the Father, who sent to us, and to us alone, his only begotten Son? Henry More, Pierre Borel, Pierre Gassendi, and Otto von Guericke illustrate some of the responses to this question.

"Infinitie of worlds! A thing monstrous if assented to, and to be startled at"—Henry More's exclamation in the preface to his poetic treatise on the infinity of worlds, *Democritus Platonissans*, reminds us how daring it still was in the midseventeenth century to place on the very title page of a book the name of one of the Atomists, who were regarded with disfavor—"hooted at, by the Rout of the learned, as men of monstrous conceits." At the same time, of course, the title indicates the frame of reference within which the revival of Atomism had to take place if it was to take place at all. The reference to Plato, or rather to the Christian neo-Platonism of the Cambridge philosophers, indicates that the existence of the infinitely many worlds formed by the configuration of the atoms is attributed not to the workings of chance but to the provident creative act of a God who is spiritually active in all the processes that operate in the universe:[73]

> For in each Atom of the matter wide
> The totall Deity doth entirely won,
> His infinite presence doth therein reside,
> And in this presence infinite powers do ever abide.
>
> (St. 69)

Thus secured against the accusation of atheism and materialism, More proceeds, in this philosophical poem of over one hundred stanzas, to unfold his vision of the innumerable planetary worlds. It is an ecstatic glorification of a Brunoesque universe, of an "immense field

---

[73]Sts. 34, 47, 48, 50. Even the idea of plenitude is to be found (sts. 50–51) in More's otherwise wholly Copernican thinking.

of Atoms" arranged in "Centreities," an infinite number of Copernican systems (st. 14):

> My nimble mind this clammie clod doth leave,
> And lightly stepping on from starre to starre
> Swifter than lightning, passeth wide and farre,
> Measuring th' unbounded Heavens and wastfull skie[.]
>
> (St. 5)

> A circle whose circumference no where
> Is circumscrib'd, whose Centre's each where set[.]
>
> (St. 8)

> Whatever is, is Life and Energie
> From God, who is th'Originall of all;
> Who being every where doth multiplie
> His own broad shade that endlesse throughout all doth lie.
>
> (St. 10)

Again and again he returns to the thrilling notion that the stars are suns, "innumerable numbers of fair Lamps," circled by planets, that is, by "worlds" like the Earth (sts. 18, 22):

> These with their suns I severall worlds do call,
> Whereof the number I deem infinite:
> Else infinite darknesse were in this great Hall
> Of th'endlesse Universe; For nothing finite
> Could put that immense shadow into flight;
> But if that infinite Suns we shall admit,
> Then infinite worlds follow in reason right,
> For every Sun with Planets must be fit,
> And have some mark for his farre-shining shafts to hit.
>
> (St. 26)

And such worlds exist not only simultaneously in the infinity of space but also consecutively in the infinity of time (st. 76). More visualizes the end of each of these worlds as a conflagration such as we witness when we see a comet (st. 93).

Precisely because this visionary cosmology has Christian overtones, it cannot sidestep the question of theological anthropocentrism. More's infinite multiplication of the Copernican model implies, and indeed he repeatedly hints at, the presence of living creatures in those other "worlds." He does not as a rule, however, speak of extraterrestrial "men" but only of "life" and "shapes," as in stanza 25:

> This is the parergon of each noble fire
> Of neighbour worlds to be the nightly starre,
> But their main work is vitall heat t'inspire
> Into the frigid spheres that 'bout them fare;
> Which of themselves quite bare and barren are,
> But by the wakening warmth of kindly dayes,
> And the sweet dewie nights, they well declare
> Their seminall virtue, in due courses raise
> Long hidden shapes and life, to their great Makers praise.

It is not clear whether this refers to praise of God actually uttered by the creatures themselves—as for instance Robert Wittie was later to envisage—in which case they must be seen as creatures at least similar to humans. Only in one stanza (which is presented as a quotation from Origen) is there mention of *men* on other planets, and then only in passing: the lines focus on the end of our world and of every world:

> To weet that long ago there Earths have been
> Peopled with men and beasts before this Earth,
> And after this shall others be again
> And other beasts and other humane birth.
> Which once admit, no strength that reason bear'th
> Of this worlds Date and Adams efformation;
> Another Adam once received breath
> And still another in endlesse repedation,
> And this must perish once by finall conflagration.
>
> (St. 76)

It is hardly to be assumed that More is speaking here of a succession of humankinds in the sense that only one exists at any one time (which might possibly have been reconcilable with Christian theology and its anthropocentrism). The whole tenor of his visionary portrayal of the universe makes it far more likely that he envisages the simultaneous existence of many worlds, each of which is destined to perish at a particular time:

> Ne can our Earth from this state standen free,
> A Planet as the rest, and Planets fate must trie.
>
> (St. 95)

But this immediately raises the theological questions of the uniqueness of man as the pinnacle of Creation and of God's justice, not to mention the dogma of the redemption by Jesus Christ. More, like Bruno, is however more interested in his grandiose conception of the universe than in dogmatic quibbles and man's search for his identity, and he evades these problems. His one gesture in this direction is a defense of God's justice in relation to our destined fate (confirmed by the Bible) of being finally consumed by fire: he argues that the Flood, too, destroyed (almost) the whole human race (st. 96). But this analogy, instead of suggesting a natural process of decline in the universe, forces one to assume that the extraterrestrial human beings have, like us, incurred God's judgment for a *moral* decline, which implies a similarity between their nature and ours. And indeed in More's final apocalyptic vision, his thoughts seem to tend this way. He speaks of the phoenixlike renewal of each world after it has been consumed by fire:

> O happy they that then the first are born,
> While yet the world is in her vernall pride:
> For old corruption quite away is worn[.]
> (St. 99)

But although More, as a scientist, claims that "what I sing" is supported by "right reason and Philosophie" (st. 102), he quickly retreats to safer ground, appealing to God's providence and goodness, which, for all man's efforts, are ultimately beyond the scope of human knowledge:

> But, ah! What mortall wit may dare t'areed
> Heavens counsels in eternall horrour hid?
> (St. 107)

The question of man and his view of himself in a universe shared with an infinite number of living beings is thus dismissed with a few pious phrases. That these did not satisfy More himself may be inferred from his poem published a year later, "Insomnium Philosophicum." It is a dream vision of another planetary world "like to our earth"; the sleeper sees not only familiar domestic animals but also what are now unmistakably human beings, as well as coaches, towns, towers, churches, and so on. But the human inhabitants have wings and are thus more like angels or devils, or rather both. For the winged beings on the dark hemisphere of this globe are ugly, deformed, and frightening in appearance, while those on the sunny side are exceedingly beautiful. Their outward appearance reflects a theological difference: the batlike creatures curse the Creator and his works; the angelic ones praise them. *Why* this should be so is once again glossed over as something beyond comprehension: "This is the mystery of that mighty Ball / With different sides" (st. 28). The significance of the two sides is hardly made clearer by their being described as the dark and the light side of Providence. Perhaps More is contrasting the view of Providence held by the saved with that held by the damned.[74] The poem is supposed to be philosophical (it first appeared in a collection entitled *Philosophical Poems*), but the questions it might be expected to tackle, about the status and nature of man, remain unanswered.

The second leading Cambridge Platonist, Ralph Cudworth, also ignores the problems raised by the subject of the plurality of worlds. He is the author of the main theoretical work produced by this circle, *The True Intellectual System of the Universe* (1678), which is similarly based on Atomism set in a theological framework. He speaks only briefly of the plurality of worlds and relates it to both the new science and the old idea of plenitude. He writes, without further commentary:

> The new celestial phenomena, and the late improvements of astron-omy and philosophy made thereupon, render it so probable, that even this dull earth of ours is a planet, and the sun a fixed star in the centre of that vortex [!], wherein it moves, that many have shrewdly suspected, that there are other habitable globes besides this earth of ours, (which may be sailed round about in a year or two) as also more suns, with their respective planets, than one. . . . Now, it is not reasonable to think that all this immense vastness should lie waste, desert, or uninhabited, and have nothing in it that could praise the Creator thereof, save only this one small spot of earth.[75]

The same combination of corpuscular mechanics with the Chris-tian Creator as "first cause" is found in the Anglican physiotheologian, experimental physicist, and member of the Royal Society Robert Boyle (1627–91), who is the leading English exponent of the attempt to reconcile Christianity and new science. He too is content to glorify God in enthusiastic terms without so much as touching upon the

---

[74]As Geoffrey Bullough suggests in his edition of Henry More, *Philosophical Poems* (Manchester, 1931), p. 235.

[75]*The True Intellectual System of the Universe* (London, 1845), 3:480–81. On Cud-worth's ideas see Emanuel Hirsch, *Geschichte der neueren evangelischen Theologie*, vol. 1 (Gütersloh, 1949), pp. 188–96.

burning theological issues. In an essay entitled "Of the High Veneration Man's Intellect Owes to God, Peculiarly to His Wisdom and
Power," he admits that God may have formed other planetary worlds,
in our solar system or in another, differently from our own world; but
he concludes a priori that a plurality of worlds, like a single one, can
only testify to the wisdom of the Creator. In many thinkers, from
Saint Thomas Aquinas on, this conclusion would inevitably raise the
question of God's justice in relation to the various humankinds—in
other words, the question of man's theological status; but Boyle ignores it, as do the Cambridge neo-Platonists More and Cudworth,
and for the same reason. As the author of a treatise on "The Excellency
of Theology Compared with Natural Philosophy," Boyle is happy to
be guided on the most central questions by revelation or to accept
God's inscrutability:

> Now if we grant, with some modern philosophers, that God has
> made other worlds besides this of ours, it will be highly probable,
> that he has there displayed his manifold wisdom, in productions very
> differing from those, wherein we here admire it. . . . Now, in case
> there be other mundane systems (if I may so speak) besides this visible
> one of ours, I think it may be probably supposed, that God may have
> given peculiar and admirable instances of his inexhausted wisdom in
> the contrivance and government of systems, that, for aught we know,
> may be framed and managed in a manner quite differing from what
> is observed in that part of the universe, that is known to us.[76]

There is a marked difference between these pious Anglican neo-
Atomists with their visionary naïveté and the no less pious but more
deeply reflective Continental neo-Atomists Pierre Borel, Pierre Gassendi, and Otto von Guericke. The latter conscientiously weigh the
problems raised by ancient Atomism in connection with plurality
(problems More, Cudworth, and Boyle were barely prepared to recognize, let alone discuss), above all, the question of man's status and
nature. They make their own task harder by courageously trying,
while dealing with this explosive subject, to be guided wholly by
reason. But unlike the Cartesian Cyrano de Bergerac, who emphasizes
the conflict between reason and Christian doctrine, they still manage

[76]*The Works of the Honourable Robert Boyle* (London, 1772), 5:139. On his conception
of God as the Creator of the atoms see for instance 2:40–43. On Boyle in general see
Kargon, *Atomism in England*, chap. 9; Richard M. Hunt, *The Place of Religion in the Science
of Robert Boyle* (Pittsburgh, 1955). Dick (*Plurality of Worlds*, p. 150) maintains that Boyle
never assumed other inhabited planets.

to enunciate the belief in plurality founded on Atomist premises in such a way that it is compatible with their strongly held Christian beliefs. Their subtle discussions mark the climax of the attempt, characteristic of the seventeenth century, to harmonize plurality and Christian convictions without the latter becoming so watered down as to be virtually meaningless, as was often to happen in the eighteenth century. Borel, Gassendi, and Guericke are much alike in their treatment of the plurality of worlds, yet there are differences between them that are worth noting because they make one aware of the less obvious ramifications of this still-dangerous subject.

Pierre Borel's *Discours nouveau prouvant la pluralité des Mondes, que les Astres sont des terres habitées, & la terre une Estoile, qu'elle est hors du centre du monde dans le troisiesme Ciel: & se tourne devant le Soleil qui est fixe, & autres choses tres-curieuses* (Geneva 1657) is a compendium of all the arguments for and against the plurality of worlds that were current around the middle of the century.[77] The very title, compared with Wilkins's cautiously formulated title of 1638, makes it clear how much more confidently it is now possible to assert that "we are not alone." For those who still cling to the belief in *one* world, Borel has only ridicule: they are as outmoded as those who doubt the existence of antipodeans (1). Surely there is an element of bluff in this apparent confidence; for although Borel speaks in his preface of the many (unnamed) scholars who share his opinion, he adds that they dare not make their views public for fear of causing offense. And it is clear that Borel himself has reason to share this fear. He cannot deny that the essence of his conception of the universe derives from Atomism (1, 18, 20, 28, 30, 34), and at the end he even admits that this work is really only a fragment of a book he plans to write on the philosophy of Democritus, adding that he himself is therefore only setting down what Democritus would have had to say about the plurality of worlds (47). Also, in the same chapter Borel repudiates the suspicion that he is irreligious, thus distancing himself from Democritus, the atheist; and he even follows the example of Campanella and Wilkins in stating his readiness to retract his opinions if both theology *and* reason should prove him to be in error.

The title also indicates that the theory of plurality as presented

---

[77]Completed as early as 1648, according to the preface. References in parentheses are to the chapters. On Borel see Dick, *Plurality of Worlds*, pp. 117–20; Paolo Rossi, "Nobility of Man and Plurality of Worlds," in *Science, Medicine and Society in the Renaissance: Essays to Honor Walter Pagel*, ed. Allen G. Debus (New York, 1972), 2:146–50; Marie Rose Carré, "A Man between Two Worlds: Pierre Borel and his *Discours nouveau prouvant la pluralité des mondes* of 1657," *Isis* 65 (1974), 322–35.

by Borel is far more thoroughly grounded in Copernican-Galilean cosmology than is the case with More, Cudworth, and even Boyle. Borel thus confirms once again the affinity between Copernicanism and Atomism, but without its atheistic aspect. The theological arguments for and against plurality take up a good deal of space in the *Discours nouveau;* Borel, a man trained in medicine and learned in many sciences, was also a committed Protestant. Science and theology meet in the question of whether and how it is possible to guarantee the uniqueness of man and his status as the chief object of the Creator's concern and love, if more than one world is assumed. The stages of Borel's reasoning lead boldly up to this question (even though the order in which he finally sets out his ideas is a muddle).

For Borel, "proof" of the plurality of worlds is supplied by reason on the basis of scientific knowledge; theology merely confirms it. The proof rests on the by-now-familiar evidence, strengthened, however, by the triumphant awareness that anti-Aristotelian science is inexorably freeing itself from the provincialism of geocentric cosmology (1). The arguments are based on the analogy, made visible by Galileo, between the Earth and the moon and planets and on that between planets and stars (4, 5, 7, etc.). (Borel believes the light of the stars to be a reflection of that of the sun, which he—like Kepler, but unlike, say, Bruno—sees as the absolute center of the universe. Thus there is no place in Borel's cosmology for planetary systems with the fixed stars as their centers; in this respect, his cosmology is even more conservative than that of the Jesuit Athanasius Kircher [5, 25, 38].)[78] Once the analogy is accepted and the Aristotelian division of the universe into Earth and heaven abandoned, it follows that if the Earth is a planet or star, then the planets and stars must also be "earths"— earths harboring life; for Borel considers it absurd to suppose that other worlds might be barren, empty, and "en vain," even if we do not know the purpose for which such bodies may have been created (4, 5, 7, 11, 13, 16, 32, 45). Here a merely formal use of teleological argument is not enough for Borel: he resorts to the protection of theology. Like most of the Atomists of his century he retains God as the Creator of the atoms, but he also reintroduces the principle of plenitude. He claims (in contradiction, for instance, to Mersenne) that to assume only one world containing life would be to restrict the Creator's omnipotence and infinite efficacy in a manner amounting to

---

[78]Rossi ("Nobility of Man," p. 146) and Carré ("Man between Two Worlds," p. 331) erroneously see Borel's conception of the universe as (like Bruno's) lacking a single center.

blasphemy (1, 18, 26, 34, 39). Conversely, the existence of a plurality or indeed infinity of inhabited worlds (in one universe) raises the glory and the greatness of the Creator to immeasurable heights (8, 18, 32, 39). This had been Bruno's belief, but whereas he had incurred the displeasure of the Curia, we know of no persecution of Borel by his church.

Borel, the scientist resorting to the theological principle of plenitude, consistently assumes that numerous worlds, if they exist, must be different from one another, as must be their inhabitants (6, 8, 13, 16, 18, and elsewhere). But how great are these differences? It is clear that for him the presence of "men" (hommes) among the creatures living in these various other worlds is implicit in the analogy between Earth and other planets (a view less circumspect than that of Campanella and Wilkins, and also that of Gassendi and Guericke). Borel seems to assume it as a matter of course (13, 16, 45). But the assumption raises problems that, for all his spirited polemics against the backwardness of those who still argue for one world, he cannot quite overcome. To put it in a nutshell, there is a contradiction between scientific analogical reasoning and the postulation, linked with the concept of plenitude, that the manifestations of God's omnipotence must be different from one another. If the human beings on all the inhabited planets are "equal" from a theological point of view—children of God who are equally prone to sin but also equally loved and destined for salvation—then it might be just possible to make the multiplication of humankinds theologically acceptable by presenting it as a multiplication of the Creator's greatness. Borel seems to have this in mind when he infers from Colossians 1:20 that there are human beings on other planets who need redemption by Christ just as we do (41) and when he counters Melanchthon's dogmatic doubts about extraterrestrial humankinds by saying they could have been redeemed from original sin by some other means unknown to us (34). It would be possible in this way to preserve a belief in the preeminence of the human species— an infinitely multiplied human species—in God's Creation and plan of salvation.

But the whole thrust of Borel's book is against this kind of metaphorical anthropocentrism in which, throughout the universe, man as we know him is the purpose of Creation. The universe is characterized by diversity: the heavenly bodies are large or small, noble or mean. But since, as Borel asserts with a vestige of medieval thinking, the Earth is the least and lowest of all heavenly bodies (7), it is natural to extend the same valuation to its inhabitants: perhaps the Earth and its human beings "serve" more perfect worlds (6)? Ten years later

Milton's Protestant Adam still allowed himself to be silenced on this subject by the Archangel Raphael; but the Protestant Borel faces up to the problem. He gives voice to the thought that worries Adam: God in his omnipotence may well have created more perfect worlds than ours (26). He invokes the authority of Campanella for the view that other worlds may have inhabitants superior to us in wisdom and insight (30); and against Melanchthon's christological scruples about plurality he also argues that the men in other worlds might perhaps be "better" than we and might therefore not need salvation by Jesus Christ (34). When speaking here of other worlds, he cannot be referring only to the paradise promised to the redeemed, although he does locate this (like hell) on a star (35, 36). For when (citing, like More, the venerable testimony of Origen) he mentions the *earlier* human inhabitants of those other bodies to whose destruction by fire the comets bear witness, he expresses the conviction, based on little-known biblical passages, that at least some of these were "better" than the descendants of Adam and Eve (42).

This casts doubt on man's position as the pinnacle of Creation or at best makes it uncertain. Like Campanella before him, Borel is able to use biblical quotations to show that the assumption of a plurality of worlds, like the heliocentric theory on which it is based, is compatible with the revealed Word of God (32, 41–43). But he cannot find in the Bible a clear answer to the question of man's status in this universe of many worlds. It is noteworthy that, unlike his Catholic colleague Gassendi, Borel makes no mention whatever of the idea that man on this Earth, even if he is not "alone" and not at the center, is singled out from all other creatures by the fact that God sent his only begotten Son to redeem him from original sin. On the contrary, Borel, like others of his day, does nothing to soften the harsh conclusion that a plurality of worlds with human inhabitants represents a loss for the sons of Adam, a diminution of the self-esteem of man for whom, according to the Creation story, the whole world including the stars had been created. And it is perhaps not too far-fetched to suggest that the glorification of God by Borel and others as a Creator whose greatness is enhanced by the plurality of humankinds is in part a reaction to this sense of loss, a way of compensating for it: man, finding himself "disinherited," at least knows that he is submitting to the decree of a Being of immeasurable greatness.

Interestingly, however, this tendency in Borel to diminish man's status is accompanied by a contrary one, an attempt to preserve a position of special significance for man even though he now has no absolute preeminence. Plato, defending the belief in *one* world (or rather one universe, which, basically, Borel and the other neo-Atomists do not disagree with), had used the argument that if there were more than one world, none of these worlds would contain "everything," and therefore each one would be imperfect. Borel counters this with Plutarch's response: man does not contain "everything" and is nevertheless "perfect" (32). This suggests that though his planet has lost its absolute status, the loss does not apply to man himself. But does that mean only that man is perfect in comparison with the other living creatures on Earth, or that he is as perfect as the human beings on other planets? Hardly, in view of what has already been said. And yet at the end Borel returns to the idea that we are no different from the human beings of "other worlds"—in respect to our status in Creation, Borel now means, not in respect to original sin and need of redemption. The principle of analogy makes it seem likely that there are plants on other planets; plants can only have been created for the benefit of animals, and animals only for the benefit of man; for man rules the stars just as he rules the Earth, "all the world was made for him, and consequently [!] there must be inhabitants on the stars" (45). The belief in the absolute preeminence of man as we know him, which just now appeared to have been destroyed, has evidently crept in again, by the back door as it were. What is more, it now has an imperialistic ring, anticipating the form it will take in the more secularized thought of Huygens.

Looked at as a whole, Borel's work does not resolve the conflict between these two opposed tendencies, the relativistic and the anthropocentric. It shows, as clearly as anything could, what ambivalent feelings the idea of the existence of many worlds aroused in educated and open-minded thinkers around the middle of the century.[79]

Similar but more complex ambivalences may be found in the writing of the priest Pierre Gassendi (1592–1655), the most significant of those who reshaped Atomism in the spirit of the new science and Christianity. He was a professor of mathematics at the Collège de France and a correspondent of the most influential philosophers of the day, and so his modernized and "theologized" version of the ancient teaching—and with it his ideas on the plurality of worlds—became widely known and appreciated, incidentally in literary as well as scholarly circles. His concern with the plurality idea was more than merely marginal, since astronomy was his chief scientific interest. That

---

[79]Carré totally fails to recognize this conflict when, on the basis of passages taken out of context, she attributes to Borel a belief in the preeminence of man, his perfection and greatness, and also "complete confidence in the destiny of modern man" ("Man between Two Worlds," pp. 323, 335).

he wrote a biography of Copernicus and was sympathetic to Galileo's beliefs and deeply concerned about his fate, indicates something of the intellectual context within which the plurality idea was developing. Gassendi's thoughts on plurality are only set out fully in the work that sums up his whole philosophy, the *Syntagma Philosophicum*, which was published posthumously in the *Opera Omnia* (1658); but it has been possible to show that the essentials had been laid down long before, in his studies of Atomist, particularly Epicurean, physics, some of which were published in his lifetime.[80]

Like the Protestants More and Borel, Gassendi too attributes the origin of the atoms, the component parts of all that exists, to God, the "first cause" (p. 130). Since for him such a creative principle can clearly only be the God of Revelation, it follows that any theory of a combination of atoms forming a plurality of inhabited worlds must be seen in the light of that revelation. At first sight this seems to cause no problem, since Gassendi rejects out of hand the classical Atomist theory of plurality. In the chapter "An-non Mundus Hic Unicus Sit pro Universo Habendus" (pp. 125–30), he reaches the conclusion that *kosmoi* outside our own visible universe, assumed on the basis of pure speculation, are mere figments of the imagination (*purum putum figmentum*) because their existence cannot be verified by observation (p. 130). He thus supports *one* world (in the sense of an Atomist cosmos), and this can readily be identified with the world of Scripture; everything the Bible says about the origin of things, about Providence, the creation of man, his "happiness," and "everything else" points to only one world (p. 127). Gassendi is of course familiar with the plenitude argument, but for him, as for Saint Augustine, the fact that God, the Infinite and Almighty, possesses the "strength" (*virtus*) and "power" (*potentia*) to create other worlds does not prove that he in fact created them; nor may one conclude that he did not: Such a question is by its very nature outside the scope of reason (and science): "praeter rationem omnino est." It oversteps the limit of what God wished man to know (pp. 127, 130). God is inscrutable; but as a scientist and as a theologian, Gassendi's belief—unless God should reveal it to be an error—is that "this world is the only one" (p. 130).

This sounds like his final word on the subject. But to stop short at this rejection of the plurality of worlds is to draw (as has been done

---

[80]Bernard Rochot, *Les travaux de Gassendi sur Épicure et sur l'atomisme* (Paris, 1944). In the following discussion, all page references in the text are to vol. 1 of the *Opera Omnia* (Florence, 1727). The *Nicolai Copernici . . . Vita* is in ibid., 5:439–56; on Gassendi's interest in Galileo see in particular his letter to him, ibid., 6:59–60.

only recently)[81] a very incomplete picture of Gassendi's role in the development of the idea in the seventeenth century. For in a later chapter of the physical section of the *Syntagma Philosophicum* he returns to the subject, and here we find that Gassendi, like Borel and the other neo-Atomists, rejects only a plurality of *kosmoi* as envisaged by the ancient Atomists, but not a plurality of planetary worlds in our solar system and also in other possible systems centered on fixed stars. He gives a clear yes to the question posed in this later chapter, "Are the heavens and the stars habitable?" (pp. 459–63), and moreover he reveals, for all his caution, that in his opinion they are indeed inhabited. But are they inhabited by *humans*? This is the question that follows on logically, and for Gassendi, the theologian, it naturally represents the crux of the matter.

More had, without much thought, taken it for granted that the humankinds in the universe must be equal; Borel had, equally impulsively, suspected they might not be. Gassendi approaches the question with all the deliberation of a critical historian of the idea. He roundly dismisses the opinion of some unnamed thinkers that the moon and other heavenly bodies are inhabited by living beings far *superior* to ourselves (*animalia nobis longe excellentiora*); also the albeit-rather-orthodox view linked with it, of the Earth as the dregs, or "scum" (*faex*), of the universe (p. 459). But this does not in any way prejudge the next question, namely whether on other heavenly bodies that resemble the Earth there are creatures which, if not superior (*nobilior*), are nevertheless similar to terrestrial beings and whether these include "some men, or certain forms of life that are similar to men" (*aliquos homines, naturasve quasdam cum hominibus proportionem habentes*, p. 459). Gassendi records the opinions of Plutarch, Epicurus, Lucian, Democritus, and the Pythagoreans. His own immediate verdict is wholly that of an anti-Aristotelian physicist: if the various continents on Earth harbor such very diverse plants and animals, then logically those on the moon and planets must be more different still. One cannot therefore maintain that the heavenly bodies are inhabited by animals and, in particular, by men. But this means only animals and men as we know them. For Gassendi immediately continues: supposing that these bodies have inhabitants, then these must be fundamentally different from the animals of our Earth and from "us."[82] Different—

---

[81]Dick, *Plurality of Worlds*, pp. 53–60. Olivier René Bloch is not explicit in his suggestion that Gassendi, in other passages that Bloch does not discuss in detail, comes close to assuming a plurality of worlds and thus to the "twofold truth" (*La philosophie de Gassendi* [The Hague, 1971], pp. 334–38). Dick devotes one sentence to it.

[82]P. 460: "Non perinde tamen probabile habemus ejusmodi ea esse, cujusmodi

but still human? Here Gassendi hesitates. Man can adapt himself to all regions of the world, but not to the wholly unaccustomed climatic conditions of the moon, with its long periods of heat and cold; just as, conversely, the moon dwellers, "if you assume there to be some on that globe," could not survive on our Earth (p. 460). "We" cannot live on the moon (let alone elsewhere in the universe), and therefore there are no humans living on the moon (p. 461).

This does not mean that the moon (used here as an example to represent all the heavenly bodies) necessarily lacks other forms of life (*aliae naturae*), but only that we cannot form any notion of them, any more than they can of us. Unlike Galileo, who first introduced this topos, Gassendi does, however, feel able to state with certainty that *if* one assumes the existence of living creatures on the planets of our solar system ("if you suppose there to be some things on them which, either because of a certain similarity or for lack of more suitable names, you call animals [*animalia*]"), then one is justified in assuming that the closer the planet's orbit is to the sun, the more perfect (*perfectiora*) the creatures on it must be, while those inhabiting the sun, "if you suppose there also to be some animals on it," would be by far the noblest (*nobilissima*; pp. 461–62). Such speculations based on analogy continue to crop up in the works of thinkers up to Kant. In the seventeenth century, at any rate, these speculations can still be seen as casting doubt on God's justice. Gassendi forestalls this objection by emphasizing that every form of life is adapted to the conditions in which it lives and therefore lives in the best possible world for it, like the fish in the water and the bird in the air.

The neatness of this solution is impaired, however, when Gassendi has to admit that, like our solar system, *other solar systems* may be inhabited. Would there not in that case be creatures corresponding to ourselves? These too cannot be supposed to be humans, Gassendi declares magisterially but without offering further elucidation (p. 462). He does admit that there is a problem in that it is difficult to see how those beings (*naturae*) can have been created, like everything else, for us, the inhabitants of this Earth. Surprisingly, however, this orthodox argument is no longer a major consideration for Gassendi. He is content with the distinction God bestowed on man in sending his only begotten Son to our planet and ensuring, by his death, our eternal salvation. Why—"to what end"—God created the other planets and

---

sunt, quae heic in terra generantur, & corrumpuntur, maximeque cujusmodi animalia, ac in animalibus homines, ut propterea ullo modo possint dici sidera ab animalibus, ac speciatim ab hominibus habitari. Esto enim habitatores habeant; at ratio postulat, ut sint penitus diversæ a nobis, animalibusque terrenis naturæ."

planetary systems, whether it was for us or for the "others," it is beyond the capacity of human beings to discover (p. 463).

Anthropocentrism, threatened by science, defended by theology and dogma, is thus maintained, but with some erosion of the Bible's absolute authority. Man on this Earth is no longer necessarily the *sole* end of Creation. But does the fact that he is singled out by Christ's mission mean that he is still its *highest* end? The creatures (if any) on the planets closer to the sun are more perfect than we, and those *on* the sun are the noblest of all. Did they not need redemption? Does that give them a higher status in Creation? And do the less perfect beings on the planets further from the sun remain unredeemed? Gassendi dismisses such questions with a quotation from Lactantius about the inscrutability of God, which is above all reason. In pursuing these philosophical problems, the theologian Gassendi has taken on more than he can manage, and he retreats to the security of his doctrine, although at other points in his work he himself has forced doctrine to bend to fit the logic of his own thought.

Theological scruples and considerations similar to Gassendi's also cause Guericke's reflections, unlike Borel's, to lead finally only to the topos of God's inscrutability. Guericke was the "founder of experimental physical research in Germany": his experiments with the "Magdeburg hemispheres" relating to the theory of the vacuum made him a prominent figure on the German intellectual scene. As a convinced Lutheran he could not tolerate a contradiction between theology and science; in this he was similar to Robert Boyle, the thinker most akin to him. The possibility of such a conflict is especially acute in connection with the plurality of worlds, which involves both disciplines. Appropriately enough, a discussion of this topic forms the culmination of his main work, *Experimenta Nova (ut Vocantur) Magdeburgica de Vacuo Spatio* (1672).[83]

Unlike Descartes (p. 241) but like Gassendi, whom he frequently quotes, Guericke sees "the whole edifice of the world" as being located in "empty space" (p. 1). Again, it is not a plurality of (Aristotelian) *kosmoi* that he envisages—Guericke cites the Scholastic position of 1277, that God "can and could create numerous worlds . . . , even if in fact he deliberately only created a single one in Time" (p. 51)—but he is referring to a plurality of planetary worlds. Like Bruno and the

---

[83]Reprint (Aalen: Otto Zeller, 1962); translated into German in a monumental scholarly edition, *Neue (sogenannte) Magdeburger Versuche über den leeren Raum*, trans. and ed. Hans Schimank (Düsseldorf, 1968); quotation, p. [216] (pages of the editorial commentary are numbered in brackets). Quotations from Guericke are based on the Latin text, and page references are to the 1962 reprint.

seventeenth-century Atomists, Guericke too believes that there are, at least potentially, as many worlds as there are planets in our solar system and in the other systems (centered on the fixed stars) that may, in the Copernican view of the universe, be assumed by analogy.[84] Guericke conscientiously acknowledges the objection that Scripture teaches "that such systems form only a single world"(p. 241), but he has already taken the wind out of its sails by rejecting the plurality of *kosmoi*. He has also subtly forestalled the objection by devoting a whole chapter to the cosmology of Athanasius Kircher's *Iter Exstaticum* (bk. 7, chap. 4). The Jesuit had postulated systems of planets circling the fixed stars, to the greater glory of God, and the Lutheran is convinced of this too:

> A plurality of created worlds not only does not detract from the majesty and immeasurable greatness of the Deity but in fact makes it the more resplendent. For in a space of unbounded height, depth, or breadth, what could a single world amount to, which cannot remotely be compared with it. . . . But it is not possible to specify the number of worlds that could fill this immeasurable space. However, just as God did not create the whole sea so that only a single fish should swim in it, or the whole domain of the air around us so that a single little bird or fly should fly about in it, so the omnipotence of God might well shine forth the more brightly from a multitude of worlds that are formed with such diversity that none is found to be the same as or even similar to another, but one is distinguished from another in many thousands of ways. (P. 243)

A plurality of *diverse* worlds, then, as with Gassendi (and, earlier, Bruno). For Kircher, whom Guericke invokes as a witness of undisputed respectability, such diversity lay above all in the fact that the Earth alone was chosen by the Divine Creator to be the abode of life. Guericke's idea of the diversity of the worlds includes, however, the presence of living creatures elsewhere than on Earth, and so, unlike Kircher, he grasps the nettle as Gassendi and Borel had done before him. Each of them, in his own way, had been prepared to countenance some modification of anthropocentrism, but not the crass anti-anthropocentrism that breaks through in Descartes and Cyrano de Bergerac. Guericke approaches this danger zone more closely.[85] But even he

---

[84]Pp. 230, 232, 240. Schimank's view that Guericke does not accept such analogous systems (p. [218]) is in contradiction to these clear statements in the *Experimenta*.

[85]Rossi ("Nobility of Man," p. 154) has insufficient justification for suggesting that Guericke is militantly anti-anthropocentric.

tries more than once to salvage some special distinction for terrestrial man. Thus this Lutheran has essentially the same aims as the Catholic Gassendi; but his strategy, despite some similarity, is different—and more risky.

While sympathizing with Galileo's strategic withdrawal on the matter of extraterrestrials (his suggestion that, if there were any, they would by definition be of a kind that human beings could not possibly imagine), Gassendi had not been content to accept this view. Neither, at bottom, is Guericke, though he quotes that very passage from the *Dialogo* (pp. 216–17). Guericke too comes to some conclusions about the nature of the extraterrestrials, but they are not always the same as Gassendi's. First he gives a decisive answer to the question Are there living creatures on the moon? (p. 181). His view is that *if* the question is answered in the affirmative, then the "creatures" can only be "of a totally different . . . kind [from the terrestrial ones] . . . and rather far removed from anything we can imagine." But unlike Galileo, he believes, with Gassendi, that one can at least establish that they are not superior to us but must be "of a far baser kind" (*longe viliora*). Like Gassendi, Guericke gives no reason for this unfavorable a priori judgment, except that he adds that the moon "lacks animate life" (*animata vita caret*) and its "so-called living creatures" would "possess no animate life, which is after all the essential characteristic of a living being." Like Gassendi, Guericke feels that this purely philosophical position is supported by physicobiologic arguments: it is impossible for life to develop without conditions of light, warmth, and humidity such as exist on Earth. Guericke concludes, then, that there can be no living creatures on the moon, for both metaphysical and physical reasons. These reasons probably do not apply, however, to the planets that are comparable to the Earth, and Guericke's answer to the question Are there living creatures on the planets? is decidedly more positive (pp. 214–17).

The reasons for this inclination to a more positive answer are derived from the analogy, seen by thinkers from Copernicus and Galileo onward, between the Earth and the other planets of the solar system. These planets must have some purpose, but despite the biblical Creation story's total clarity on this point, particularly as the story is understood by the literalist Lutherans, Guericke believes the planets "appear not to have been created for the illumination of our Earth, as they themselves have no light of their own, and no one would, for the use of the occupants of a house, light a torch that is larger than the house itself" (p. 216). Here, going beyond purely analogical reasoning, Guericke presupposes a purposeful Creator, but at the same time he disputes that man is his sole concern. (The same doubt is voiced by Borel and implied by Gassendi). For, Guericke says, pursuing this train of thought, to believe that such a Creator should have created life only on the Earth, a tiny dot in the universe, is contrary to the idea of his omnipotence. In such general terms Guericke, the physicist, admits the competence of theology. Whereas for instance in Galileo's writings, analogical reasoning had stood on its own feet, Guericke here backs it up with the plenitude argument. God's complete power theoretically makes anything possible, including humans on other planets, and in principle analogical reasoning sets up no obstacles to such an assumption; but Guericke, concerned now about his orthodoxy, at once protects himself from any suspicion of the heresy of out-and-out anti-anthropocentrism by adding: "However, if anyone were to think that there are human beings [the pinnacle of Creation] on the other planets, he would be making an error as great as the distance of such a star from us" (p. 216). Why the idea is so absurd, he does not explain. He simply states that "all beings that are there are fundamentally different from ours." Galileo and Gassendi had said the same. Guericke even uses the same comparison with America as Gassendi: "Anyone who would deny the presence of living creatures on the planets because he is not capable of imagining any creatures other than those that he sees here on Earth should know that in America there is no wild animal of exactly the same kind as in Europe, Asia, or Africa" (p. 216). But America, one might remind Guericke, does have its own human beings—different, perhaps, but still human: so if this aspect of the analogy applies to the regions of the world, does it also apply to the heavenly bodies?

After the quotation from Galileo, Guericke makes only the cautious statement that "there are therefore on other planets neither men nor other creatures of the same kind as those on Earth" (p. 217). Indeed, it is with apparent satisfaction that at the very start he had introduced a quotation, perhaps meant to serve as a *captatio benevolentiae*, from Nicholas of Cusa (whom, admittedly, one could quote in support of almost any position), to the effect that "the spiritual being that lives here on this Earth . . . seems [*videtur*] to have no equal in nobility and perfection, even if there are inhabitants of another kind on other stars" (p. 214). This shores up the defenses against the relativistic view of man demanded by scientific analogical thinking. But just as Gassendi thought there might possibly be more perfect beings than ourselves on other planets, so Guericke is at least convinced, as Kepler was, that analogy "necessarily" (p. 217) leads to the assumption of "rational beings" on other planets. And these beings that Guericke envisages must surely be at least similar to human beings! They venerate God—what could more clearly set them apart from all nonhuman creatures?—but in a different way, a way "appropriate to their understanding and intelligence," just as on Earth, God "gave the Jews different laws from the other peoples and also permitted and indulgently granted to them what he did not grant to others. Rather, he treats different people quite differently" (p. 217).

So are there, after all, humans on other stars? Perhaps even "nobler" ones? Guericke admits that "among objects which differ from one another, one is always more noble or more outstanding than another." Therefore "there will necessarily be beings there that are superior to and more rational than the others (we will say nothing at all of the stars themselves) and occupy a more privileged position than they" (p. 217). Superior to and more intelligent than man on this Earth? For Guericke is obviously not speaking of differences within the "population" of a planet, but of differences between the various inhabited planets. Gassendi, when faced with this suspicion, managed to salvage man's special status, under threat from science, by means of theology and dogma, namely by reference to Jesus Christ's unique mission of salvation. It is noteworthy that Guericke, like the Protestant Borel, attempts no such rescue. Certainly he retains God as the Creator of *all* living creatures in the universe: we may "suppose," he says, that they all, including the extraterrestrial beings, serve to "manifest his glory, his omnipotence, and his honor" (p. 217). But in the same breath with which Guericke the experimental physicist stresses that *our* only knowledge of "the actions of God" is ignorance, the theologian in him concedes that one must "nevertheless" not suppose that "God has kept his infinite omnipotence secret; rather, what he has not permitted to some creatures he has allowed others to discover" (p. 217). Others—not us. But does this not sabotage his own attempt, supported by the quotation from Nicholas of Cusa, to uphold the preeminence of man (which he himself undermines by his scientific arguments)? Whereas Gassendi trusted in the dogmatic teaching that Christ was the Savior of man on this Earth, and only on this Earth, Guericke expresses only a vague confidence in the wisdom of the Protestants' hidden God, who may have shown his mercy in other worlds too, "but in a manner different and most different from ours" (p. 217).

Guericke does not enter into further particulars. In the final section of his book (a book that, one almost forgets, is an account of his experiments relating to the vacuum), he returns like a good Protestant to the inscrutability of God. As we have seen, his Catholic colleague had done the same. But whereas Gassendi had cited Lactantius, who was one of the church fathers and thus above suspicion, Guericke quotes Descartes, whose orthodoxy was at least in doubt. What is more, Guericke looks to Descartes not only for corroboration as to the impenetrability of God's decrees ("as if the power of our cognition could reach beyond that which God has created") but also as an authoritative voice against the view that "all things were created for us" (p. 231). In Descartes this polemical anti-anthropocentrism represents a genuine conviction, which reaches its full fruition in the radical and cynical skepticism of his pupil Cyrano de Bergerac. For the moment, Guericke simply gives the quotation without commenting on it. The closing words of his book, however, make his doubts about the central position of man in God's plan of salvation appear rather as a cipher for the inscrutability of God, to whom anything at all is possible: knowledge of "the unerring truth" (p. 243) is something that can only be hoped for. "'For in this mortal state,' as the Apostle says, 'we know in part; but when that which is perfect is come, then that which is in part shall be done away. For now we see through a glass, darkly; but then face to face'" (p. 244).

5. The Risk of Reasoning: Descartes and Cyrano de Bergerac

When Guericke invoked the support of Descartes for anti-anthropocentrism and also for the inscrutability of God—Descartes, of all people, who was suspected of atheism—he did not take over the philosophical context of Descartes's statements but assimilated it to his own ideas that were rooted in his unshakable Lutheran faith. Descartes's own thoughts would hardly have culminated, like Guericke's, in the quotation from the first letter to the Corinthians (13:9–12). *His* anti-anthropocentrism is not merely a cipher for the inscrutability, established as a dogma, of the Christian God, but a necessary consequence of his philosophical physics; and in the context of Descartes's thought, the inscrutability of God itself is little more than a phrase referring simply to the areas that still remained blank on the map of scientific discovery. The truth Guericke has in mind, which is "above all reason," has no meaningful place in the system of thought of a man for whom reason is the sole criterion of truth and the only means of access to it.[86] Not only that, but his trust in this reason

[86]*Principia Philosophiae*, pt. 3, § 1 in *Oeuvres*, ed. Charles Adam and Paul Tannery,

invites, by its own inner logic, inferences that bring it into conflict with that revealed and institutionalized truth "above all reason," whereas More and Borel, Gassendi and Guericke were always ultimately secure in their belief in that truth. For this very reason, when Descartes's thinking brings him even within hailing distance of the theory of the plurality of worlds, he proceeds with far more caution than the other thinkers just mentioned (who also stand opposed to him in positing that space is empty, not full). The whole tenor of Descartes's thought (whatever his personal beliefs may have been) is such that a retreat into the bosom of religion is not open to him: what appear to be pious statements, and are taken to be such by Guericke, are only prudently adopted camouflage.

Descartes is so wary of the powerful influence of theology that at the various stages of his exposition of his cosmology and cosmogony— the atomistic "vortex" theory that provided the physicophilosophical basis for the most influential books of his century on the plurality of worlds[87]—he does not even touch upon the idea that other heavenly bodies might be inhabited by rational beings, or even by any living creatures at all. Even so, he held back publication of the first draft of the vortex theory, Le monde (written 1629–32), although it still assumed "God" as the creator of matter and of its laws of motion (in other words, as the originator of the Copernican planetary system) because, as he himself said, he was afraid of suffering the fate that overtook Galileo in 1633.[88] And whereas in this draft he was able to write "if God had created many worlds . . ." (11:47), in the main work containing his anti-Aristotelian physics and philosophy, the Principia of 1644, which develops further the basic cosmologic ideas put forward in Le monde, he anxiously avoids even this harmless conditional mode. In the Principia, Descartes, like all the neo-Atomists, explicitly rejects a plurality of kosmoi such as the ancient Atomists imagined (pt. 2, § 22), on the grounds that it is incompatible with his fundamental conviction of the "indefinite" extent of matter.[89] But in the same work, Descartes accepts the analogy between our sun and the fixed stars (pt. 3, §§ 9, 13, 46)—

"nouvelle présentation," vol. 8:1 (Paris, 1964), p. 80. Page references in the text are to this edition (1964–74).

[87]See chap. 4, sec. 2 below; Dick, Plurality of Worlds, chap. 5; Arthur O. Lovejoy, The Great Chain of Being (1936; New York: Harper Torchbooks, 1960), pp. 124–25 on Descartes's philosophy as the main pillar of the belief in plurality in the seventeenth century.

[88]See the editors' comment, Oeuvres 11:702, and Descartes's letter to Mersenne, ibid., 1:270–72 (late November 1633).

[89]"Indéfini" is only a diplomatic euphemism for "infini"; see Descartes's letter to Pierre Chanut, 6 June 1647 (Oeuvres 5:51).

by now a familiar idea, though Borel was still to reject it in 1657—and in the light of his cosmology this logically implies that our planetary system is not the only one. But Descartes himself never directly mentions what incontrovertibly follows from the law of analogy, namely (Earth-like) planets orbiting the fixed stars (physically necessary phenomena associated with the "vortex"); still less, as I have said, does he mention their inhabitants, whether human, humanlike, or of any other kind. Also, although at all stages of the development of his thought he retains the conviction, derived directly from Copernicus, that the Earth is "one planet among others" in the solar system ("inter Planetas . . . posse numerari" [pt. 3, § 13]), he never explicitly draws from this the analogical conclusion that had long since become commonplace, that there is a plurality of "earths," that is, of inhabited worlds in the solar system.

This caution is all the more striking, as at the opening of the cosmologic (third) section of the *Principia* (Guericke quotes from this famous passage), Descartes refers not only to the inscrutability of God but also to his omnipotence. For when he asserts in this context that this God by no means created "everything" for terrestrial man, he comes close to using the teleological argument, or the plenitude argument, in favor of the plurality of worlds and of humankinds without actually formulating them. He writes:

> 1. *That the works of God cannot be thought too great.* We have discovered certain principles concerning material things; and there can be no doubt about the truth of these principles, since we sought them by the light of reason and not through the prejudices of the senses. We must now consider whether we are able to explain all the phenomena of nature by these principles alone; and we must begin with those phenomena which are the most universal and on which the rest depend, namely, the general structure of this whole visible world. In order to reason correctly about this matter, we must pay special attention to two things. First, remembering God's infinite power and goodness, we must not be afraid of overestimating the greatness, beauty, and perfection of His works; rather, we must beware of accidentally attributing to them any limits of which we do not have certain knowledge, and of thus seeming to have an inadequate awareness of the Creator's power.
>
> 2. *That we must beware, lest, thinking too highly of ourselves, we suppose that we understand for what ends God created the world.* Second, we must beware of overestimating ourselves. We would be doing so if we were to attribute to the universe limits of which we had not been assured either by reason or by divine revelation; for this would be to assume

that our minds can conceive something which is greater than the world which God actually created. We would be overestimating ourselves still more if we were to imagine that God created all things solely for us, or if we were to consider our intellect powerful enough to understand His ends in creating the universe.

3. *In what sense it may be said that all things were created for man.* For although, from a moral point of view, it may be a [good and] pious thought to believe that God created all things for us, since this may move us all the more to love Him and to give thanks to Him for so many blessings; and it is true in a sense, since there is no created thing which we cannot put to some use, even if it is only a matter of exercising our minds by contemplating it and, by means of it, being moved to praise God: it is, however, in no way likely that all things were made for us in the sense that God had no other purpose in creating them. And [it seems to me that] it would be clearly ridiculous to attempt to use such an opinion to support reasonings about Physics; for we cannot doubt that there are many things which are currently in the world, or which were formally here and have already entirely ceased to exist, which no man has ever seen or known or used.[90]

The *reasons* given for this anti-anthropocentrism are purely scientific. Descartes maintains (in contrast for instance to Kepler) that from the time when Copernicus convinced adherents of the new science of the "immense distance" between the Earth and the fixed stars, only those who have remained behind the times regard the Earth as the "most important part of the universe" (*praecipuam partem universi*) because it is the abode of man, "for whose benefit all things were created" (pt. 3, § 40), including for instance the sun, created solely in order to give man light (3:431–32).

If on the basis of this conviction, Descartes had felt able to take the obvious next step of expressing support for the plurality of worlds, he would of course have been contradicting the biblical Creation story, according to which even the faintest of the fixed stars had the sole function of illuminating the night for man. At this time, both Catholics and Protestants still upheld the authority of Genesis in this matter with as much obstinacy as Melanchthon a century earlier. True, Descartes could confidently have defended himself against the accusation of irreligiosity by arguing that the plurality of worlds followed perfectly logically from the omnipotence of God, which he had so force-

---

[90]*Principles of Philosophy*, trans., with explanatory notes, Valentine Rodger Miller and Reese P. Miller (Dordrecht, Boston, London, 1983), pp. 84–85. Brackets in orig.

fully invoked, and not merely from his version of the new science. This was the dilemma of the plurality theory in the seventeenth century. Some, particularly anti-Cartesians such as Gerhard de Vries, chose the literal understanding of the Creation story and its *one* world; others, including avowed Cartesians such as Huygens, opted for the omnipotence of the Creator and his *many* worlds, without feeling they were crossing the threshold of unbelief. In contrast Descartes, and some of his followers, dare not even enter this arena.[91] In the very paragraph asserting God's omnipotence (quoted above) Descartes had also declared reason to be autonomous; and he must have sensed, with the wisdom of the burnt child, that the conclusions to which his reason led him could not convincingly be incorporated into a theology of omnipotence. His atheistic disciple, Cyrano de Bergerac, was soon to demonstrate this quite openly in both his theory and his imaginative treatment of the plurality of worlds. Unlike that notorious freethinker, Descartes allowed himself to be intimidated by the power of the One True Church, which had after all been able to make even a man of Galileo's stature recant.

Strictly speaking, however, all this applies only to those statements of Descartes's that were intended for publication. In private letters he was prepared to speak more freely and revealed quite clearly that he himself recognized the irrefutable logic of the inference his followers drew from the "vortex" theory, namely that the Earth could not be the only abode of intelligent life in the universe. In 1629 he had half-jokingly said in a letter to the lens grinder Jean Ferrier that he dared to hope that improvements to the telescope would soon make it possible to see "whether there are animals [*des animaux*] on the moon" (1:69). After the publication of the *Principia* he adopts a more serious tone. His friend Pierre Chanut, the French ambassador in Stockholm, had on 11 May 1647 told him of certain reservations Queen Christina of Sweden had, on dogmatic grounds, against his assertion of the infinite, or "indefinite," extent and duration of matter. They made her fear that man's theologically guaranteed status as the pinnacle and highest purpose of Creation was being called into question: for the assumption that the universe was spatially and temporally unlimited tempted, indeed forced, one to think in terms of more than one inhabited world—inhabited, perhaps, by beings more intelligent and better than we. Theological anthropocentrism was thus endangered:

[91]On this dilemma see Dick, *Plurality of Worlds*, p. 117; on de Vries see ibid., pp. 120–23.

And it is certain that, if we imagine the world as having the vast extent which you attribute to it, it is impossible for man to preserve that honorable rank within it; on the contrary, though he inhabits the whole earth, he will see himself as dwelling in a little corner, out of all proportion to the immoderate size of the rest. He will very probably consider that all those stars have inhabitants, or rather other earths all around them all filled with creatures more intelligent and better than he; certainly he will at the very least cease to be of the opinion that this world of infinite size is made for him, or could be of any use to him.[92]

In his reply, dated 6 June 1647, Descartes employs a maneuver that enables him to hold fast to his beliefs about physics without directly challenging the orthodox interpretation of the Bible. The Bible, he says, teaches that all things (*toutes les choses*) were made for man only in the sense that all of them are in some way useful to us. For Descartes, this does not mean that man is also the purpose (*la fin*) of Creation. The purpose of Creation is, rather, God himself, who as the "efficient cause" is also the "final cause." Moreover, Descartes continues, the idea that "everything" was made for man derives from the teaching of priests rather than from the Bible itself. Admittedly the Bible creates that impression, but only because it addresses itself to men on this Earth. If, therefore, one cannot exclude the possibility that other heavenly bodies too possess not only living creatures but also intelligent, indeed human, beings, then equally the doctrine of the redemption of man by the Son of God is no longer a compelling argument for the uniqueness of man on this Earth. Just as Christ by his blood did not redeem only the thief on the cross but "a very great number of other men, so I do not see at all that the mystery of the Incarnation and all the other benefits which God has bestowed on man prevent his having bestowed an infinity of other very great ones upon an infinity of other creatures" (5:53–55). This not wholly logical analogy does not make it clear whether Descartes believes that the Crucifixion has the same significance for the stellar humankinds whose existence, he says, the "new" astronomers lead us to assume as it has for us; or whether he supposes that they have received *other*, comparable proofs of God's mercy. He does not enter into details of

[92]*Correspondance*, ed. Charles Adam and Gérard Milhaud, vol. 7 (Paris, 1960), pp. 313–14. The passage is quoted from this edition as it is not included in the *Oeuvres* (see *Oeuvres* 5:19–22). The passages I have quoted from the correspondence are also used by Rossi, "Nobility of Man," pp. 152–54.

dogma and immediately begins to retreat, saying, "And although I do not on this account infer that there are intelligent creatures in the stars or elsewhere, I equally do not see that there is any argument by which one could prove that there are none; but I always leave questions of this sort undecided, rather than deny or assert anything about them" (5:55).

In fact, however, he does not leave the question open. For he adds that our merits are not diminished by the fact that other intelligent beings on other heavenly bodies "have similar ones." This ignores the point made by Queen Christina that those beings might be *more* intelligent and better than we. Not long afterward, however, he did freely admit—again in private—that this was a logical assumption. In his discussion with the Dutch philosopher Frans Burman about his philosophical writings, Descartes, on 16 April 1648, made the following comments (according to Burman's protocol) on the second paragraph of the cosmologic part of the *Principia* (quoted above): Men are in the habit of considering themselves to be the creatures dearest (*carissimos*) to God, who created everything for them. But how do we know that God did not use his creative power (*potentia*) to form other creatures, on other heavenly bodies, "and other 'men', so to speak— or at least beings analogous to man" (*ut ita dicam, homines, aut saltem homini analogos*)? The phrase "or at least beings analogous to man," which could be taken as a partial retraction, is not in fact a subtle way of suggesting that he is not challenging man's theological position of preeminence. This is clear from the surprising conclusion that follows: We should not be so arrogant as to assume that everything centers upon ourselves. "For an infinite number of other creatures far superior to us may exist elsewhere" (*cum forsan infinitae aliae creaturae nobis longe meliores alibi existant*).[93]

This conclusion explicitly confirms the worst fears of the Swedish queen and of Descartes's orthodox opponents.[94] His philosophy and physics proved indeed to form the basis of a theory of the plurality of worlds which no longer necessarily upheld man's special theological status and the greatness of the Creator-God. In this, Descartes differed from the other Atomists, especially Gassendi, his antagonist in both

[93]*Oeuvres* 5:168; *Descartes' Conversation with Burman*, trans. with Introduction and Commentary by John Cottingham (Oxford: Clarendon, 1976), p. 36.

[94]Descartes's works were placed on the index in 1663. He was, however, already suspected of atheism in his lifetime; Pascal was one of the first to doubt the sincerity of his pious phrases; see Francisque Bouillier, *Histoire de la philosophie cartésienne*, vol. 1 (Paris, 1854), esp. chaps. 10, 11, 21, 22; on Pascal see p. 553. After Descartes's works had been placed on the index, the suspicion of atheism was voiced particularly by the Newtonians (Philipp, *Werden der Aufklärung*, p. 89).

physics and philosophy. The essentially un-Christian conception of the plurality of worlds implicit in his physics was in fact a major one of the matters to which his opponents objected. In addition to the Dutch Protestant Gerhard de Vries, already referred to, another opponent worthy of mention is the French Jesuit Gabriel Daniel. In his *Voiage du monde de Descartes* (1690), a polemical account of the journey of a soul to the moon and stars, he rejects Descartes's "vortex" cosmology and the view based upon it that there may be "men" on the moon or on other heavenly bodies. It is no coincidence that the soul's guide on this journey is Père Mersenne.

Of course Daniel cannot accuse Descartes himself of publicly advocating this view; instead he lays the heresy at the door of Cyrano de Bergerac, whose thinking was similar to Descartes's.[95] Cyrano was a very obvious choice, as this famous, or notorious, duelist, freethinker, and man of letters, one of the circle surrounding Molière and Chapelle (Claude Emmanuel Lhuillier), liked nothing better than to flaunt his heterodoxy in a scandalous manner. He is the author of two "utopian" or satiric novels, written around the middle of the century but published only some years after his death, *Les estats et empires de la lune* (1657) and *Les estats et empires du soleil* (1662)—also known jointly as the *Histoire comique* or *L'autre monde*. In them he draws the most radically anti-anthropocentric and also antitheological conclusions from Descartes's Atomist cosmology.

Or perhaps from Atomist cosmology in general. Even as a novelist and "creator of the philosophical novel,"[96] Cyrano stresses his commitment to Atomism; but in both of these accounts of space journeys, it remains unclear whether he inclines more toward Gassendi or Descartes, toward the theory of empty or of "full" space—especially as the second story breaks off before a promised discussion of Descartes's physics. The novels make respectful mention of both Gassendi, whose private lectures Cyrano attended as a young freethinker in Paris in the 1640s, and Descartes, whose corpuscular physics he exactly reproduces in his *Fragment de physique*, written about the same time as the novels and published, together with the second novel, in the *Nouvelles oeuvres* (1662).[97] In any case, Cyrano's precise position within seventeenth-century Atomist thinking (and evidently he himself had not

---

[95]*Voiage du monde de Descartes* (Paris, 1690), p. 152.

[96]E. Hönncher, *Fahrten nach Mond und Sonne: Studien insbesondere zur französischen Literaturgeschichte des XVII. Jahrhunderts* (Jena and Leipzig, 1887), p. 51 ("Schöpfung").

[97]On Cyrano's relationship to Gassendi and Descartes see Erica Harth, *Cyrano de Bergerac and the Polemics of Modernity* (New York and London, 1970), pp. 99–108; Alcover, *Pensée philosophique et scientifique*, pp. 58–67, 93–107, 137–39, 167–68.

finally decided where he stood) is of less significance than the unmistakable fact, evident in the novels too, that Cyrano bases his theory
of the plurality of worlds firmly on the scientific insights of his age.
Not only is there biographical evidence of his intensive study of the
new philosophy; the text of the *Histoire comique* itself clearly shows the
influence of Copernicans such as Galileo, Jacques Rohault, and of
course Descartes, and also that of scientifically oriented "pluralists"
such as Borel, Gassendi, Campanella, and Godwin, some of whom
are actually referred to by name. The decisive difference between
Cyrano and all his predecessors—Copernicans and neo-Atomists—is
that he does not assume his worlds, his atoms of matter, to be the
work of a divine Creator. Accordingly, he no longer concludes his
reflections with the pious topos of God's inscrutability, even as a mere
formality, but instead consistently submits to the sole guidance of
reason, his "queen," wherever she may lead.[98] The views implicit in
Cartesian thinking but masked by Descartes, at least in public, with
professions of loyalty to the Catholic church are openly expressed by
Cyrano with a positive delight in scandalizing the clergy. Reason,
freed from any constraint, leads from the premise of Copernicanism
to the idea of the plurality of worlds in the form that inflicts on human
self-esteem the greatest possible anxiety and disillusionment; and it
leads to uncompromising scientific atheism and the rejection of all
metaphysics. But this courageous willingness to follow the scientific
premises through to their logical conclusion obviously caught the
mood of many of Cyrano's contemporaries: though toned down somewhat by the first editors, the two space novels were extraordinarily
successful and influential all over Europe.[99]

The fact that Cyrano's imaginary accounts of "other worlds" are
scientifically inspired makes it possible to arrive at a more precise
assessment of their character than is conveyed by the usual classifica-
tion of them as utopian, satiric, or parodistic (parodying Christianity
or other things). Far from being merely playful fantasies, or indeed

---

[98]"La raison seule est ma reine" (*Oeuvres diverses*, ed. Frédéric Lachèvre [Paris, n.d.],
p. 47). See also *Oeuvres libertines*, ed. F. Lachèvre (Paris, 1921), 1:36. Page references in
the text are to this latter edition. English renderings of longer quotations are taken from
Cyrano de Bergerac, *Other Worlds: The Comical History of the States and Empires of the Moon
and the Sun*, trans. and introduced by Geoffrey Strachan (London, 1965); the page
reference to this translation is given first. Harth (*Cyrano de Bergerac*) and Alcover (*Pensée
philosophique et scientifique*) agree on Cyrano's atheism. Only once does Cyrano name
God as the Creator of the atoms (*Oeuvres libertines* 1:127). Alcover regards this as merely
a formula intended to give the appearance of religious belief (pp. 160–63; see also pp.
132, 171).
[99]See Lachèvre, *Oeuvres libertines* 1:xcvii–cix.

stories poking fun at the genre of the cosmic voyage itself,[100] they represent a serious exploration of ideas by an empirical scientist. Undoubtedly they contain utopian, satiric, and polemical elements, but above all they portray the imagined meeting between man and a counterpart in space. As philosophical novels set in the framework of modern cosmologic thought, the two parts of *L'autre monde* are among the earliest examples of serious "encounter" science fiction. One factor that contributes markedly to their intellectual stature and appeal is the fact that traditional theological reasoning involving the idea of plenitude, which still cropped up here and there even among Cyrano's neo-Atomist predecessors, is here entirely discarded.[101] Cyrano's conception of an inhabited universe derives wholly from the identity of substance between the celestial bodies that is inherent in Copernicanism and Atomism. Also he assumes, unlike Descartes, that there are as many Copernican planetary systems as there are suns (1:14) and that these planets are, like the moon and the sun, inhabited by *rational* beings. This in turn means that Cyrano's attention is focused on the question of man's nature and status. His response to this question, which forms the central theme of the novels, is a view of man more virulently anti-anthropocentric (and critical of theology) and philosophically misanthropic than the views of any other seventeenth-century thinker.

For the sake of their structure as novels, the two parts of the *Histoire comique* deal only with the inhabited moon and sun, but Cyrano's conception of the universe is such that the moon and sun can each be assumed also to represent other, analogous bodies.[102] The implication is that there are very many inhabited worlds. But whereas cosmic thinkers like Bruno and More were excited by the infinite number of worlds, Cyrano is less interested in this than in the fact that he can only imagine the aliens as being *superior* in intelligence to the rational beings on Earth, if not necessarily exemplary. Such an idea had been merely a transient nightmare for Kepler and many of his successors; others, like Borel, had suggested it rather casually, as a possibility they found plausible but had not fully thought through. But for Cyrano's novels this belief is the very precondition and basis; he treats it seriously and, having introduced it, does not then in any way retract or invalidate it.

In *Estats et empires de la lune*, the narrator, Dyrcona (an anagram

---

[100]As suggested by Nicolson, *Voyages to the Moon*, p. 159.

[101]This is also the view expressed by Harth, *Cyrano de Bergerac*, pp. 96–98; she argues against Lovejoy's general thesis.

[102]See ibid., p. 87.

of d[e] Cyrano), undertakes his journey after a dispute among a group of friends in Paris on the question—a topic of urgent debate since the time of Galileo's discoveries, if not before—of whether the moon is a world or earth like ours. The best evidence could be obtained by actually seeing it. Dyrcona's first attempt to fly to the moon (with the aid of bottles filled with dew, which he ties to his body) is unsuccessful, but when, after a promising start, he lands in French Canada, it at least gives him the opportunity of discussing the Copernican basis of his project with the viceroy there. What could be more natural? The very fact that he has landed in the New World proves that during his ascent and descent, the Earth has turned on its own axis. And as the viceroy is also convinced by the other Copernican theorems, the narrator is able to introduce the theme, deduced from them, which forms the leitmotif of both the moon voyage and the sun voyage: that the universe cannot exist for our benefit. On the contrary—he hints at, rather than spells out, what this implies:

"Sir," I replied to him, "the majority of men, who only judge things by their senses, have allowed themselves to be persuaded by their eyes, and just as the man on board a ship which hugs the coastline believes that he is motionless and the shore is moving, so have men, revolving with the earth about the sky, believed that it was the sky itself which revolved about them. Added to this there is the intolerable pride of human beings, who are convinced that nature was only made for them—as if it were likely that the sun, a vast body four hundred and thirty-four times greater than the earth, should only have been set ablaze in order to ripen their medlars and to make their cabbages grow heads!

"As for me, far from agreeing with their impudence, I believe that the planets are worlds surrounding the sun and the fixed stars are also suns with planets surrounding them; that is to say, worlds which we cannot see from here, on account of their smallness, and because their light, being borrowed, cannot reach us. For how, in good faith, can one imagine these globes of such magnitude to be nothing but great desert countries, while ours, simply because we, a handful of vainglorious ruffians are crawling about on it, has been made to command all the others? What! Just because the sun charts our days and years for us, does that mean to say it was only made to stop us banging our heads against the walls? No, no, if this visible god lights man's way it is by accident, as the King's torch accidentally gives light to the passing street-porter." (p. 10; 1:13–14)

Descartes had said the same, though only in the confidence of a private letter. For Dyrcona, the correctness of his deduction is forcefully confirmed soon afterward in a way he himself finds most surprising. He takes off in a flying machine, which is not described in detail but which is thrust into the air by firework rockets, and he actually lands on the moon. This turns out to be the Paradise from which Adam and Eve were banished. (Ironically, Cyrano, the mocker of religion, here finds himself in agreement with the speculations of many doctors of the church,[103] which Campanella had touched on in his *Apologia pro Galileo* in connection with the theory of the plurality of worlds.) What is more, as the traveler learns after biting into an apple from the Tree of Knowledge, the moon is in every sense a "world" similar to Earth. Like Godwin's moon, it is the home of giants. They are twelve ells tall and walk on all fours but are otherwise "men" (*hommes*) who look just "like us." They cannot believe that the new arrival, this miniature two-legged creature, is a man ("que je feusse un homme"). After all, even animals use the limbs with which nature has equipped them in the same way as the (lunar) "men" do (1:32)! Here we have a pointer to the central theme, which despite the wealth of other satiric and polemical elements superimposed on it, unifies *Les estats et empires de la lune* and also links it with *Les estats et empires du soleil*.

It is no accident that animals (*les bestes*) are mentioned here. Exploring the question of terrestrial man's status in the context of "more than one world," Cyrano queries not whether we are comparable with the "men" on the moon but—provocatively—whether in reason and humanity we can even compare ourselves with the *animals* of the moon (and later of the sun). Cyrano does not even bother to refer to man's theological preeminence as the masterpiece of Creation; for any such suggestion would immediately be seen to be absurd. The comparison with the extraterrestrial animals suffices to reveal the arrogance of earthly man's claim to sole possession of reason and immortality (and as Cyrano at least in the moon novel never allows us to forget, the animals are of course far inferior to the extraterrestrial humans).

The episodes of the story provide a series of illustrations of this downgrading of man. An entertainer teaches the visitor (whose astonishment has rendered him speechless) to do tricks, just as we train

---

[103]This is noted with some satisfaction by Camille Flammarion, *Les mondes imaginaires et les mondes réels*, 12th ed. (Paris, 1874), p. 358.

monkeys, and makes him perform for the Lunarians' amusement. (Alexander Pope later uses the same comparison to suggest that in the cosmic context man's intelligence is only relative: in the *Essay on Man* he considers the possibility that extraterrestrial races, to whose abode a man of Newton's genius had strayed, might show him in public as we show apes at fairs—2, line 34.) The mountebank "began pulling at my rope harder than ever, so as to make me leap about, until the spectators had debauched themselves with laughing and insisting that I was almost as clever as the beasts in their country, and each returned to his home" (p. 36; 1:38). Dyrcona soon meets Gonsales (from Godwin's novel), who is being kept as a comical and exotic pet by the queen. He is thought to be a baboon and is displayed in a private zoo— while having scientific discussions with the newly arrived Frenchman! Hardly has the rumor that the two bipeds are defective *humans* begun to spread than the church intervenes with an official declaration that to compare the terrestrial creatures to "men" or even animals amounts to blasphemy. Traditional theological arguments in man's favor are now used against him: it is more likely that the animals share "humanité" and immortality with "men" (i.e., moon people) than that the earthlings do. God has obviously shown not the slightest concern for them, since he has left them, with their clumsy upright posture, exposed to certain destruction. Even the birds, admittedly also two-legged, are better equipped to survive (1:52–53). The earthlings can only be regarded as monsters (*Monstres*), at best perhaps as "plucked parrots," but certainly not as humans (1:53). The worst blow to man's self-esteem comes, however, when the priests also officially deny that the French visitor possesses reason, supposedly the gift of the Creator by which man alone is distinguished from the rest. Cyrano, like the anti-Cartesian thinkers of his day, chooses to claim, if only to make his point, that the animals possess reason, at least to a certain degree (1:52–54).[104]

As in the writings of Montaigne and Charron, such philosophical onslaughts obviously contain much contemporary polemic against the exaggerated self-confidence of reason intent on emancipating itself. But Cyrano turns again to the cosmic aspect of his questioning; that is to say, he continues to pursue the theme of the extraterrestrial encounter. When Dyrcona has to stand trial for having maintained that his own planet is an earth and not a "moon," as the Lunarians believe, the argument used in his defense is that this "animal without

---

[104]On this subject, which was much discussed at the time, see George Boas, *The Happy Beast in French Thought of the Seventeenth Century* (Baltimore, 1933).

reason" is so totally insignificant *in the universe* that his views and actions are not worthy of the moon dwellers' attention:

> "Truly, gentlemen, if you encountered a grown man policing an anthill, now cuffing an ant for knocking down its comrade, now imprisoning one for stealing a grain of corn from its neighbour, now bringing another to justice for abandoning its eggs, would you not consider him mad to be busying himself with these things so far beneath him and striving to make creatures submit to reason which have not its use? Then what name, venerable pontiffs, would you give the concern you show about the caprices of this little animal? Just men, I have spoken." (Pp. 61–62; 1:59–60)

It is with sheer irony, therefore, that the court officially declares the accused to be a "man" (*sensé homme*)—simply in order to be able to sentence him to the humiliating penance (*amende honteuse*) of recantation. Dyrcona's punishment is that he must call out at every street corner: "People, I declare unto you that this moon here is not a moon, but a world; and that the world down there is not a world but a moon. Such is what the priests deem it good for you to believe!" (p. 62; 1:60). Even speaking the truth—saying what he himself has thought from the outset to be the truth, namely that the moon is a "world"—here serves to degrade terrestrial man, to reduce him to a "plucked parrot"!

In the further course of *Les estats et empires de la lune* this definition of man's status is reinforced by discussions with, in particular, two professors and a "Demon" from the world of the sun. These discussions firmly establish the book as a "philosophical novel" but seriously detract from the narrative interest of this part of it. Here Cyrano also gives a full account of his Atomist conception of the origin of the heavenly bodies and of life, taking a markedly materialistic and atheistic line (e.g., 1:95), which, among other things, strikes once again at the "image" of man. Man's belief in a Creator, we are told on the one hand, derives only from a defect in his reason (1:75–76). Atheism, on the other hand, is apparently simply one of the superhuman moon-dwellers' modes of behavior, customs, habits, and attitudes—which to earthly eyes seem perverse but which Cyrano portrays as rational, refined, and cultured. The Lunarians feed on smells, sleep on blossoms, pay their bills in poems, view death as the fulfillment of life, venerate youth, know no prudery, conduct even their wars with scrupulous fairness and humaneness, and being highly intelligent, are of course convinced that the stars are inhabited (1:71). The fact that they are, naturally, atheists thus appears in this favorable light.

True, at the end of the book, Cyrano does let the Devil carry off one such Lunarian philosopher of Atomist persuasion in the midst of his blasphemy and drag him down to hell (thereby conveying the space traveler, who clings to the philosopher, back to Earth). But of course this is just an added touch of ironic polemic, not a refutation of the philosopher's beliefs: *this* Devil is no logician.

And in no way is the transport-to-hell episode a retraction of the downgrading of earthly man by comparison with other beings. Such a recantation by this notorious freethinker would hardly carry conviction (though he is said to have been converted to Catholicism on his deathbed), even if *Les estats et empires de la lune* were not followed by *Les estats et empires du soleil*, where man is still further downgraded in comparison with the animals. The second part in fact adds little to the thematic content of the work. Only a few comments need therefore be made, before we consider the precise manner and hence the meaning of this demotion of man.

The early part of *Les estats et empires du soleil* is filled with action and adventure, as if to make up for the excessive weight of philosophy in the last third of the moon novel. But even when Dyrcona finally ascends to the sun in a new flying machine driven by solar reflectors (taking off from the tower of a fortress in which he has been held captive and passing on the way Earth-like planetary worlds with their own moons), Cyrano continues to let lively action and description of the exotic and fantastic features of the sun take precedence over philosophy. After a flight lasting four months, the aeronaut lands first on one of the "sunspots"; this is in reality one of the many "worlds" circling the sun. Here a small, naked man (*un petit homme, tout nu*, 1:129) explains to him how not merely life but man himself came spontaneously into being (through the action of the sun's rays on mud). On the sun itself, which Cyrano describes in brilliant colors as a wonderland of "burning snowflakes" (1:137) and vegetation of overpowering beauty, the traveler sees the repeated metamorphosis of living creatures into trees, birds, human beings—a fairy-tale happening in which he can witness the miracle of Atomism. The creatures all speak Dyrcona's language—for in this world of marvels nothing is to be marveled at—and they explain to him, among other things, that they are not spirits but "animals more perfect" than man (1:145).

Such fairy-tale motifs do not so much diminish man as confront him with wondrous things; science fiction of high literary quality has continued to derive much of its appeal from such elements. The lowering of man's status is, however, the focus of attention in the central portion of *Les estats et empires du soleil*, the "Histoire des oiseaux"

(Story of the birds). The sun, Dyrcona is told, contains the realms and provinces of lovers, of the just, of the souls of creatures of all the "worlds" in the universe, the regions of truth and of peace, and the state of the philosophers. (The novel is about to focus on this latter state, having introduced the souls of Campanella and Descartes, when it breaks off.) But the sun is also, and most particularly, the home of the birds, and this is a cue for Cyrano to take up again, and to emphasize even more pointedly, the theme of man as the defective animal handicapped by nature, the plucked bird (1:150, 158). Dyrcona is taken prisoner by the birds and placed on trial quite simply for being a man. He has to submit to being radically disabused of the then-current self-definition of man, whether in a theological or in a secular framework, on the basis of his own assumed superiority.

In the context of an idyllic nature seen with pre-Darwinian eyes, man appears to the birds as an instigator of conflict and a brutal destroyer (1:158–59). To ascribe to him reason and immortality such as the wise and cultured birds themselves possess would be the height of absurdity. There is no more compelling proof of his lack of reason than the delusion of this featherless monster ("Monstre" once again) that the whole of the animal world is meant to be subservient to him (1:150). This mockery reaches the level of the grotesque when Dyrcona tries to evade the condemnation of the court by pretending he is not a man but an ape—and the court is not deceived. In short, the birds' tribunal confirms what is acknowledged even by a magpie who shows pity for the accused: that man is a pest "of which every well-policed state should be rid" (p. 175; 1:155). It condemns him to death, but he is reprieved at the last minute in a highly dramatic manner.

Certainly at first sight such accusations against the human species read as a satire on the arrogance of man's desire to preserve himself at the expense of all other life—at the expense, we might say, of his environment. It is usual to try to define the thesis of the two novels in some such terms, and this is logical enough, bearing in mind the anti-physicotheological theories that set out to prove, in a playfully polemical manner, that God created everything for the benefit not of the highest but of the lowest being.[105] But can we accept that the whole point of the work lies in this condemnation of "the human race . . . from the standpoint of strictly objective justice, when the creatures that have for so long and so often been oppressed by man can at last exercise the right of retaliation"? Are the demand for the "democratic

---

[105]See Clarence J. Glacken, *Traces on the Rhodian Shore* (Berkeley and Los Angeles, 1967), pp. 504 ff., esp. p. 534. His examples are, however, from the eighteenth century.

ideal" of justice for all living things, including cabbages, and the hope that the destruction of anthropocentrism may bring about a new era of "understanding and tolerance among men," really the main ideas that it conveys?[106]

It hardly seems possible that the lasting appeal of L'autre monde rests on nothing more than this somewhat banal and sentimental message. This interpretation presupposes that man is being judged (and found wanting) against a standard of truly rational and humane behavior. But is this the case? To start with the "Story of the Birds": The birds on the sun lay claim to such a standard. But do they themselves measure up to it (as, later, Lasswitz's Martians, for instance, were to be justified in claiming that theirs was the higher civilization)? Is not the portrayal of the birds' thinking and behavior calculated to show that, on the contrary, they are just as narrow minded, irrational, and "inhumane" as the humans? Their brutality and cruelty is easily comparable to that of human hunters and bird trappers; they condemn Dyrcona not simply to death but to being eaten alive by flies. They are just as "manlike" in the sneering arrogance with which they believe themselves to be the highest, most rational, and most cultivated beings in the whole of Creation, not to mention the naïve simplicity of the reasons they give for believing in this superiority in relation to man: "As for the common mob, they cried out that it was horrible to believe that a beast without a face like theirs should be possessed of reason. 'What?' they murmured to one another. 'It has neither beak, nor feathers, nor claws—and yet its soul is supposed to be spiritual! O gods! What impertinence!' " (p. 171; 1:150).

In other words, their world is just like ours. Image and counter-image are the same. This is clearly not satire or utopia in the usual sense, where reality is confronted with an ideal (whether implied or actually depicted). Instead, the two worlds are both out of joint and simply mirror each other, instead of presenting an ideal and a distorted image. Each world dislocates the other, with the result that *everything* becomes relative, and nothing of absolute validity remains. Certainly man is downgraded, but only as part of a process in which *all* values are exposed for what they are—not in the context of a *hierarchy* of many worlds "of which our earthly dunghill is the worst."[107]

If we now look back at Les estats et empires de la lune, we find that here too, each society is shown to fall short of the ideal, and this

[106]Hönncher, Fahrten nach Mond und Sonne, p. 40; Howard G. Harvey, "Cyrano de Bergerac and the Question of Human Liberties," Symposium 4 (1950), 128; Winter, Compendium Utopiarum 1:90.
[107]John Dryden, "Eleonora," line 82.

result is achieved more cleverly still. The moon dwellers who treat the human who arrives on the moon as a comical circus animal are behaving exactly as humans would behave if a Lunarian who considered *his* species to represent perfection were to land on Earth. This is not only stated explicitly (1:33) but also conveyed implicitly by the Gonsales episode. The Lunarians, who dress their pet monkeys in Spanish costume for comic effect, take Domingo Gonsales for a monkey— while he, with the same blind arrogance, proclaims on the moon that all other men are created only in order to provide the Spaniards with slaves or objects of amusement (1:45).

Another of Cyrano's targets in *Les estats et empires de la lune* is the claim of theological dogma to speak with authority on the natural world. Here Cyrano makes his own original contribution to the theme that dominated the discussion of plurality in the seventeenth century—to the question of whether the belief in plurality was compatible with Christian teaching. Dyrcona's assertion that he comes from the Earth and not from the moon, and that not the Earth but the world of the Lunarians is the moon, appears absurd to the moon dwellers. This is a further demonstration of the irony that the moon's inhabitants are just as intellectually limited as the Earth's—they believe the Earth to be uninhabited just as we believe the moon to be. But, more to the point, it is above all the priests who take exception to Dyrcona's views, which thus constitute not merely a scientific error but a religious heresy, for which the penalty is death by drowning (1:58). Institutionalized religion on the moon may differ from that on Earth in preferring water to fire as the instrument of death, but it is just as fanatic, just as mentally lazy, just as attached to established dogmas for reasons of mere convenience, and just as inhumane. The religious view of the world is shown to be, by definition, unenlightened and absurd. And of course the sentencing of Dyrcona publicly to recant his view that the moon (our Earth) is a world is a sly contrafact of the Inquisition's action against Galileo and his proof that the moon was an "earth." In each case, the religious authority, judging from the standpoint of sacrosanct dogma, condemns scientific truth as false.

Scientific truth is in fact the one thing Cyrano leaves intact while undermining the validity of everything else. But it does nothing to restore man's "image." Cyrano has, with devastating effect, dismissed man's confident claim to preeminence and hence to uniqueness in "creation," and the truth of science confirms his picture of man. For in the dispute about whether the moon is an earth and the Earth a moon (which stands representatively for this whole complex of questions on plurality), the scientific truth that emerges is that there *is* more than one "world"—with beings who are just as narrow minded and mean as we are.

Any sort of accommodation with Christian dogma—the main issue in the seventeenth-century debate about the plurality of worlds—is of course out of the question; and Cyrano has not the slightest desire to find one. But even an undogmatic and generalized conception of a god who has, in his unbounded omnipotence, created the many worlds and "humankinds" is ruled out. This solution becomes common only in the last decades of the century, when thinkers become less interested in specifically dogmatic matters relating to the plurality of worlds but are still concerned to some extent with religious issues. A couple of examples from the end of the century have already been cited in order to give an overview of the whole range of responses that were possible in the seventeenth century (see sec. 1 above). Before the middle of the century, this conception of divine omnipotence—just hinted at, for instance, by John Wilkins (see sec. 3 above)—is very much the exception. It is as though it was first necessary for the emotional and intellectual impact of the baroque "shock" to be felt in its most extreme form,[108] and this extreme is represented by Cyrano's treatment of the philosophy of the plurality of worlds in *L'autre monde*. At any rate, it is only *after* this option had been explored—an option that constantly presented itself to intellectually alert Europeans in the age of the wars of religion—only after this consistently atheistic rejection of any higher Being, that we find (even if only as the understandable psychological reaction of turning from disillusionment to the opposite extreme) widespread acknowledgement of that divine "Guardian" whose praises are sung by all humankinds in the universe. This development continues from the late seventeenth century until well into the eighteenth—a century that in spite of its "coming of age" (Kant's *Ausgang aus der Unmündigkeit*) did not find it easy to be without this "Guardian." On this basis, belief in the plurality of worlds becomes general among educated people in the Enlightenment. The heresy, having passed through the transitional stage of universal despair represented by Cyrano, becomes a new, nondenominational gospel. This, too, severely challenges man's view of himself, but the problems are now philosophical rather than theological and dogmatic.

[108]On the baroque "shock" see Wolfgang Philipp, *Werden der Aufklärung*.

CHAPTER 4

The Enlightenment: Man as "the Measure of All Things"?

1. "Coming of Age": Triumph and Trauma

During the Renaissance and baroque periods the presumption was that the idea of the plurality of worlds was heretical, and so those who did bring an open mind to this sensitive topic were much preoccupied with the question of whether, and how, the concept might nevertheless be reconciled with theological dogma. But as the Enlightenment gained ground, from the closing decades of the seventeenth century onward, the belief that "we are not alone" came increasingly to be accepted by the educated as a matter of course. This was due chiefly to the impact of two works that made the idea accessible to the educated public, Fontenelle's *Entretiens sur la pluralité des mondes* (1686) and Huygens's *Kosmotheoros* (1698). The ground was of course prepared for both these works by the triumphal advance of Copernicanism, which in turn received further impetus from their success; from 1757 onward *De Revolutionibus Orbium Coelestium* no longer figured on the Catholic church's index.[1]

As early as 1705 a treatise on geomagnetism by Edmund Halley, the discoverer of Halley's Comet, contains the passing remark that "it

---

[1]On the popularization of science and in particular the process by which Copernicanism became widely known and accepted, see Walter Schatzberg, *Scientific Themes in the Popular Literature and the Poetry of the German Enlightenment, 1720–1760* (Bern, 1973); Christof Junker, *Das Weltraumbild in der deutschen Lyrik von Opitz bis Klopstock* (Berlin, 1932); Karl Richter, "Die kopernikanische Wende in der Lyrik von Brockes bis Klopstock," *Jahrbuch der Deutschen Schillergesellschaft* 12 (1968), 132–69; idem, *Literatur und Naturwissenschaft: Eine Studie zur Lyrik der Aufklärung* (Munich, 1972), pp. 131–39.

is now taken for granted that the Earth is one of the Planets, and they are all with reason suppos'd Habitable."[2] The writer adds, "though we are not able to define by what sort of Animals"; but such caution is soon thrown to the winds by many, even in apparently orthodox circles. Only two decades later, Johann Christoph Gottsched, whose popularization of science formed a significant part of his work as an "educator of the German nation,"[3] published his translation of Fontenelle's *Entretiens* (*Gespräche von mehr als einer Welt*, 1726) and in his preface was able to cite with approval the *Journal de Trévoux* (1722) (a Jesuit publication!) to the effect that "all who wished to be thought intelligent" (*aufgeweckte Köpfe*) agreed with Fontenelle's thesis of the plurality of worlds; and Fontenelle had expressed very definite views on the nature of the inhabitants of the other heavenly bodies. A century and a half after the concept of plurality had been denounced by the first *praeceptor Germaniae*, Melanchthon, the Lutheran minister Andreas Ehrenberg did not feel it incompatible with his office to lay before the public, "rationally and clearly," his *Curiöse und wohlgegründete Gedancken von mehr als einer bewohnten Welt* (Curious and well-founded ideas on a plurality of inhabited worlds, c. 1710); and the treatise by the Frankfurt rector Johann Jakob Schudt totally endorsing the theory of life on other planets (*De Probabili Mundorum Pluralitate*, Frankfurt 1721) was actually presented in 1720 as a course of lectures to his students. The difference between this age and the preceding one is sharply brought out in 1746 by the scientist and poet Daniel Wilhelm Triller when, commenting on the Christian anthropocentrism of Hugo Grotius, he refers to the work of Fontenelle, Huygens, Christian Wolff, and William Derham (a disciple of Newton) and observes, "Nowadays, however, it is considered fairly probable that the other planets, like our Earth-planet, are inhabited by certain, albeit unknown, creatures."[4] As the "century of philosophy" draws to its close, this assumption has become so widespread that to set about proving it would be an insult to common sense: In 1787, in the preface to his edition of Chevalier de Béthune's *Relation du monde de Mercure* (1750), the dramatist Charles Garnier writes, "The opinion that leads us to believe that all the planets are inhabited like our Earth has become so familiar to us, since we have seen M. de Fontenelle's ingenious description of the worlds, that one has no anxiety lest the account of Mercury should be considered an absurd idea; for it seems, on the contrary, that one would

---

[2]*Miscellanea Curiosa*, vol. 1 (London, 1705), p. 56.
[3]See Walter Schatzberg, "Gottsched as a Popularizer of Science," *Modern Language Notes* 83 (1968), 752–70.
[4]Martin Opitz, *Teutsche Gedichte*, ed. D. W. Triller (Frankfurt, 1746), 4:748.

have more difficulty in seriously denying the plurality of inhabited worlds than in speaking very convincingly in support of it."[5]

Of course such a change of attitude does not take place overnight or everywhere. Even in the Age of Enlightenment the idea that there were other species of men continued to come under attack now and again (as indeed it has even in the twentieth century, in Jean Guitton's staunchly Catholic *Dialogues avec M. Pouget sur la pluralité des mondes*, 1954). This resistance, on dogmatic and especially christological grounds, came for instance from strict Reformed Churchmen such as Herman Witsius and from Lutherans such as Johann Franz Buddeus and Valentin Ernst Löscher;[6] in England it came from John Wesley or again from anti-Newtonian scientists such as John Hutchinson (*Moses' Principia*, 1724–27), among not a few others. Catholic Spain showed particular reluctance to embrace the belief in plurality, and this was true even of those Spaniards, like the Benedictine scientist and father of the Spanish Enlightenment Benito Jerónimo Feijóo, who willingly accepted the Copernican-Newtonian revolution.[7]

In contrast, the century of Enlightenment also saw theologians attempting to demonstrate that the scientifically grounded belief in the existence of more than one world was not necessarily contrary to Scripture and in particular to the New Testament. One view, voiced again in our own day by C. S. Lewis, was that the people on other planets were free from sin and were therefore not in need of redemption by Christ. This was the conclusion reached by men such as Joachim Böldicke, in his *Abermaliger Versuch einer Theodicee* (Another attempt at a theodicy, 1746), and the Baptist minister Andrew Fuller, in *The Gospel Its Own Witness* (1799).[8] An alternative premise—adopted

---

[5]Charles Garnier, ed., *Voyages imaginaires*, vol. 16 (Amsterdam, 1787), pp. 158–59.

[6]Otto Zöckler, *Geschichte der Beziehungen zwischen Theologie und Naturwissenschaft*, vol. 2 (Gütersloh, 1879), p. 62.

[7]*Cartas erúditas y curiosas*, vol. 2 (1745), carta 26: "Si hay otros mundos?" See John Browning, "Fray Benito Jerónimo de Feijóo and the Sciences in Eighteenth-Century Spain," in *The Varied Pattern: Studies in the Eighteenth Century*, ed. Peter Hughes and David Williams (Toronto, 1971), pp. 353–71. Eighteenth-century Spanish literary space voyages remain within the boundaries set by Kircher; see Monroe Z. Hafter, "Toward a History of Spanish Imaginary Voyages," *Eighteenth-Century Studies* 8 (1975), 270–72, on Diego de Torres Villarroel (1724) and Cándido María Trigueros (1777). Lorenzo Hervás y Panduro is the first to join the mainstream of European development with his book (whose title contains an echo of Kircher's), *Viage estático al mundo planetario* (1793–94). Following the by-now-considerable body of literature on the subject, this work—more a textbook on astronomy than a novel of fantasy—maintains that the celestial bodies are inhabited.

[8]On Böldicke see Zöckler, *Geschichte* 2:63; on Fuller see Ralph V. Chamberlin, *Life in Other Worlds: A Study in the History of Opinion*, Bulletin of the University of Utah 22:3 (1932), p. 24.

by Johann Heinrich Becker in *De Globo Nostro* (1751), by Emanuel Swedenborg, and by James Beattie in *Evidences of Christianity* (1786)— was that Christ's mission of salvation embraced the inhabitants of other planets too.[9] The by-now-familiar passages from Saint Paul (Eph. 1:10, Col. 1:20) formed the chief evidence for this view.

Not surprisingly, the antitheologians could play at this game too. They too could cite biblical passages in support of their case, and they could even make capital out of the line of argument used by the Inquisition against Galileo, by turning it on its head. This ploy was used as late as 1793 by Thomas Paine, when he made his sensational frontal attack on revealed religion in *The Age of Reason*. For him the plurality of worlds was scientifically irrefutable as well as philosophically attractive, and it seemed to him that Christianity could be most effectively pilloried by stressing the incompatibility of the two beliefs. This he did with polemical gusto, while not contributing any new thoughts of his own to the by-now-familiar theme:

> What are we to think of the Christian system of faith that forms itself upon the idea of only one world, and that of no greater extent . . . than twenty-five thousand miles? . . .
>
> From whence, then, could arise the solitary and strange conceit that the Almighty, who had millions of worlds equally dependent on His protection, should quit the care of all the rest, and come to die in our world, because, they say, one man and one woman had eaten an apple?
>
> And, on the other hand, are we to suppose that every world in the boundless creation had an Eve, an apple, a serpent and a redeemer? In this case, the person who is irreverently called the Son of God, and sometimes God Himself, would have nothing else to do than to travel from world to world, in an endless succession of deaths, with scarcely a momentary interval of life.[10]

And the young Shelley, commenting on a passage of *Queen Mab* about the "innumerable systems" of the universe, provides the finest example of the strong influence that such ideas, current in popular philosophy, could have on a specifically literary imagination:

[9]On Becker see Zöckler, *Geschichte* 2:64; on Swedenborg see sec. 3 below; on Beattie see Chamberlin, *Life in Other Worlds*, p. 25.

[10]*The Complete Writings of Thomas Paine*, ed. Philip S. Foner (New York, 1945), 1:504. See Joseph V. Metzgar, "The Cosmology of Thomas Paine," *Illinois Quarterly* 37 (1974), 47–63.

The plurality of worlds,—the indefinite immensity of the universe, is a most awful subject of contemplation. He who rightly feels its mystery and grandeur is in no danger of seduction from the falsehoods of religious systems, or of deifying the principle of the universe. It is impossible to believe that the Spirit that pervades this infinite machine begat a son upon the body of a Jewish woman; or is angered at the consequences of that necessity, which is a synonym of itself. All that miserable tale of the Devil, and Eve, and an Intercessor, with the childish mummeries of the God of the Jews, is irreconcilable with the knowledge of the stars. The works of His fingers have borne witness against Him.[11]

Thus the theological theme so prominent in the baroque age still has its echo in the Age of Enlightenment, but for all the protagonists' fervor, this is only a faint, distant rumbling, the skirmishing of an army in retreat. Such theological quibbling had had its day. Paine engaged in it in order to attack Christian doctrine; the other side used it in defense of the faith, seeking to demonstrate that the words of Holy Writ were compatible with the assumption of more than one world and more than one species of intelligent beings. Such arguments were advanced once again by Christian apologists in response to the publication of Paine's *Age of Reason*.[12] But all this disputation had long since become a virtual anachronism in an age when the subject of the "myriad" worlds had moved from the arena of theology to that of philosophy. It has been claimed, even by nonchurchmen,[13] that the eighteenth century was even more intensely preoccupied with the dogmatic problems than the seventeenth century had been; but this is not borne out by the evidence. True, in the history of ideas, one must expect to find phenomena characteristic of different periods

[11]*The Complete Poetical Works of Percy Bysshe Shelley*, ed. Neville Rogers, vol. 1 (Oxford, 1972), p. 296. It was in fact his studies in astronomy that led Shelley to reject Christianity; see B. Ifor Evans, *Literature and Science* (London, 1954), pp. 66–71.

[12]E.g., Andrew Fuller's *The Gospel Its Own Witness* (Clipstone, 1799), already mentioned; Uzal Ogden, *Antidote to Deism* (Newark, N.J., 1795), and Edward Nares, Ἐις Θεις, Ἐις Μεσιτης: *or An Attempt to Shew How Far the Philosophical Notion of a Plurality of Worlds is Consistent, or Not so, with the Language of the Holy Scriptures* (London, 1801). Nares cites dozens of biblical passages that may be interpreted as favoring the plurality of worlds, and he aligns himself with those who assume that not only the terrestrial human race but all humankinds need redemption and have been redeemed (on other planets, and in a manner unknown to us). See also Marjorie Nicolson, "Thomas Paine, Edward Nares, and Mrs Piozzi's Marginalia," *Huntington Library Bulletin* 10 (October 1936), 103–33.

[13]Arthur Jack Meadows, *The High Firmament: A Survey of Astronomy in English Literature* (Leicester, 1969), p. 127. In support of this assertion Meadows cites only a single passage from a poem.

occurring side by side; but as we trace the development of the theme of plurality, reiterations of the old arguments by theologians committed a priori to the orthodox view are of less interest than new emphases of interpretation. The chief supporters of plurality in the Age of Enlightenment—Fontenelle and Huygens, and also Leibniz and Wolff and the English Newtonians—give only very limited consideration to the biblical and christological problems that threaten to disturb the very foundations of Christianity; often, indeed, they mention them only to brush them aside.

It is also revealing to see how the theological difficulties are handled by those writers who disseminate the ideas of the "classic" exponents of the plurality concept to a public assumed still to be bound by the conventions of church doctrine. For instance Andreas Ehrenberg hastens to stress in his preface to *Curiöse und wohlgegründete Gedancken* (Jena, c. 1710) that the case he makes for the existence of many inhabited worlds will give "all Christians" cause to extol God's wisdom, power, and majesty still more highly; but he reaches his conclusion that "we are not alone" purely by the scientific method of analogical reasoning and by arguing from the belief that the Creator's activity must be purposeful. This demonstration is followed by a second section: "How Holy Scripture possibly can and must be taken, understood, and elucidated in order not to be contrary to our opinion" (pp. 7–8). Some other theologians, even long after Ehrenberg—for instance Hutchinson, whom I have mentioned, and Alexander Maxwell (*Plurality of Worlds*, 1817)—adduce particular biblical passages referring to cosmology (taking them in their literal sense) as conclusive evidence *against* the existence of more than one world and expect science to conform to theology. But for Ehrenberg, the Bible is no longer an independent source of knowledge about the natural world: on the contrary, *it* must be read in the light of scientific knowledge. In other words, it is necessary to show that, correctly interpreted, the words of the biblical passages that speak of a circling sun and a stationary Earth (Josh. 10:12–13; Ps. 19:6–7) are in accord with the findings of science. Viewed in the proper light, as a description of the mere appearance of things, or as a version of the truth adapted to unsophisticated minds, the Bible only *appears* to express a geocentric conception of the universe. In reality its cosmology is that of Copernicus—which Ehrenberg, like so many others, sees as the basis for the plurality of worlds (see, e.g., pp. 164, 200).

As for the christological problems that for writers from Melanchthon onward were the strongest doctrinal arguments against the plurality of worlds, these present even less difficulty to this enlightened parson. In the *Curiöse und wohlgegründete Gedancken* he quickly dismisses them. No doubt there are intelligent beings on other planets, but:

> Since the rational creatures [*die Vernünfftigen Creaturen*] have without doubt been created in the image of God in holiness, justice, and the necessary degree of perfection, the question arises whether they have remained in this excellent state or whether, like us poor humans, they too have been led astray and have fallen from it. But no certain answer can be given, for we cannot know the extent of their holiness, perfection, and mastery [*Herrschafft*]. It may vary from one globe to another; and, as an *ens liberrimum* [a being enjoying full freedom] it may use its godlike freedom to act either in one way or in another. It is very possible that in one world the rational creatures have remained in their state of holiness and innocence, while in another they too have been led astray and have fallen, as we have. Satan, a swift-moving spirit, is capable, with the rebellious wicked angels as [his attendant] spirits, of having traveled quickly to all places, including the planets, and there is no doubt that they will have mounted their attack on the rational creatures created in a state of holiness and will have tried to lead them astray and make them, like themselves, rebel against God. Whether Satan did this, how, and where he set about it; also whether God gave a commandment as a test, and if so, what it was; whether Satan led all astray, or only some—all these things we cannot know or guess. After the dissolution of the flesh we shall learn and come to know wondrous things about these matters. It may well be that in one world or another the created rational creatures remained in their state of perfection, while in yet another they have lost these qualities of theirs.
>
> If, however, it did happen again that one or another world was led astray and rebelled against God, did God once again take pity on it and give to it a Savior and Redeemer, as he did to us? God's great and infinite mercy and compassion make it hard to doubt that he did; but this too we cannot know for certain, but must trust in the almighty Creator to do, in his supreme wisdom and mercy, what pleases him. (Pp. 152–54)

Here Ehrenberg is not resorting to the topos of God's inscrutability, like Gassendi or Guericke, as a genuine expression of piety but expressing his disdain for all such sterile sophistries indulged in by the dogmatists. Ehrenberg, himself a theologian, disposes of them in barely three pages of his 270-page work. The only theological element in his book is the physicotheological reasoning according to which the magnificent spectacle of the multitude of inhabited worlds is a proof of God's wisdom, power, and goodness (pp. 258–61). But in the intellectual context of the time, this way of thinking must be classed as philosophical rather than strictly theological.

Schudt, in his *De Probabili Mundorum Pluralitate* (Frankfurt 1721), argues along similar lines and dismisses the christological problems in an even more forceful manner (pp. 51–52). Questions about the theological status of the extraterrestrial beings—whether they are descended from Adam, whether they have been redeemed by Christ, and so on—can only be answered by conjectures (*meris conjecturis*). So Schudt does not enter into a discussion of them—any more, he laughingly adds, than he proposes to discuss the ravings (*deliramenta*) of Lucian or Fatouville's satiric farce, *Arlequin empereur dans la lune*. There could hardly be a more unequivocal indication of the status of christological questions in the Enlightenment. Indeed, from here it is only a short step to parody, with sophistic argument used to ludicrous effect as in the Cartesian *Traité de l'infini créé* (1769), published anonymously but ascribed to the Abbé Jean Terrasson, where each of the innumerable worlds is accorded a Christ of its own.[14] In this work, as in Schudt's, the old dogmatic scruples about the idea of the plurality of worlds are impatiently waved aside: by now, the idea is accepted as a matter of course. The questions now under discussion are whether the intelligent beings on the various planets are the work of God or of Nature, whether they were created on the principle of uniformity or of diversity,[15] whether among the intelligent creatures of the universe there are differing degrees of perfection and happiness, and so on. Obviously these not radically new questions contain an element of theology, in the broadest sense, especially where they enter the realm of theodicy. Also, more often than not in the Enlightenment, the idea of plurality is justified by teleology (thus, the planets were created not for us but for their inhabitants, who sing hosanna to the greatness of God), and of course this too is a theological or at any rate metaphysical mode of thought (on teleological thinking in the Enlightenment, more in sec. 2). Nevertheless, there is no mistaking the fact that this way of

[14]See Aram Vartanian, *Diderot and Descartes* (Princeton, N.J., 1953), pp. 73–75. The argument set out in the treatise is taken seriously by R. R. Palmer, *Catholics and Unbelievers in Eighteenth Century France* (Princeton, N.J., 1947), pp. 110–12.

[15]See sec. 2 below, in particular the difference of opinion on this point between Fontenelle and Huygens. The two possibilities—that the worlds are either different or alike—had of course been pointed out long before by Saint Thomas Aquinas, who had rejected them both as absurd (see the end of chap. 1 above). Little emerges from David Knight's article "Uniformity and Diversity of Nature in Seventeenth Century Treatises on Plurality of Worlds," *Organon* 4 (1967), 61–68.

thinking, which in many cases, as I have said, is philosophical rather than theological, is far removed from the strictly theological, *dogmatic* approach that characterized the sixteenth and seventeenth centuries. This is true even of writers like, say, Swedenborg or Johann Kaspar Lavater, whose treatment of the idea of plurality incorporates specifically Christian elements because of the authors' personal beliefs; it applies equally to a work like Friedrich Gottlieb Klopstock's *Messias* (Messiah, 1748–73), where the plurality of worlds is carefully situated in the blank spots on the map of dogma (rather an anachronism in the intellectual climate of the day) and is then imaginatively elaborated in detail.

Clearly, what began as a "new heresy" in the wake of the Scientific Revolution has now been absorbed into the consciousness of Christians in general, a consciousness that for its part has, in the spirit of the Enlightenment, become secularized and liberalized, has freed itself from dogmatism and begun to think for itself. Moreover, in the Age of Enlightenment the idea of a multiplicity of inhabited worlds takes on the character of a new, up-to-date gospel that even finds its way into devotional literature.[16] Thomas Paine himself, radical as he is, slips into the language of religiosity when, arguing against the one world of orthodox Christianity, he expresses his belief in the plurality of humankinds. Like Ehrenberg and Schudt, Paine declares that precisely this thought of the infinite number of worlds with human inhabitants in the universe increases our admiration for the Creator and our gratitude toward him: "Our ideas, not only of the almightiness of the Creator, but of His wisdom and His beneficence, become enlarged in proportion as we contemplate the extent and the structure of the universe" (*Complete Writings*, 1:503). This is by now, at the end of the eighteenth century, a familiar leitmotif of that more or less secular piety which finds its support in the book of Nature rather than in the Book of Books (see chap. 3, sec. 1 at n. 31). In this, even Paine's declared opponent, the minister Edward Nares, is on his side.[17] As early as 1703, David Russen, in his commentary on Cyrano, *Iter Lunare: or, A Voyage to the Moon* (London), summed up this religious attitude shared by believers in plurality in the memorable polemical statement "I confess it savours more [of] Religion to admit a plurality of Worlds, than of Pride to deny it" (p. 67). The theological dogmatism that admits of only one world is now merely a function of man's all-too-human egocentricity; in the plurality idea the new science has provided the

---

[16]Zöckler, *Geschichte* 2:425.
[17]Nares, Εἰς Θεις, Εἰς Μεσιτης, p. 171.

Age of Enlightenment with a contemporary symbol to aid man's understanding of himself—with a "myth."[18]

As the Enlightenment proceeds, so the scope of the "new gospel" expands to take in ever more of the universe. It becomes less and less realistic, except in satires and utopias, for extraterrestrial humankinds to be assumed to live on the moon. With Huygens's *Kosmotheoros* (1698) and Erik Engman's *Dissertatio Astronomico-Physica de Luna non Habitabili* (1740), which confirmed Galileo's and Riccioli's assumption that the moon had no water and no atmosphere, the idea of life on the moon had finally become scientifically untenable. If there was life in outer space, it was more likely to exist on the planets of the solar system and of other systems around a fixed star.[19] Thus the legacy of Giordano Bruno is taken up after a lapse of a hundred years. In the Enlightenment philosophy of the plurality of worlds, his attempt to liberate human consciousness from the straitjacket of dogmatism comes triumphantly into its own. One of the spokesmen of the Enlightenment, Gottsched, provides forceful evidence of this liberation in a footnote to his translation of Fontenelle. Like Huygens before him, he makes fun of the dogmatic hairsplitting of the theologians by recalling how Kircher, in his *Iter Exstaticum* (1656), had pondered the question of "whether someone could receive a valid and effective baptism by the water that flows on Venus."[20]

This sounds like the sort of rationalistic arrogance with which the Enlightenment has for a long time been associated. In fact, however, it was precisely their willingness to acknowledge a plurality of worlds and, consequently, a plurality of humankinds which faced enlightened thinkers with a series of new problems that were not dogmatic or strictly theological, but philosophical in the sense indicated above.

Of course, as the scientific "truth" of the plurality of worlds became, in the course of the eighteenth century, a part of the accepted view of things, it easily won over even those who remained insistently Christian; man's understanding of himself entered upon a new phase: in the name of Enlightenment, he triumphantly broke with the Middle

---

[18]Ira O. Wade, *Voltaire's "Micromégas"* (Princeton, N.J., 1950), p. 59: "Science had at last created a myth, and the Enlightenment had adopted it."

[19]See Marjorie Nicolson, *A World in the Moon*, Smith College Studies in Modern Languages 17:2 (1936), pp. 53–57; Steven J. Dick, *Plurality of Worlds: The Origins of the Extraterrestrial Life Debate from Democritus to Kant* (Cambridge, 1982), pp. 176–83; idem, "Plurality of Worlds and Natural Philosophy: An Historical Study of the Origins of Belief in Other Worlds and Extraterrestrial Life" (Ph.D. diss., Indiana University, 1977), pp. 319–72; chap. 3 above, at the end of sec. 1.

[20]*Gespräche von mehr als einer Welt*, 3d ed. (Leipzig, 1738), p. 107. On Huygens see n. 49 below and chap. 3, n. 22.

Ages. But this is not the whole story. If pluralism brought about, or contributed to, a momentous shift in thinking, indeed nothing less than the "crisis of European consciousness" (*la crise de la conscience européenne*),[21] then it did so in part in the sense that the chorus of jubilation at the new cosmic freedom and the praise of God as the Lord of untold humankinds were sometimes drowned out by the descant of new doubts and anxieties.

Significantly, it was increasingly the literary writers, with their acute sensitivity, who formulated the philosophical problems that became crucial after the strictly theological questions had been left behind. This was because of the way literature viewed itself and its role at the time. It was not simply that in this, the "century of philosophy," literature too aspired to be philosophical, to take hold of the great themes of the momentous epoch (just as, conversely, philosophy, as the "secular wisdom" of the *beaux esprits*, tended toward literary articulation). In the terms of Leibniz's theory of knowledge, literature made a point of presenting itself as the expression of *anschauende Erkenntnis* (cognition through the senses, or experience), which afforded an access to reality no less valid than "clear" or purely rational cognition (*klare Vernunfterkenntnis*); this view of literature had been current at least since the publication of Alexander Baumgarten's *Meditationes Philosophicae de Nonnullis ad Poema Pertinentibus* (1735) and was the view held by writers such as Lessing and Moses Mendelssohn. But beyond this, much literature in the Age of Enlightenment also had specifically scientific aspirations; it aspired to be abreast of, and validated by, current scholarship, particularly in the natural sciences. This approach was applied early on by Barthold Hinrich Brockes and was advocated by Gottsched in his *Critische Dichtkunst* (Critical poetics, Leipzig 1730, pp. 88–89), most programmatically by John Aikin in his *Essay on the Application of Natural History to Poetry* (1777), and also by Schiller in his review of the poetry of Gottfried August Bürger (1791), by Friedrich Schlegel in his discussion of mythology ("Rede über die Mythologie") in *Gespräch über die Poesie* (Conversation on poetry, 1800), and by Wordsworth in the preface to the second edition of the *Lyrical Ballads* (1800).[22] And for a literature concerned to reflect the current

[21]Alexandre Calame in the foreword to his edition of Fontenelle's *Entretiens sur la pluralité des mondes* (Paris, 1966), p. li.

[22]A particularly early instance of this attitude appears in John Dryden: "A man should be learn'd in severall Sciences, and should have a . . . philosophicall . . . head to be a compleat and excellent poet" (*Notes and Observations on "The Empress of Morocco"* [London, 1674], p. 70). See Ralph B. Crum, *Scientific Thought in Poetry* (New York, 1931); Ria Omasreiter, *Naturwissenschaft und Literaturkritik im England des 18. Jahrhunderts* (Nuremberg, 1971).

state of scientific knowledge, a particularly apt subject was the "possible worlds" that were a major issue not only in contemporary cosmology but also in poetics, where especially in the circle of Johann Jakob Bodmer and Johann Jakob Breitinger, they were the subject of much discussion in relation to the works of Milton and Klopstock.

But what were the difficulties posed by these possible worlds? These worlds seemed at first sight simply to call forth praise of the Creator and of man emancipated from dogmatism; yet when they feature in that "literary" philosophy and philosophical literature, it is sometimes in a very problematic light. Why, when the dogmatic problems have been overcome, do philosophical problems make their appearance?

Briefly, the point at issue is the nature and status of man in relation to other intelligent beings in the universe. This question presents itself sooner or later—at least sometimes and to some authors—once the dogmatic viewpoint with its emphasis on divine control has been abandoned. But now, the question of man's position is no longer—or no longer exclusively, or even primarily—considered in the light of God's attitude toward him, but instead in the light of man's claim to autonomous reason.

Ernst Cassirer said that the one thing that united the diverse movements within the European Enlightenment was their common renunciation, in the name of reason, of the Christian dogma of original sin[23]—the very doctrine that had so dogged earlier discussions about the plurality of worlds. This being so, the step in the history of thought that decisively marks the emergence of the Enlightenment is that, in renouncing this belief, man has now come to see himself as the "measure of all things." (There is a parallel here with the first, the Greek, Enlightenment, when Protagoras coined the key phrase that has become familiar to us in its Latin form, *homo mensura*.) This elevation of man to represent the standard by which everything is measured is still quite commonly used as the defining characteristic of the seventeenth- and eighteenth-century Enlightenment.[24] It is also implied in Kant's definition of Enlightenment as man's readiness to make use of his own reason, as the coming of age when man emerges from his self-incurred tutelage ("der Ausgang des Menschen aus seiner selbstver-

[23]*Die Philosophie der Aufklärung* (Tübingen, 1932), p. 188.
[24]For instance Kurt Schilling, *Geschichte der Philosophie*, vol. 2, 2d ed. (Munich, 1953), pp. 114–17, 249; Wolfgang Philipp, *Das Werden der Aufklärung in theologiegeschichtlicher Sicht* (Göttingen, 1957), pp. 169, 177–78, 181; Horst Möller, *Aufklärung in Preußen* (Berlin, 1974), p. 46.

schuldeten Unmündigkeit") and above all from the God-dominated view of the world. It is implicit, too, in Pope's polemical admonition that the proper study of mankind is man, and not the stars or God. (The same point is made earlier by Charron, and later by Lessing.) It was indeed the conventional wisdom of the Enlightenment that man, endowed with reason and now able to think and speak for himself, was the highest form of life, even if he was still undergoing a process of perfection; he was, therefore, the measure of all living things. How strange, then, that extraterrestrial beings in eighteenth-century literature and philosophy are so very often *superior* to the humans of Planet Earth in virtue, wisdom, and even reason. Seeking an answer to the question What is man? the writers typically did not compare him with animals or savages, or for that matter with angels; instead they compared him with imaginary creations of philosophical anthropology, believably delineated creatures resembling man, but located on other planets. And the result is not that man, the proud possessor of reason, measures all things against himself but that instead it is he who is measured—and often found wanting.

Can man's world still be "the best of all possible worlds," if in other planet-worlds "superior beings . . . shew'd a Newton as we shew an Ape"?[25] Is the supposition that there are higher beings in the countless worlds of the firmament perhaps a secret admission that the Enlightenment, not yet quite secure in its vaunted independence, is succumbing to the need for a "guardian"? Is it regressing from emancipation to an undogmatic but still quasi-religious dependence? The psychoanalyst Robert Plank thinks that man, when freed from institutionalized religion, generally has just such a need and that this accounts for the conviction that there are higher beings on other planets. Werner Krauss, the Marxist literary historian, speaks of the "wish-dream of an encounter with more highly evolved beings" as a "relapse into the inexperienced infancy of man, when a life beset by dangers could not be endured without the concept of a protective higher world [*Überwelt*], without the compensations of religion."[26] Renewed interest is being shown in what has been called the pessimism of the Enlightenment, and much of what is meant by that phrase may perhaps also be linked to this idea of a failed attempt to achieve

[25]Pope, *Essay on Man* 2, lines 31–34.
[26]Robert Plank, *The Emotional Significance of Imaginary Beings: A Study of the Interaction between Psychopathology, Literature and Reality in the Modern World* (Springfield, Ill., 1968); Werner Krauss, *Perspektiven und Probleme: Zur französischen und deutschen Aufklärung und andere Aufsätze* (Neuwied, 1965), p. 366. See chap. 1, sec. 2 above.

maturity. Such pessimistic traits include both the feelings of anxiety (as described by Pascal) to which man without God is prey, and which Peter Gay in particular has seen as a feature of the European Enlightenment, and also the sense of disorientation (identified too by Pascal) as man, acquiring new knowledge through the microscope and telescope, finds himself poised midway between the infinitely small and the infinitely large.[27]

If it was indeed the search for a guardian figure that led to the assumption that planet-worlds were inhabited by higher beings, the phenomenon would be a manifestation of a theme that occurs quite frequently in the literature and philosophy of the time: the anxiety caused by man's isolation in the universe. Even if a Creator-god is assumed, man, being the only representative of intelligent life in the cosmos, regards himself as an "alarming anomaly," as an exception; whereas only a short time before, he confidently saw himself as the cherished only child of God. Ernest Troeltsch suspected that belief in a "pluralité des mondes" was always, at bottom, inspired by this fear.[28] Even in Thomas Paine, the confident destroyer of the one world of Christianity, we find, mingled with exultation over the indisputable existence of many worlds, relief that "the solitary idea of a solitary world rolling or at rest in the immense ocean of space gives place to the cheerful idea of a society of worlds so happily contrived."[29] In the naturalist Comte de Buffon's Époques de la nature, European civilized man is praised without reserve as the sovereign species in nature; yet even Buffon adds to this the reflection—at first sight a contradiction— that this view of man is only tolerable if we assume by analogy that other planets are inhabited by humans, or beings resembling humans, including some more highly developed ones. For otherwise, man remains frighteningly unique: "Is it not more noble [plus grand], more worthy of the conception that we ought to have of the Creator, to think that everywhere there are beings that can know him and celebrate his glory, than to depopulate the universe, with the exception of the Earth, and to empty it of all sentient beings, reducing it to a profound emptiness in which one would find nothing but the desert of space,

---

[27]See—as well as Peter Gay's The Enlightenment, 2 vols. (New York, 1967–69)—his concise formulations in The Party of Humanity (New York, 1964), pp. 124–27; Lester Crocker, An Age of Crisis: Man and World in Eighteenth Century French Thought (Baltimore, 1959); Karl S. Guthke, Die Mythologie der entgötterten Welt (Göttingen, 1971), chap. 2; Ira O. Wade, Voltaire's "Micromégas," pp. 61–77 ("Great and Small").

[28]Der Historismus und seine Probleme, Gesammelte Schriften, vol. 3 (Tübingen, 1922), p. 86, whence also the phrase (erschreckende Anomalie) cited in the previous sentence.

[29]Complete Writings 1:503–4.

and dreadful masses of wholly inanimate matter?"[30] This fear of isolation is also present as an undercurrent in a very similar celebration of the nonuniqueness of man by Charles Garnier, who was by no means an intellectual like Paine or Buffon, passionately dedicated to a pet theory. In the foreword to Béthune's *Relation du monde de Mercure*, mentioned above, he asks rhetorically: "Is it possible that God could have used so much art to make such a great number of useless things? Can one imagine that in forming those prodigious masses of matter, he did not see fit to create anything but vast deserts and frightful wildernesses?"[31]

The idea of a multitude of inhabited worlds can dispel the sense of cosmic isolation and provide a feeling of security, but it can also have the opposite effect. As Wolfgang Philipp says, "The countless instances of human life in the universe left terrestrial man isolated as a solitary individual in a boundless collective. The single human being within the terrestrial human race became an exposed, solitary individual in the mass of humanity."[32] One reaction to the thought of a plurality of inhabited worlds is an eagerness to "embrace the millions," not just on Earth but in the whole universe, and a sense of relief at not being alone; but this reaction may suddenly turn into its opposite, a terror of being lost precisely in such an infinitely populated universe. In an essay in the *Spectator*, Joseph Addison vividly captures this feeling of terror inspired by the Enlightenment conception of plurality. Contemplating the starry sky, he feels crushed by the immeasurable expanse of space with its countless inhabited worlds; he recalls Huygens's suspicion that some stars may be so distant from the Earth that their light has not yet reached us. "I could not but look upon myself with secret horror, as a being that was not worth the smallest regard of one who had so great a work under his care and superintendency. I was afraid of being overlooked amidst the immensity of nature, and lost among that infinite variety of creatures, which in all probability swarm through all these immeasurable regions of matter" (no. 565, 9 July 1714).

This dialectic of isolation and the sense of being lost in the anonymity of an infinitely populated universe may have contributed—

[30]Comte de Buffon, *Histoire naturelle*, Supplément, vol. 2 (Paris, 1775), p. 528. See (ibid., p. 516) Buffon's "enlightened" mockery of the anthropocentric arrogance of the "highest" being. See also, in general, Michael Winter, "Utopische Anthropologie und exotischer Code: Zur Sprache und Erzählstruktur des utopischen Reiseromans im 18. Jahrhundert," *Zeitschrift für Literaturwissenschaft und Linguistik*, suppl. 8 (1978), 169–73.
[31]*Voyages imaginaires*, 16:160.
[32]In his anthology, *Das Zeitalter der Aufklärung* (Bremen, 1963), p. l.

alongside the downgrading of man by comparison with higher intelligences on other planets—to the revival during the Enlightenment of an idea that perfectly compensates for the crushing blow inflicted by man's new knowledge. This is the idea of the transmigration of souls, in the form of palingenesis. According to this idea, the soul would, after death, inhabit a succession of other planet-worlds, becoming progressively more perfect. Such a concept of perfectibility linked to migration between planets offers a perspective in which the fears of inferiority, of isolation, or of being lost in the cosmos are transformed into ultimate triumph for the immortal part of man's being. This idea is embraced by such diverse thinkers as Kant, Johann Gottfried Herder, Thomas Wright, and Charles Bonnet (see sec. 3).

There is another way in which the Enlightenment's belief in superior forms of life in the universe afforded metaphysical consolation: it answered an unacknowledged desire for a nondogmatic form of theodicy. Kant made this connection when he hinted that only the thought of the perfection and happiness of the beings on other planet-worlds could make the indubitable imperfection of man tolerable. We would then represent "merely one level [*Stufe*] in the hierarchy of intelligent beings, one which, *with* its faults, could not be left out."[33] This idea was developed further by, among others, George Berkeley, the classic proponent of total philosophical subjectivity. In his literary-cum-philosophical dialogues titled *Alciphron, or, The Minute Philosopher* (1732), one speaker argues that it is an error to conclude from the shortcomings of our world that God is not all-powerful, benevolent, and wise; the Earth is, after all, the prison house in God's kingdom, and both revelation and common sense lead us to conclude that "there are innumerable orders of intelligent beings more happy and more perfect than man, whose life is but a span, and whose place, this earthly globe, is but a point, in respect of the whole system of God's creation." Rightly understood, therefore, the universe gives us every reason to praise the wisdom of the Creator, despite the undeniable imperfections to be found on this Earth. Characteristically, however, Alciphron, the freethinker and thus the representative of the new age, reacts with some skepticism to this pious apologia for theism based on the assumption of a plurality of worlds.[34] True, the argument of the dialogues develops in such a way as to refute such skepticism; but Alciphron's philosophical hesitation is nevertheless instructive as a

[33]B. Erdmann, ed., *Reflexionen Kants zur kritischen Philosophie*, vol. 2 (Leipzig, 1884), p. 494: "nur eine Stufe der vernünftigen Geschöpfe, die zusammt ihren Mängeln nicht fehlen durfte."

[34]*Works*, ed. A. A. Luce and T. E. Jessop, vol. 3 (London, 1950), pp. 172–73.

symptom of the age. Two years later we encounter a similar hesitation in exactly the same context in Albrecht von Haller's poem "Über den Ursprung des Übels" (On the origin of evil). Here too the consolation offered by the idea of happier beings on a higher rung of the cosmic ladder is by no means unproblematic (see sec. 5 below). And as late as 1757, the idea is vigorously rejected by Samuel Johnson, opposing Berkeley on this matter as on others. In his brilliant review of Soame Jenyns's *A Free Inquiry into the Nature and Origin of Evil*, he categorically declares that it is absurd to make the evils of our world tolerable by means of a theodicy that depends on the assumption of a hierarchy of beings and of cosmic worlds. For Johnson, as for Voltaire, the "Great Chain of Being" is nothing but a product of fantasy.[35]

Not only Kant and Berkeley, but also, among others, Addison (*Spectator*, no. 519) and, earlier, John Locke (*An Essay Concerning Human Understanding*, bk. 3, chap. 6, § 12) refer to the Golden Chain in the context of the vindication of God's wisdom and goodness; but it is an awkward image to use. For the idea of a ladder leading up from the simplest living creature to man and then continuing beyond him to the angels in the aura of the Creator is a strictly vertical mode of thought that unmistakably remains attached to the geocentric world view and the theological hierarchy associated with it. But when thinkers and poets from Bruno onward speak of inhabited worlds in the universe and their various degrees of perfection, what they have in mind is ubiquitous life in a post-Copernican universe containing an innumerable multitude of planets. The Golden Chain of life-forms, deriving from neo-Platonism and linked from the start with theological argument based on the idea of plenitude, is shattered by the view of the world revealed by the new science. Lovejoy's standard work *The Great Chain of Being* (1936) has tended to obscure the discontinuity between the neo-Platonic and the modern scientific conceptions of plurality (see chap. 2, sec. 1 above). Along the linear, vertical length of the Golden Chain held in God's hands, every link—every form of life and thus every "world"—has its own allotted, fixed place between "above" and "below," in an unalterable hierarchical order; after Bruno, if not even earlier, the worlds are scattered all around in a space often seen as without limits and without structure.

Strictly speaking, then, it is inappropriate to speak of "levels" in the context of the new science, as Kant did, and the incompatibility of the two views is clearly evident in Alexander Pope's *Essay on Man*.

---

[35]Samuel Johnson, *Works* (New York, 1811), 8:32, 41. On Voltaire see Arthur O. Lovejoy, *The Great Chain of Being* (1936; New York: Harper Torchbooks, 1960), p. 252.

For Pope, despite his now-proverbial admonition that "the proper study of mankind is man" (*Essay on Man* 2, line 2), had a keen interest in the new cosmology, fueled in particular by his attendance at Whiston's Astronomical Lectures.[36] When he carries this new conception of the universe over into his philosophical poetry, however, one can see how hard it is for him, while accepting the new ideas, to let go of the old hierarchical order. The result is a new disorientation that obviously disturbs him. The most telling evidence of this disquietude is to be found in the passage from the opening of the *Essay on Man*. These lines urge that now, in the post-Copernican age, an understanding of man must be sought in the perspective of the plurality of worlds, not in that of the one world. This exhortation is set, however, in a context that betrays his uncertainty in the conflict between the old and the new:

> Say first, of God above, or Man below,
> What can we reason, but from what we know?
> Of Man, what see we but his station here,
> From which to reason, or to which refer?
> Through worlds unnumbered though the God be known,
> 'Tis ours to trace him only in our own.
> He, who through vast immensity can pierce,
> See worlds on worlds compose one universe,
> Observe how system into system runs,
> What other planets circle other suns,
> What varied Being peoples every star,
> May tell why Heaven has made us as we are.
> But of this frame the bearings, and the ties,
> The strong connexions, nice dependencies,
> Gradations just, has thy pervading soul
> Looked through? or can a part contain the whole?
>   Is the great chain, that draws all to agree,
> And drawn supports, upheld by God, or thee?
>   Presumptuous Man! the reason wouldst thou find,
> Why formed so weak, so little, and so blind?
>   (1, lines 17–36)

These lines begin "realistically": Pope places man unmistakably in a pluralistic universe with its myriad inhabited worlds in solar system upon solar system. How these worlds and systems may be

---

[36]Marjorie Nicolson and G. S. Rousseau, *"This Long Disease, My Life": Alexander Pope and the Sciences* (Princeton, N.J., 1968), pp. 147–49.

interconnected, and what is the Creator's purpose, man cannot possibly perceive or understand. It is clear that Pope's teeming universe is no longer the neo-Platonic one with its hierarchically ordered worlds and forms of life that, anachronistically, still underlay the theodicy of Kant and Berkeley. Yet, astonishingly, Pope still uses the metaphor of the Great Chain. Is he transforming the traditional image, with its association of vertical movement, into a poetic symbol for gravitation and its circular, or rather elliptic, law of motion?[37] But if so, can it still refer to a cosmic order? The image would appear to allude to the one solar system, but as soon as it is introduced, the context requires that it be related instead to the countless systems of modern cosmology, which would, if anything, call for the image of an inextricable tangle of chains. Is it not more likely that Pope is using the familiar image in its traditional sense, based on the Aristotelian cosmos of spheres— that this is the vertical chain that God holds in his hands? In that case there is indeed a total contradiction here.[38] But precisely this dichotomy in Pope's outlook may explain why, in this very passage, Pope turns away from all such dizzying "Milky Way speculation" (Thomas Mann), preferring to focus attention on man himself. The exhortation to study man in the perspective of innumerable worlds and species of beings is therefore issued, only to be instantly rejected as being unfitting for us.

Not all of Pope's readers, in the age of Newton, were prepared to follow him in this hasty retreat, and he himself was not wholly happy with it (see n. 160 below). Some of his contemporaries did attempt to understand the order of the new whole—Pope had not denied the possibility of such an order—and the status of terrestrial man within it. This is, of course, the bold undertaking of all those who have taken up the challenge of the idea of the plurality of worlds. So it almost seems as if a polemical point is being made when Kant takes the middle section of the passage from Pope, with its exhortation that man should be viewed in the light of the plurality of worlds, separates it from the context that negates it, and places it as a motto at the head of the section of his *Allgemeine Naturgeschichte und Theorie des Himmels* (General natural history and theory of the heavens) in which he seriously reflects on the nature of manlike beings on other planets. And in lyric poetry with a cosmologic theme, in both England and Germany, the "worlds on worlds"—the "thousand suns"—of the "new

---

[37]Bernhard Fabian, "Pope und die Goldene Kette Homers," *Anglia* 82 (1964), 150–71.

[38]See F. E. L. Priestley, "Pope and the Great Chain of Being," in *Essays in English Literature from the Renaissance to the Victorian Age*, ed. Millar MacLure and F. W. Watt (Toronto, 1964), esp. pp. 223–24.

astronomy" are one of the most prominent leitmotifs.[39] Man's consciousness in the age of Enlightenment has reached a point where his attempts to define himself can virtually only take place within the framework of the many worlds of modern cosmology, whatever philosophical problems this entails.

Recalling the Great Chain of Being, Pope shrank from this bold enterprise, saw the notion of plurality as a temptation to be resisted, and took refuge in the view of man as the measure of all things. In those writers who were not afraid to follow through the idea of defining man in relation to a plurality of worlds, the image of the chain appears in a very dubious light, a light that also falls on yet another philosophical problem arising from plurality: the question of whether the many worlds have their origin in "God" or in "Nature." This can be seen for instance in Johann Gottfried Herder.

He states his position in the very first sentence of the *Ideen zur Philosophie der Geschichte der Menschheit* (Ideas on the philosophy of the history of mankind, 1784):[40] "Heaven must be the starting point of our philosophy [not "theology"] of the history of mankind, if it [i.e., the philosophy] is to be at all worthy of the name." He had already, in the preface (p. 8), criticized those "who, because they can see no design, actually deny that there is one." (One writer to throw in the towel in this way had been Pope, though Herder does not mention him by name.) Instead, according to Herder, one must "seek to find order" in the "apparent disorder" created by modern cosmologists from Copernicus to Kant (p. 8). Instead of "losing" oneself in the "sea of immeasurableness" and finding in it neither beginning nor end (*nirgend Ausgang und Ende*), one must try to track down the "laws that operate in the immeasurable" and to "ask what I am meant to be in this particular place and presumably can be only in this place" (pp. 14–15). Herder expressly states that he is referring to a place within a universe containing countless worlds, all inhabited, some by "creatures perhaps more splendid" than ourselves. It is in this context, then, that Herder takes up the neo-Platonic image of the "chain" (p. 17) but quietly detaches it from its original meaning. For Herder, there is no doubt that "the organization and the destiny" of the "inhabitants"

---

[39]See Junker, *Weltraumbild*, e.g., pp. 37, 39, 89; William Powell Jones, *The Rhetoric of Science: A Study of Scientific Ideas and Imagery in Eighteenth Century English Poetry* (London, 1966). On Pope's uncertainty, see his letter to Caryll, 14 August 1713, in *The Correspondence of Alexander Pope*, ed. George Sherburn, vol. 1 (Oxford, 1956), pp. 185–86.

[40]*Sämmtliche Werke*, ed. Bernhard Suphan, vol. 13 (Berlin, 1887), p. 13. Page references in the text are to this edition.

of the planets are matters of which we "can know nothing and are intended to know nothing" (p. 17). At most, we may assume "that everywhere, as here, unity and diversity reign" and that therefore there may well be "more perfect creatures" elsewhere in the universe. But—and this is central to man's view of himself at this time in the light of the plurality of worlds—man as we know him is not, and does not possess, the yardstick by which these life-forms in the cosmos may be measured. In the Great Chain of Being, man at least had a set place about halfway between animals and angels, and according to the anthropocentric Christian teaching associated with the chain, one thing that man knew for certain was that he was the apple of God's eye. Herder, a clergyman in an enlightened age, knows that, on the contrary, "the measure of our intelligence . . . by no means provides us with a yardstick. We are not at the center but among the throng; like other earths, we sail about in the current [*schiffen . . . im Strom umher*] and have no measure for comparison" (p. 18). For Herder, the notion of the chain no longer serves to orient man.

Despite all this, Herder does not brand Milky Way speculation as misguided: he focuses his questioning on God, though without entering the domain of theology as such; for God only appears in this context as the originator of the universe of the new science, as the creator of the physical laws and consequently also of man's destiny (*Bestimmung*, p. 17). "My destiny," Herder concludes, "is not tied to the Earth's dust, but to the invisible laws that govern the Earth's dust," that is, to the "world of God" (p. 16). "Everywhere the great analogy of Nature has led me to religious truths" (p. 9). The initial overwhelming of intellect and imagination, and the dizzying effect of the "astonishment, pure and simple, that annihilates us" (p. 14), is mitigated by trust that a divine order governs the whole. This is the basic position of the physicotheology of the day: the shock of Copernicanism, which could lead to the brink of nihilism, is overcome by the intuition of a wise Creator continuously at work.[41]

Unlike most physicotheologians, however, Herder does not demonstrate the existence of a divine "order" but simply postulates it. Indeed, it appears as little more than a hope, conceived in the vaguest terms (just as the symbol of the chain as used by Herder no longer offers any fixed points of orientation). But this means that Herder, even more than others who accept the plurality of worlds at the period of the Enlightenment, could easily be suspected of seeing not "God"

[41]Philipp, *Werden der Aufklärung*, pp. 61, 78–87.

as the underlying creative principle (as the Lutheran pastor Ehrenberg did, around 1710, as a matter of course) but instead an autonomous, mechanical "Nature."[42] In that case, reference to the supreme being in religious terms would be a mere phrase, an act of lip service perhaps, but at all events unnecessary. This is the risk run by any philosophy that, like Herder's (p. 8), wishes to eliminate chance. Not only Descartes and Leibniz had been accused by contemporaries of having a mechanistic conception of the world (in which the world-machine was seen as being something like a clock); the same charge was leveled at Newton, who hotly denied it, as did his followers Samuel Clarke and Richard Bentley. It was only later that Laplace, in his now-famous conversation with Napoleon, felt able to assert candidly that "that hypothesis," namely God, was no longer necessary to him. How precarious is the hold that belief in a creative power conceived in religious terms maintains in the late eighteenth century can be seen, appropriately enough, in a paragraph of Herder's preface to the *Ideen*. The casual and relaxed tone is deceptive:

> Let no one therefore be misled by the fact that I sometimes use the name of Nature personified. Nature is not an independent being; on the contrary, *God is everything in his works;* however, by mentioning his name often, without being able each time to accord it the full measure of holiness, I might have taken in vain that most holy name which no grateful creature ought to pronounce without the deepest reverence, and this at least I wished to avoid. Let anyone for whom the name of "Nature" has been rendered meaningless and base by some writings of our time substitute for it in his mind *that all-powerful strength, goodness, and wisdom,* and let him name in his soul that invisible Being whom no earthly language can name. (Pp. 9–10)

We have already seen various developments in man's view of himself that took place once the plurality of humankinds had become a theoretical possibility or indeed an inescapable conclusion; to these, this uncertainty as to whether the supreme power in the universe is "God" or merely "Nature" adds a new and threatening dimension. The jubilant feeling of cosmic brotherhood, the praise of God's omnipotence extended to infinity, can easily give way to the horror of being lost in the *uncreated*, merely "natural" Pascalian infinite spaces of

---

[42]See in general E. J. Dijksterhuis, *Die Mechanisierung des Weltbilds* (Berlin, 1956); also Thomas P. Saine, "Natural Science and the Ideology of Nature in the German Enlightenment," *Lessing Yearbook* 8 (1976), 72–75.

the post-Copernican universe, or indeed to fear of the superiority of intelligent extraterrestrial beings that are now no longer subject to the wise surveillance of a Creator-God.

Paradoxically, however, this realization of man's insignificance also points the way toward an unexpected elevation of man. This is foreshadowed in Pascal's image of the "thinking reed" and even in Copernicus's and Kepler's feeling of satisfaction in the greatness of the human mind that is capable of grasping the laws governing the universe (and which thereby raises the Earth, the proverbial "dung heap," to the level of a star among stars).[43] Man is elevated by the fact that the majesty of the edifice of the universe finds its equal in human reason, or the human soul, that is capable of such grand conceptions of the universe: "That, as I am moved by such greatness, I myself am great."[44] Louis Sébastien Mercier's *L'an 2440* (1771) states, looking back in time, that astronomy is a science which has become important for the understanding of man "because with a magnificent voice it makes known the glory of the Creator and the dignity of the thinking being" (chap. 40). The "thinking reed" has by now become a self-confident individual. Mercier's individual still knows himself to have issued from "God's hands"; but in this context too the Creator often proves to be a mere pseudonym for Nature. But even where mention of "God" is more than formulaic, man—who is worthless before God, lost to God's sight in the mass, unnoticed by God—man, with his ability to perceive and to know, is triumphantly extolled.

Where this perception takes the form of a primarily emotional and imaginative experience, it gives rise to the feeling of cosmic sublimity such as we frequently find in eighteenth-century nature poetry from Brockes to Schiller. The sense of humiliation and of being lost in the immeasurable spaces of the new science with its countless "worlds" is transformed into the consciousness of a heightened sense of self, into increased self-assurance. This is well exemplified by Klopstock, the first writer in the German-speaking countries who made "the discoveries of modern astronomy" and hence "the scientific world view the basis of his own philosophy and the subject of his poetry"—

[43]On Pascal see chap. 3, sec. 1 above; on Copernicus see Karl Richter, *Jahrbuch der Deutschen Schillergesellschaft* 12 (1968), 135; on Kepler see his letter to Maestlin, 9 April 1597, in *Johannes Kepler in seinen Briefen*, ed. Max Caspar and Walter von Dyck (Munich, 1930), 1:44–45 (*Ges. Werke* 13:113). On the new science's view of itself see also Hans Blumenberg, *Die kopernikanische Wende* (Frankfurt, 1965), pp. 122 ff.

[44]Barthold Hinrich Brockes, *Irdisches Vergnügen in Gott*, vol. 1, 6th ed. (1737; Bern: Peter Lang, 1970), p. 454: "daß, da mich solche Grösse rühret, / Ich selber etwas grosses bin."

and thus "carried through the Copernican revolution in the domain of poetry."[45] In his odes we constantly encounter those many "worlds" and "earths" in the heavens whose presence the telescope has enabled us to see or at least to infer. At first this expansion of the consciousness of man on his "little grain of sand" (which, like the "myriad worlds," becomes one of the leitmotifs of the new self-definition) is overwhelming. But then the greatness of the universe is transmuted into the greatness of the soul. The cosmic dimension is internalized, for instance in the ode "Die Allgegenwart Gottes" (The omnipresence of God, 1758):

> Hier steh ich Erde!
> Was ist mein Leib
> Gegen diese selbst den Engeln
> Unzählbare Welten!
> Was sind diese selbst den Engeln
> Unzählbare Welten
> Gegen meine Seele![46]

Here I stand, Earth! What is my body compared to these worlds which even the angels cannot number! What are these worlds, which even the angels cannot number, compared to my soul!

This imaginative leap had, long before Klopstock, become tremendously widespread in popular philosophy through the numerous editions of Gottsched's *Erste Gründe der gesammten Weltweisheit* (First principles of all branches of philosophy, 1733–34). The flyleaf bore, as a sort of motto, a copperplate engraving showing the heavens with many randomly arranged radiant suns circled by planets, with the accompanying explanatory text:

> Hier starret Sinn und Witz, der Geist verliert sich gantz,
> In aller Welten Heer, Pracht, Ordnung, Lauf und Glantz.
> O! was ist hier der Mensch? Er wäre nichts zu nennen,
> Könnt er am Wercke nicht, des Meisters Größe kennen.

Here sense and wit are numbed, the mind quite loses itself in the multitude, splendor, order, motion, and brilliance of all the worlds.

---

[45]Karl Viëtor, *Geschichte der deutschen Ode* (Munich, 1923), p. 112; Robert Ulshöfer, "Friedrich Gottlieb Klopstock: 'Die Frühlingsfeier,' " in *Die deutsche Lyrik*, ed. Benno von Wiese (Düsseldorf, 1957), 1:173–74.
[46]*Ausgewählte Werke*, ed. Karl August Schleiden (Munich, 1962), p. 81.

Oh, what is man in comparison? He would be accounted as nothing, were he not able to recognize the greatness of the Master in his work.

The classic exponent of this idea, however, who actually made it a literary fashion, was Edward Young, in his *Night Thoughts* (1742–45). The countless worlds are as constant a leitmotif in Young as in Klopstock. As man contemplates the night sky he feels himself to be "enclos'd by these innumerable worlds" (bk. 9, line 1400). "A thousand systems" (bk. 9, line 1750) of such worlds surround him,

> Orb above orb ascending without end!
> Circle in circle, without end enclos'd!
>   . . . . .
> What involution! what extent! what swarms
> Of worlds, that laugh at earth! immensely great!
> Immensely distant from each other's spheres!
> What, then, the wondrous space through which they roll?
> (Bk. 9, lines 1097–1105)

The first reaction is a feeling of being overwhelmed:

> At once it quite engulfs all human thought;
> 'Tis comprehension's absolute defeat.
> (Bk. 9, lines 1106–7)

This is a cliché of lyric poetry about the cosmos. But the feeling is relieved not, as in Brockes's much-quoted poem "The Firmament," by a sense of the greatness of God—who brings meaning and order to the confusion ("Allgegenwärt'ger Gott, in Dir fand ich mich wieder"— Omnipresent God, in Thee I found myself again)—but instead by a sense of the greatness of the human being who apprehends it:

> As yet thou know'st not what it is: how great,
> How glorious, then, appears the mind of man,
> When in it all the stars, and planets, roll!
> And what it seems, it is: great objects make
> Great minds, enlarging as their views enlarge;
> Those still more godlike, as these more divine.
> (Bk. 9, lines 1061–66)

This idea of Young's was given wide currency by the *Meditations and Contemplations* (1746–47) of James Hervey, a Methodist preacher and admirer of Newton. This work went through many editions right into the nineteenth century and was a sort of best seller in the field of scientific devotional literature. The sight of the numberless suns with their planets, "which . . . are supposed to be abodes of *intellectual life*,"[47] brings home to him both the omnipotence of the Lord *and* "my own utter insignificancy" (pp. 259–60)—until, with direct reference to the ninth book of *Night Thoughts*, the soul discovers how great is its own inner expanse, evidenced by its ability to perceive God's glory in his works: "Under the influence of such considerations I feel my sentiments expand, and my wishes acquire a turn of sublimity. . . . My soul, fired by such noble prospects, weighs anchor from this little nook, and coasts no longer about its contracted shores, doats no longer on its painted shells. The *immensity* of things is her range, and an *infinity* of bliss is her aim" (pp. 261–62).

The writings of Young and Hervey reveal that this feeling of sublimity associated with the plurality of worlds gives rise, however, precisely in its soaring enthusiasm, to its own set of problems: "Nothing can satisfy but what confounds" (*Night Thoughts*, bk. 9, line 838). Maybe; but when Hervey finds himself writing, without any subversive intent, "perhaps nobler systems" (p. 259), or when Young speaks in passing of planet-worlds "where nobler natures dwell" (bk. 9, line 1607), then even the enthusiast is faced, at moments of sober "meditation," with the question of man's status, posed in philosophical rather than dogmatic terms. Instead of following Pope's exhortation to define himself "anthropocentrically," using his own measure, terrestrial man now compares himself to the inhabitants of the "unterrestrial sphere[s]": "how unlike the lot of man! . . . But what are we?" (*Night Thoughts*, bk. 9, lines 1754, 1797–98, 1804). The reflections on the nature of the cosmic humankinds and thus also on that of man are therefore an important element in the *Night Thoughts*. Dogmatic questions are still marginally present, but they are not given greater weight than those relating to the Golden Age, and above all they are not answered in a way that accords with dogma. Typically they are submerged in a whole catalog of philosophical questions, which are the same questions as those asked by modern science fiction: it too is ultimately asking the question, "What are we?" (H. G. Wells).[48] Young's questions are already those of an "age more curious than devout" (bk. 9, line 1852). They concern above all the intellectual status of the extraterrestrial beings, their way of thinking and feeling,

[47]*Meditations and Contemplations* (London, 1856), p. 257. Page references are to this edition.
[48]*The War of the Worlds*, bk. 1, chap. 13.

their social and moral behavior—in other words, they suggest alternative "yardsticks" in the light of which those of our planet become questionable or need to be justified; clearly, at the height of the Enlightenment man is no longer the measure:

> How can man's curious spirit not inquire,
> What are the natives of this world sublime,
> Of this so foreign, unterrestrial sphere,
> Where mortal, untranslated, never stray'd?
>   "O ye, as distant from my little home,
> As swiftest sunbeams in an age can fly!
> Far from my native element I roam,
> In quest of new, and wonderful, to man.
> What province this, of His immense domain,
> Whom all obeys? Or mortals here, or gods?
> Ye bord'rers on the coasts of bliss! what are you?
> A colony from heav'n? or, only rais'd,
> By frequent visit from heav'n's neighbouring realms,
> To secondary gods, and half divine?—
> Whate'er your nature, this is past dispute,
> Far other life you live, far other tongue
> You talk, far other thought, perhaps, you think,
> Than man. How various are the works of God!
> But say, what thought? Is Reason here enthron'd,
> And absolute? or Sense in arms against her?
> Have you two lights? Or need you no reveal'd?
> Enjoy your happy realms their golden age?
> And had your Eden an abstemious Eve?
> Our Eve's fair daughters prove their pedigree,
> And ask their Adams—'Who would not be wise?'
> Or, if your mother fell, are you redeem'd?
> And if redeem'd—is your Redeemer scorn'd?
> Is this your final residence? If not,
> Change you your scene, translated? or by death?
> And if by death; what death?—Know you disease?
> Or horrid war?—
>   .   .   .   .
>   . . . In our world, Death deputes
> Intemperance to do the work of Age;
> And hanging up the quiver Nature gave him,
> As slow of execution, for despatch
> Sends forth imperial butchers; bids them slay
> Their sheep (the silly sheep they fleec'd before),

And toss him twice ten thousand at a meal.
Sit all your executioners on thrones?
With you, can rage for plunder make a god?
And bloodshed wash out every other stain?—
But you, perhaps, can't bleed: from matter gross
Your spirits clean, are delicately clad
In fine-spun ether, privileg'd to soar,
Unloaded, uninfected; how unlike
The lot of man! how few of human race
By their own mud unmurder'd! how we wage
Self-war eternal!—Is your painful day
Of hardy conflict o'er? or, are you still
Raw candidates at school? and have you those
Who disaffect reversions, as with us?—
But what are we? You never heard of man;
Or earth, the bedlam of the universe!
Where Reason (undiseas'd with you) runs mad,
And nurses Folly's children as her own;

. . . . .

But this, how strange to you, who know not man!
Has the least rumour of our race arriv'd?
Call'd here Elijah in his flaming car?
Pass'd by you the good Enoch, on his road
To those fair fields, whence Lucifer was hurl'd;
Who brush'd, perhaps, your sphere in his descent,
Stain'd your pure crystal ether, or let fall
A short eclipse from his portentous shade?

. . . . .

But this is all digression: where is He,
That o'er heav'n's battlements the felon hurl'd
To groans, and chains, and darkness? Where is He,
Who sees creation's summit in a vale?
He, whom, while man is man, he can't but seek;
And if he finds, commences more than man?
O for a telescope His throne to reach!
Tell me, ye learn'd on earth! or blest above!
(Bk. 9, lines 1752–1835).

2. *Authorities in Conflict: Fontenelle and Huygens*

It is noteworthy that these and similar problems that arose in the eighteenth century once the "doctrine" of plurality had become widely accepted already crop up in the two works most influential in spreading the idea of the plurality of worlds at the very beginning of the Enlightenment. These are Fontenelle's *Entretiens sur la pluralité des mondes* (1686) and Christian Huygens's *Kosmotheoros* (1698). Both authors are proudly conscious that, as Copernicans and Cartesians, they have developed a *philosophy*, not a theology, of plurality. The intellectual climate in which the discussion now takes place is radically different from that of the baroque period, and this is well illustrated by a passage at the beginning of the second book of the *Kosmotheoros*. Huygens ridicules Athanasius Kircher as a representative of the (negative) theology of plurality; conscious of the philosophial emancipation of his own thinking, he adds that had Kircher been free to express his true opinions, he might have come to quite different conclusions.[49] Fontenelle's "Bible" of enlightened pluralism is the product of thinking that is equally emancipated from theology; on the very first page of his preface he makes a point of referring to himself as a "philosopher." The fact that their hypotheses about extraterrestrials still raise problems—of a philosophical nature—is not consciously acknowledged by either of the two, but it is nevertheless apparent in both, if one reads between the lines. These two classic exponents of the idea of plurality, whose conclusions are so very different, have at least this in common.

The *Entretiens sur la pluralité des mondes* contributed more than almost any other publication to bringing about that shift in thinking among the educated public which has been called "the crisis of European consciousness."[50] Friedrich Melchior Grimm, in his *Correspondance littéraire*, gives Fontenelle the credit for having been the first to make philosophy popular in France and for having spread the spirit of philosophy (*l'esprit philosophique*) and the love of the truth, and so extended "the empire of reason";[51] and above all this is true of the *Entretiens*, the astronomical best seller of the Age of Enlightenment and a book that imparts its scientific content in an irresistibly chatty style. By the time of the author's death in 1757, thirty-three French

---

[49]As late as 1771, Kircher is supported, against Huygens, in François Xavier de Feller's *Observations philosophiques sur les systèmes de Newton, de Copernic, de la pluralité des mondes, etc.* (Liège, 1771), pp. 164, 171–72.

[50]Calame, in his edition of the *Entretiens*, p. li. Quotations in English in the text are from the 1688 trans. by John Glanvill, *A Plurality of Worlds*, repr. with a prologue by David Garnett (London: Nonesuch Press, 1929). Page references to this translation are followed by the corresponding references to Calame's edition. (Translator's note: one passage, as indicated, was not translated by Glanvill, as it was added by Fontenelle only in a much later edition.)

[51]*Correspondance littéraire*, ed. Maurice Tourneux, vol. 3 (Paris, 1878), p. 338, 1 February 1757.

editions had already appeared; several English and German translations went through a number of editions. For decades these evening
conversations between the galant scientist and his eager pupil the
marquise, in the rococo gardens of a French chateau, featured in the
everyday talk of educated people, and not least of women. In Addison's moral weekly, *The Guardian*, we learn that "women of quality or
fortune" read Fontenelle's dialogues aloud to each other even while
preserving fruit: "It was very entertaining to me to see them dividing
their speculations between jellies and stars, and making a sudden
transition from the sun to an apricot, or from the Copernican system
to the figure of a cheese-cake" (no. 155, 8 September 1713). This might
suggest a very bland popularization of the subject, but in fact the
very real philosophical problems make themselves felt in spite of the
engaging style of the *Entretiens*.

These are already evident in the preface. Fontenelle (like Arthur
C. Clarke,[52] the philosopher among science fiction writers of our own
day) is firmly convinced of the philosophical importance and topical
relevance of this general subject: "For what can more concern us than
to know how this World which we inhabit, is made; and whether
there be any other Worlds like it, which are also inhabited as this is?"
(pp. 3/5). For this very reason he stresses that he has scrupulously
avoided building metaphysical castles in the air and has instead adhered strictly to rational conclusions drawn from science ("vrais rai-
sonnements de Physique," p. 6). In a manner symptomatic of the
intellectual climate of the time, he clears the ground for such a discussion of the subject by nonchalantly disposing of the more narrowly
theological questions: in order to mitigate the theological danger inherent in the assumption of living creatures in an "infinity of Worlds"
(pp. 6/8), he assumes that the planet-worlds have inhabitants who are
"not Men" (pp. 6/9)—without, however, giving any closer description
of them at this strategic point. Of course the theological authorities
were not fooled by this ploy but fully recognized the serious antitheological tendency of this little book of philosophy with its air of harmless
playfulness. It was on the index of the Catholic church from 1687 until
well into the nineteenth century, and then again in the twentieth.[53] In
the present context, however, this is less important than the fact that
while explicitly avoiding theological (for instance christological) sore
points, Fontenelle unwittingly created philosophical ones. For, as for

[52]*Report on Planet Three and Other Speculations* (1972; New York: Signet Books, 1973),
p. 79.
[53]Calame, *Entretiens*, pp. xxxix–xl.

instance the *Journal des savants* immediately pointed out,[54] he himself implies that his planet dwellers are indeed "human" in nature. He does so in the very way he justifies the assumption of a plurality of worlds. His main argument, constantly repeated, is the scientific inference by analogy, on the basis of the Copernican system, that the moon is of "the same nature" as the Earth, the Earth is a planet among planets, and the fixed stars are suns that in their turn must each have a planetary system:

> You grant that when two things are like one another in all those things that appear to you, it is possible they may be like one another in those things that are not visible, if you have not some good reason to believe otherwise: Now this way of arguing have I made use of. The Moon, *say I*, is inhabited, because she is like the Earth; and the other Planets are inhabited, because they are like the Moon; I find the fix'd Stars to be like our Sun, therefore I attribute to them what is proper to that. (Pp. 119/138)

"What is proper to that"—does this, on the basis of analogy, include rational beings on their planets, indeed (despite all the assertions to the contrary) human beings? "Why not?" For why should just we, and our grain of sand, be an exception in the universe (pp. 79–80/91–92)—in Troeltsch's phrase, an "alarming anomaly"? Fontenelle therefore does not hesitate to transpose the European/Indian schema to the cosmic arena. The analogy has of course long since ceased to be original, but what is new, and offers a startling challenge to the self-image of Enlightenment man, is the fact that Fontenelle abandons the Eurocentric chauvinism of the colonial masters: it is immaterial to him whether in this discovery of cosmic new worlds we Europeans play the part of the Europeans or that of the Indians, immaterial whether we make the journey to the moon dwellers or they to us, immaterial who colonizes whom—presumably "enlightening" or subjugating them: " 'Tis no great matter whether we go to them, or they come to us, we shall then be the *Americans*, who knew nothing of Navigation, and yet there were very good Ships at t'other end of the World" (pp. 61/72).

What matters is the encounter with unknown life-forms from elsewhere in the universe, and that would be an encounter between humans and humans. Fontenelle ironically underlines this when, referring to the analogical case of the discovery of America, he imagines

[54]Ibid., p. xxxvii.

the Indians' reaction to the conquistadors, who are clothed in iron, ride on monstrous animals, and hurl deadly flashes of lightning: "Are they Sons of the Sun? for certainly they are not Men" (pp. 59/70). Fontenelle carries the humanness of his moon dwellers and planet dwellers to extremes. On the fourth evening he offers various reflections on the influence of the climates of Venus, Mars, Jupiter, and so forth on the physiology and temperament of their inhabitants (such reflections were common in the Enlightenment in relation to *Homo sapiens*); what is more, he does not spare his readers an account of the all-too-human occupations of his planet dwellers, which include observing the Earth through telescopes and writing reports about it in learned journals the populace at large cannot understand (pp. 104/123). Anthropocentric thinking (which assumes the universe was created for our happiness) is shown to be a "folly" with which men on Earth flatter themselves (pp. 20/23–24), just as—Gottsched adds in his translation, giving homespun expression to his enlightened views—the maggots in a cheese believe it to have been created for them.

So there are human beings everywhere in the universe; life elsewhere is just the same as chez nous. But that is not the whole story. Fontenelle is not content to "abate the Vanity of Mankind" by the thought that our human race, as it has seemed so far, is merely one of many that are identical to us (pp. 25/30) and leave it at that. The devastating impact of this thought promptly makes the marquise feel that man's struggle with life is "ridiculous" (pp. 35/43). Nihilism threatens at this point, but Fontenelle clearly does not want to let matters rest there. As a popular philosopher of the Enlightenment he has too great a sense of responsibility. Despite the new assumptions, somehow the preeminent status of man on Earth must be confirmed or at least redefined. Fontenelle sets out to do this, but to begin with he only exposes man's view of himself to even more serious doubts.

Let us return to the encounter between the Indians and the Spaniards in America. The ironizing of this meeting *between man and man* is Fontenelle's contribution. For the Indians (and indeed for the Europeans, who had to be convinced by papal decree that the Indians were humans), it was an encounter with beings who were quite different and yet familiar: a shock of recognition of the kind that defines one's own identity, but only after first making it uncertain. Accordingly, however much Fontenelle stresses the humanness of the extraterrestrials, there is no mistaking how very different they are. Nature loves "infinite diversity" (pp. 7/9); she hates repetition (pp. 82/95). The inhabitants of the cosmic worlds are therefore credited with a diversity that reads like a catalog of science fiction anthropology; and the diversity explicitly extends to temperament and ways of thinking. On one planet the inhabitants use their voices, on another they use sign language for communication, and on a third they do not communicate at all. In one world they form judgments wholly on the basis of experience, in another by reasoning. In one place in the universe they are concerned more with the future than with the past, in another the reverse is true, and in yet a third they do not worry about either, "which by the way, I think, is much the better." Perhaps other humankinds have the sixth sense that we lack, but they may lack one of the senses we have. Perhaps—a favorite speculation of the day—those in other worlds have a large number of senses of which we know nothing. Their knowledge may not be bounded by the same limits as ours, yet they may be ignorant of what is familiar to us. Here war predominates; elsewhere perhaps peace reigns but love is unknown and "time lies on their hands" (pp. 83–84/96–97). For the marquise, this picture of the diversity of the products of Nature is still too vague, and so her chivalrous instructor assiduously supplies further details:

On a planet whose name I will not tell you as yet, there are inhabitants who are very lively, very industrious, very skillful; they live entirely by pillage, like some of our Arabs, and that is their sole vice. Apart from this, there is perfect understanding between them, they constantly work in concert and with zeal for the good of the State, and above all their chastity is incomparable; to be sure, this shows no great merit on their part; for they are all sterile, they have no sex. (Trans. H. A.; Calame, p. 98)

I find, *says the Countess,* it is easie enough to guess at the Inhabitants of *Venus;* they resemble what I have read of the *Moors* of *Granada,* who were a little black People, scorch'd with the Sun, witty, full of Fire, very Amorous, much inclin'd to Musick & Poetry, and ever inventing Masques & Turnaments in honour of their Mistresses. Pardon me, Madam, *said I,* you are little acquainted with the Planet; *Granada* in all its Glory, was a perfect *Greenland* to it; and your gallant *Moors,* in comparison with that People, were as stupid as so many *Laplanders.*

But what do you think then of the Inhabitants of *Mercury?* They are yet nearer to the Sun, and are so full of Fire, that they are absolutely mad; I fancy, they have no memory at all, like most of the *Negroes,* that they make no reflections, and what they do is by sudden starts, and perfect hap-hazard; in short, *Mercury* is the Bedlam of the Universe. (Pp. 89/105)

The Saturnians are at the greatest distance from the sun and are in every way the opposite of the Mercurians: phlegmatic and lacking any sense of humor, they have to spend a whole day thinking before they are ready to give an answer to the simplest question put to them; they would have found Cato of Utica too waggish and playful (pp. 110/130).

Such diversity among the "humans" of the worlds even within the limited range of our own planetary system imposes an unexpected readjustment of man's status (just as the Indians studying the appearance of the conquistadors reduced the latter's status to that of nonhumans—and vice versa). The dignity and self-assurance of the civilized European, indeed his very identity as a member of the human *genus* is urgently called into question. Fontenelle has an answer, and one that is interesting above all because it revives, however discreetly, the Earth-based anthropocentrism he had supposedly rejected. Just as the theologians faced with the heliocentric system were able to salvage terrestrial man's centrality in the history of salvation, so this representative of the Enlightenment surreptitiously preserves his pride by attempting to secure for earthly man a different but equivalent position of centrality. The physical form, temperament, and way of life of the beings, so very similar to humans, on the other planets of the solar system represent different extremes (conditioned by climate), while we embrace them all: we resemble now the people of Mercury, now those of Saturn, and our world alone is a miniature universe: "It is, I suppose, because our Earth is plac'd in the middle [!] of the other Worlds, . . . we are a mixture of the several kinds that are found in the rest of the Planets. . . . One would think we were collected out of different Worlds; we need not travel, when we see the other Worlds in Epitome at home" (pp. 110–11/130).

Quite clearly this is a secularized form of anthropocentrism that has crept in again by the back door and that must be accounted "vanity." Fontenelle's own bravado in destroying theological anthropocentrism (which believed "all things in Nature design'd for our use"—pp. 20/23) is now causing him anxiety. And is it not striking that he does not even hint, anywhere in these conversations, that some extraterrestrial beings might be *superior* to man? He accepts pluralism, but only in the form of "enlightened" self-flattery, congratulating the denizens of Earth on not being like the others (pp. 111–12/ 131). The expression used here, "our proportion of happiness" (*la mesure de bonheur*, pp. 112/131) clearly refers not merely to our good fortune in this respect but also to that happiness which the Enlightenment, robbed of the consolation of Christianity, sought more avidly than any other age. And of course it was the conviction, growing ever stronger, that other stars possessed different and possibly happier beings, that most persistently obstructed this all-too-human demand for happiness. Fontenelle himself had alluded to this problem in his catalog of the differences between the planetary human races: Nature has distributed happiness and talents unequally among men on our Earth, and among the human species in the universe the same will be true (Calame, p. 97; Glanvill, p. 84, speaks only of "Gifts"). Obviously, even in a secularized context in which God's justice is no longer a matter of concern, the subject of man's happiness in comparison to that of other forms of intelligent life in the universe remains a burning issue—if only, perhaps, as an aspect of what Fontenelle calls "the Vanity of Mankind." He expressly exposes this vanity in his attack on traditional anthropocentrism, but in fact he himself cannot renounce it: despite all appearances, he is far removed from "dethroning man as . . . the crowning glory of Creation"[55] but rather, attempts to secure for man the "happiest" situation and the rank of greatest perfection. In a sense, this is also an attempt to perpetuate man's role as the measure of all things despite the changed view of the universe.

The marquise, always a practical thinker, brings these reflections on philosophical anthropology down to a more trivial level, pointing out that man should also be grateful to Nature for having spared him both intolerable heat and intolerable cold (pp. 111/130). The Berlin astronomer Johann Elert Bode, in a comment on this passage in his translation, *Dialogen über die Mehrheit der Welten* (1780), informs her that indeed we need "neither pity nor envy the other planets in the matter of climate." Interestingly, however, Bode, a professional astronomer who pursues Fontenelle's lines of thought further, also tries to preserve a privileged position for man, one that in Bode's case is supported by science. Like Kepler he sees the Earth's location as the best possible position in the solar system for astronomical observation. He quotes from an essay by Lessing's teacher of geometry and philosophy, Abraham Gotthelf Kästner (to whom Herder also refers at the beginning of his *Ideen*):

I conclude with the observation that, of all planets, our dwelling occupies the most convenient position from which to determine the order of the universe and the relationships between the orbits of the

[55]Michael Winter, *Compendium Utopiarum*, vol. 1 (Stuttgart, 1978), p. 121. The same view also in William C. Heffernan, "The Singularity of Our Inhabited World: William Whewell and A. R. Wallace in Dissent," *Journal of the History of Ideas* 39 (1978), 97.

planets. The decreasing and increasing light of Mercury and of Venus shows us conclusively that there are planets which revolve around the sun. Mercury can draw no such conclusion, since it can see no planet below itself. Mercury and Mars have, as far as we know, no companions from whose darkening, spots, or other phenomena they might deduce anything about the nature of the heavenly bodies. To Jupiter and Saturn, it would seem, the small globes that are nearer to the sun are lost from view. We alone can recognize that we, in a company of fourteen worlds, revolve around the sun. Few of them can know anything of our tiny dot, a little piece of which is often divided among many peoples by fire and the sword. We alone can establish truths about the motions and properties of the planets. If the Creator had not wished us to do this, he would not have given us such a convenient observatory.[56]

Compared with this scientific demonstration, Fontenelle's rescue of the "centrality" of man on our planet is of course more philosophical—suggesting, indeed, that the enlightened author is pulling himself out of the bog by his own bootstraps. For, faced with man's sense of being lost and insignificant in the universe, Fontenelle attempts to extricate himself by means of an explanation involving only our neighbors in our own solar system and says nothing about how this explanation would stand up in the wider context of inhabited systems based on the fixed stars. This makes his effort all the more revealing.

Immediately afterward, however, on the fifth evening, Fontenelle has occasion to speak of the countless fixed stars and their planetary systems, and at once he has to admit the double-edged effect of the post-Copernican and post-Galilean awareness of infinity: on the one hand, liberation from the oppressive narrowness of the pre-Copernican world; on the other hand, the sense of disorientation induced by the thought of unimaginably numerous worlds. Indeed, this feeling of terror in the face of the universe is so marked in the *Entretiens* that Hans Blumenberg's studies of the philosophical significance of Copernicanism, which have received much attention in German-speaking countries, present the shock of disorientation as a new phenomenon first encountered in Fontenelle.[57] This is not quite true, as Pascal and Donne at least were Fontenelle's forerunners in interpret-

---

[56]*Dialogen über die Mehrheit der Welten*, 2d ed. (Berlin, 1789), pp. 269–72. Kästner's text first appeared in the *Hamburgisches Magazin* 1:2 (1748), 221–22. The article is entitled "Das Lob der Sternkunst" (pp. 206 ff.); its author is described as a teacher "of geometry and philosophy" (p. 206). Quoted from the *Hamburgisches Magazin*.
[57]For instance in his edition of Galileo, *Sidereus Nuncius* (and other works, Frankfurt, 1965), p. 26.

ing the plurality of worlds in this way. Still, Fontenelle presented the twofold emotional impact of the awareness of plurality in a way that had an enormous influence on eighteenth-century literature.

The marquise's learned companion identifies himself with one of the two possible reactions, the triumphant one:

> When the Heavens were a little blue Arch, stuck with Stars, me-thought the Universe was too strait and close, I was almost stifled for want of Air; but now it is enlarg'd in heigth [sic] and breadth, and a thousand & a thousand Vortex's taken in; I begin to breath with more freedom, and think the Universe to be incomparably more magnificent than it was before. (Pp. 115/135)

But the marquise has the opposite reaction:

> You have made the Universe so large, *says she*, that I know not where I am, or what will become of me; what is it all to be divided into heaps confusedly, one among another? Is every Star the centre of a Vortex, as big as ours? Is that vast space which comprehends our Sun and Planets, but an inconsiderable part of the Universe? and are there as many such spaces, as there are fix'd Stars? I protest it is dreadful. (Pp. 114–15/134–35)

Fontenelle had just reestablished anthropocentrism—in relation to one solar system, our own—but this confidence is now shaken once again. This passage provides the cue for Bodmer's aesthetic of the sublime.[58] But it is characteristic of Fontenelle that here the over-whelming effect of the many worlds is *not* overcome (as it is in the aesthetics of the sublime). On the contrary, the marquise now, more than ever, loses all faith that human action or thought can have any meaning in such a universe:

> We our selves . . . must confess, that we scarce know where we are, in the midst of so many Worlds; for my own part, I begin to see the Earth so fearfully little, that I believe from henceforth, I shall never be concern'd at all for any thing: That we so eagerly desire to make our selves great, that we are always designing, always troubling & harassing our selves, is certainly because we are ignorant what these Vortex's are; but now I hope my new lights will in part justifie my

[58]*Critische Betrachtungen über die poetischen Gemählde der Dichter* (Zurich, 1741), pp. 213–14.

laziness, and when any one reproaches me with my carelessness, I will answer, *Ah, did you but know what the fix'd Stars are!* (Pp. 116/136)

It is also significant that the ambivalence in Fontenelle between feeling liberated and feeling overwhelmed—like the ambivalence, indicated earlier, between the destruction and reintroduction of anthropocentrism—presupposes the abandonment, or at least questioning, of the religious view of the universe. Such a mentality can no longer find security in an omnipotent God. Addison, an enlightened Christian, was able (in the 565th issue of the *Spectator*) to overcome his cosmic "horror" by retreating to the safe haven of God's omnipresence and omniscience; the *philosophe* Fontenelle pointedly seeks no such refuge. It would accord ill with the concealed antitheological tendency of his writing; and indeed the quoted passage about the liberating effect of the boundless cosmic perspective is immediately followed by an allusion to the creative power, not of God, but of Nature: "Nature hath spar'd no cost [in producing the universe]." There are many other such allusions: in one instance, the marquise states on the third evening that it is "Nature" who has made the inhabitants of the various planets quite different from one another (pp. 82/95); a second instance occurs in the preface (pp. 7/9). Yet Fontenelle declares in the same preface that he does not wish to offend against "religion" (pp. 6/8); is he, then, not challenging the idea of the Creator-God and his Providence? Fontenelle's declaration may be a hypocritical one, made with the Catholic censor in mind. But this shows all the more clearly that adopting a pluralistic conception of the universe no longer merely leads to conflict with specific articles of Christian dogma, as was generally the case in the seventeenth century, but threatens *any* metaphysical world view: it opens the door to philosophy of a mechanistic and naturalistic kind.

In Gottsched's translation of Fontenelle, which went through many editions, this third ambivalence (God or Nature) stands out clearly: in the text, Gottsched allows Fontenelle to refer to the creative power of Nature, but in the notes he praises God as the Creator of the plurality of worlds.[59] An early English translator of the *Entretiens*, Aphra Behn, was more consistent in this respect. In the foreword to *A Discovery of New Worlds* (1688), she remarks that Fontenelle "ascribes all to Nature, and says not a Word of God Almighty, from the Beginning to the End; so that one would almost take him to be a Pagan." Thus, while Gottsched saw it as his duty to claim Fontenelle for

---

[59]*Gespräche von mehr als einer Welt*, e.g., pp. 119, 141.

the metaphysical view of the world, the naturalistic tendency of the *Entretiens* is sufficient cause for Aphra Behn to reject the idea of inhabited planetary worlds altogether: she does this both in the preface to her Fontenelle translation and in her dramatic satire *The Emperor of the Moon* a year earlier. Her compatriot and colleague Horace Walpole could not bring himself to do this, and—as Alexander Maxwell was admonishingly to recall even in the nineteenth century[60]—it was Fontenelle's *Entretiens* that destroyed Walpole's faith in revealed truth. These small sidelights vividly illustrate how the idea of the plurality of worlds, which for many was the gospel of God's omnipotence and wisdom, could imperceptibly turn into a breeding ground for atheism and secularization. Anyone who, like Fontenelle, justifies the plurality of worlds on scientific grounds and not on grounds of theological teleology or the theological principle of plenitude (which Lovejoy claims, wrongly, to find in Fontenelle)[61] runs the risk of setting up a purely mechanistic cosmology, as Cyrano had already done, not ambivalently but openly and exultantly. A recent major study of the plurality idea in relation to the history of science takes the view that for Fontenelle, the diversity of the beings inhabiting the many worlds redounds to the greater glory of God; but this is wholly unsupported by the text.[62]

It is significant, however, that Fontenelle disguises his leaning toward a mechanistic view of the universe—which follows only too naturally from his Copernican-inspired analogical reasoning—by using teleological expressions. Teleology—thinking in terms of "design"—is of course the typical means of justifying belief in the plurality of worlds in the Age of Enlightenment. Thus Andreas Ehrenberg, mentioned above, says in his *Curiöse und wohlgegründete Gedancken* (Jena, c. 1710) that while it has often been "stated that men and other creatures are to be found on the planets," the "motivating reasons" for this view have been "thought out and brought to light only recently." And when he himself sets them forth (p. 19) we find that these reasons are the teleological ones relating to God's purpose in creating the planetary bodies (and not, for instance, arguments based on plenitude). The Christian physicotheologians also thought in teleological terms. They believed that everything that God created, whether blades of grass or insects, volcanoes or planets, was solely

[60]See *Monthly Magazine* (1798:2), 116, and Alexander Maxwell, *Plurality of Worlds*, 2d ed. (London, 1820), p. 21 (1st ed., 1817).

[61]Lovejoy, *Great Chain of Being*, p. 132. He takes Fontenelle's phrase "fécondité de la Nature" (Calame, p. 161) to be a reference to plenitude.

[62]Dick, *Plurality of Worlds*, p. 124.

for the benefit and use of man on "Earth." Writers from Descartes and Cyrano onward—indeed from Galileo onward (see chap. 2, sec. 6 above)—had ridiculed this anthropocentric form of teleology. The teleology of the pluralists, however, does not lay itself open to such mockery: according to Ehrenberg and Schudt, and before them Kepler, Wilkins, and Borel, God created the heavenly bodies for the benefit of the humankinds dwelling on them, not for us. "It is more suitable to the wisdom, power, and greatness of God," says John Locke, "to think that the fixed stars are all of them suns, with systems of inhabitable planets moving about them, to whose inhabitants he displays the marks of his goodness as well as to us; rather than to imagine that those very remote bodies, so little useful to us, were made only for our sake."[63] In the course of the century, the Christian physicotheologians did gradually accept this pluralistic and nonanthropocentric teleology; but they did not find it easy, as is evident from Abbé Noël Antoine Pluche's *Spectacle de la nature* (1732–42), probably the most popular work of the physicotheological school in the eighteenth century. He sees Copernicanism as a temptation to "irreligion," because man can no longer believe that the heavens were created for him, when there is the "magnificent suspicion" that other worlds have their own inhabitants. Pluche extricates himself, however, by what amounts to sleight of hand: even if the stars are not made for us, we still make use of them, "and that is the same thing"![64] God's acts of beneficence are still for us, even if others share in them—just as the Parisians are favored by the kings of France, who open the Tuileries and Luxembourg gardens to them, even though these are also enjoyed by the inhabitants of those palaces and even by people from farther afield (4:520). In this sense, man still remains at the "center" of God's concern, and he is a "false philosopher" who "from suspecting the existence of the plurality of worlds straight away concludes that he is no longer the center of the excellent arrangement of this one, and who in multiplying them imagines that he can lose himself in the crowd, escape from the goodness of God, and discharge himself of the burden of gratitude" (4:522).

So Fontenelle too uses teleological language. But what he has in mind is not Pluche's "geocentric" reinterpretation of the plurality of worlds, nor the usual pluralistic teleology as put forward for instance by Ehrenberg. Instead, as one would expect in view of his ambivalence with regard to God and Nature, it is Nature and not God who carries

[63] *Elements of Natural Philosophy*, chap. 3, in *Works* (London, 1823), 3:309.
[64] Vol. 4 (Paris, 1787), pp. 519–20. Cf. Descartes, chap. 3, sec. 5 above.

out purposes or intentions (pp. 80/92, 161), and Nature who has a plan, a "great design" (pp. 23/27–28). The question is to what extent such semipoetic formulations really express a purely mechanistic conception. Calame judges correctly that "beneath the teleological language [*la langue finaliste*], Fontenelle is essentially a mechanist" (p. xxxix).

Fontenelle, the amateur scientist, and Christian Huygens, the specialist in practical and theoretical astronomy, meet on the common ground of the belletristic philosophy of the plurality of worlds. There could be no clearer indication of how broad the appeal of this idea had already become by the end of the seventeenth century. Like the *Entretiens*, the *Kosmotheoros* (The Hague 1698) is primarily a work of literature rather than of science, though science provides the basis for both. The second part of the book with its imaginary journey to the inhabited planets of the solar system is introduced very much in the manner of science fiction (pp. 85–86); and the astonishing number of editions and translations suggests that the *Kosmotheoros*, with its sometimes almost chatty style, was probably read as science fiction rather than as a scientific study, which could not have attracted such a large readership. Lessing read the book in his school days, and he will not have been alone in this.[65]

Huygens's readers will have found many elements in the book reminiscent of Fontenelle, in particular the basic premise that he shares with him: the idea that the planets are inhabited is a reasoned inference from the analogy that, in the Copernican world view, exists between the Earth and the other planets and can also be assumed to exist between the sun and the fixed stars. Huygens too believes that any objection to this analogical reasoning can be disarmed by the "why not?" line of argument (pp. 16–17, 67), and for him, too, teleology—present on almost every page of his book—provides the ultimate support for the analogical reasoning. (This is teleology in its anti-anthropocentric form: it is not for the descendants of Adam that the planets and stars with their plants, animals, and so on were created—most certainly not those invisible to the naked eye—but for their own rationally endowed inhabitants; see for instance pp. 32–33.) Unlike Fontenelle, however, Huygens only very occasionally represents the

[65]On the printing history of *Kosmotheoros* in Latin, English, French, and German see Dick, *Plurality of Worlds*, pp. 134–35 and 221, n. 13. See David Knight, "Celestial Worlds Discover'd," *Durham University Journal* 58 (1965), 23–29; Dick, *Plurality of Worlds*, pp. 127–35, discusses Huygens's position in the history of science. On Lessing's reading of Huygens, see *Sämtliche Schriften*, 3d ed., ed. Karl Lachmann and Franz Muncker, vol. 5 (Stuttgart, 1890), p. 66.

creative principle that acts with intention and gives a purpose to the whole as a mechanical "Nature" (e.g., p. 10); in the overwhelming majority of instances it is God or Divine Providence whose wise ordinance Huygens emphasizes in opposition to the belief in chance of Atomists such as Democritus and Descartes (e.g., pp. 11, 18–19). The *Kosmotheoros* has for this reason even been called Huygens's most religious work.[66] It certainly is the one that most fully expresses the undogmatic piety of the Enlightenment that makes the plurality of worlds into a new gospel which says that the wisdom and goodness of the Creator are venerated not only by us on Earth but also by the rational beings on other planets; for, by the way he has arranged the universe, God has provided for the happiness of them all (pp. 127–28).

Another departure from Fontenelle, with more far-reaching consequences, is the way in which Huygens visualizes the rational creatures in space. Fontenelle imagined them to be of almost infinite diversity and claimed to regard them as nonhuman. Huygens takes exactly the opposite view (though, as we shall see, this does not enable him to avoid philosophical problems). For Huygens, the principle of the diversity of nature is decidedly overshadowed by that of its essential unity. He therefore assumes—thus illustrating the wide range of opinion within the scientifically inspired philosophy of plurality—that the creatures on other planets in our solar system and in systems based on the fixed stars correspond exactly to those familiar to us. In particular he imagines the creatures endowed with reason to be as much like us "as one egg is to another."[67] If the planet Earth is not unique in the universe, then neither, according to his rigorously precise analogical reasoning, are its rational inhabitants. But Huygens—the representative, among the pluralists of his day, of the land of democracy—adds another and if anything more convincing argument, a sort of cosmic egalitarianism: why should God in any way give us preferential treatment (p. 33)? Huygens repeatedly deflates the "pride" of mankind's collective egocentricity: we are not more worthy of any privileges than other humankinds (e.g., p. 45). The ultimate effect of this is to attribute to the Creator a sort of egalitarian sense of justice that is seen as the essence of his constantly extolled wisdom: God is made to appear as an enlightened constitutional monarch who does not grant a monopoly to one group. There is a conflict

[66] A. E. Bell, *Christian Huygens and the Development of Science in the Seventeenth Century* (London, 1947), p. 202.
[67] Gottsched, *Gespräche von mehr als einer Welt*, p. 116n.

between the suggestion that the nature of God can be wholly grasped in this way and the hidden God of Calvinist Protestantism Huygens also represents, with personal conviction.

Be that as it may, it follows for Huygens that although the planet-worlds differ so greatly in climate and physical conditions, those of their inhabitants that are endowed with reason are all human beings like ourselves, with practically all the same achievements (p. 85). Huygens knows that his assessment of the relationship between "us" and "the others" differs from that of all his predecessors; they all believed that any inhabitants of the planet-worlds would be "not, indeed, men like ourselves" but some other kind of creatures possessing reason (p. 32). Huygens, on the contrary, postulates that in this respect there is no difference between the "regions of the planets" and here ("eodem modo se habere in Planetarum regionibus atque hic"—pp. 40–41). The complete similarity between the terrestrial and other human beings follows from the equality of the reason with which the Creator, unfailingly just, must have equipped the highest beings in every place (pp. 37–38). This premise leads to the assumption that all humankinds share not only the same moral ideas but also essentially the same senses; for without these, reason would be useless, life would be unattractive, and there would be an inconceivable inequality of pleasure, benefits, and happiness. Without the sense of touch, for instance, the human beings on a particular planet would constantly be injuring themselves (p. 43). And possession of the same senses by all humankinds has all sorts of other consequences. Since the sense of sight—in concrete terms, the eye—is made for the purpose of observation (p. 44), it follows that not only do all extraterrestrials have upright posture, which facilitates observation particularly of the heavens, and thus have the same form as we (p. 63), but they are also versed in astronomy and as far advanced in it as we are (pp. 52–59). This again presupposes the ability to make the instruments necessary for observation, and also the art of writing to record the findings made—so they must have hands, and so forth. Huygens is so intent on demonstrating that the rational beings on all planets are alike that he rejects as implausible even the speculation (familiar for instance from Nicholas Hill) that other rational beings are larger or smaller than we, according to the size of their planet (pp. 48, 66). That in other worlds a rational mind could inhabit a body different from ours cannot, he says, be logically disproved, but it is an idea that he cannot entertain without a feeling of horror (non sine horrore) and that he therefore repudiates (p. 65). In short, in science and inventions, social institutions and aesthetic sensibility, morality and mathematics, music and architecture—and, of course, happiness—the Martians, Venusians, and the rest are not inferior to us, and we are not "better" than they.

This cosmic "égalité," carried to extreme lengths—the extraterrestrials even have "theater" (*spectacula*, p. 68)—not surprisingly met with resistance even (or especially) from the pluralists among Huygens's critics, for instance Ehrenberg, Schudt, and Bode.[68] But problems are apparent not only in Huygens's conclusions but in the logic of his argument. These are, as with Fontenelle, philosophical problems, but of a quite different kind. Precisely for this reason, Huygens's attempt affords a valuable insight into the view of man that lies behind his spirited support for the plurality of inhabited worlds.

Fontenelle explicitly humbled the anthropocentric pride of the Christian who thinks that the whole universe was created for him, but he still took refuge in a secular kind of anthropocentrism, by declaring the extreme life-forms on other planets to have fewer advantages and by flatteringly presenting man on our planet as the golden mean. This solution is not open to Huygens, as he has from the outset rejected the idea of *diverse* human creatures in the universe. When he makes fun of theological anthropocentrism (the Almighty, he says, created the stars so that the descendants of Adam and Eve should be able to observe them through the telescope, p. 8), and when he also rejects Fontenelle's arrogance (the suggestion that we are in any way "better," "happier," or more privileged than the extraterrestrials), his brand of anti-anthropocentrism runs the risk of conveying precisely that impression which will be a nightmare to Addison and Pluche— a picture of man lost in the anonymity of the cosmic collective.

In fact, however, Huygens's anti-anthropocentrism, too, contains its own dialectic, but of a different kind from that which appears in Fontenelle. Huygens's cosmology in reality amounts to a celebration of the greatness of man. For of course man—enlightened man—is for Huygens the highest conceivable form of life. The corollary of his oft-repeated formula that we are not more highly organized or advanced than the "others" is naturally that they are not superior to us either, that there can in fact be no higher form of intelligent life! Huygens's theory, in sharp contrast to those of some eighteenth-century pluralists, means that man cannot feel threatened and downgraded by the higher degree of reason possessed by other planet dwellers (p. 36).

[68]Ehrenberg, *Curiöse und wohlgegründete Gedancken von mehr als einer bewohnten Welt* (Jena, n.d., c. 1710), pp. 149–50; Schudt, *De Probabili Mundorum Pluralitate* (Frankfurt, 1721), p. 53; Bode in a note to his Fontenelle translation, *Dialogen über die Mehrheit der Welten*, p. 188.

Homo sapiens is still the glory of the universe. The peopling of the universe with beings of *this* kind thus reveals itself as the expression of an imperialistic urge for expansion on the part of self-confident human reason as "we" know it: it proves itself by its universal dominance, which is unchallenged and cannot be challenged by any higher beings. A subtle form of anthropocentrism and "pride" is reestablished under the guise of the humbling of that same pride (e.g., pp. 53–54). There could hardly be a more grandiose way of overcoming man's fear of his uniqueness in the universe and his other fear of sinking from view in the anonymity of the cosmic collective: man is the measure of all things in the whole universe! But it is questionable whether this maneuver is any more convincing than Fontenelle's attempt to pull himself out of the bog by his own bootstraps.

We are also bound at this point to consider what we are to think of Huygens's teleology. Clearly he is not to be suspected of bad faith in his expressions of belief in the wisdom of the Creator of the many worlds. Nevertheless, there is no hint as to what might be the point of an infinite multiplication of essentially identical worlds. Writers from Saint Thomas Aquinas onward had maintained over and over again, especially in the seventeenth century, that it was an absurdity.[69] Even the eighteenth century still felt that this was impassable territory, and an astronomical textbook actually points this out. In his monumental *Astronomie*, Jérôme de Lalande argues in favor of the plurality of worlds and—*as a scientist*—specifically aligns himself with Huygens.[70] As a philosophically minded thinker, however, he feels compelled to consider the implications. The dogmatic objections are easily disposed of with a reminder that the omnipotence and glory of God are infinitely enhanced by the many worlds. But the philosophical complications remain. Does a teleology that is applied to an infinite multiplicity of planets and that postulates the *same* kind of intelligent creatures everywhere make any sense, or is it invalidated precisely by that multiplication?

> The only problem that a philosopher may have with the existence of the inhabitants of so many million planets is the obscurity of the final causes, which it is very difficult to accept when one sees the errors into which the greatest philosophers—Fermat, Leibniz, Maupertuis, etc.—have fallen when they have tried to use these final causes, or metaphysical suppositions of claimed connections between the effects

[69]Grant McColley, "The Seventeenth-Century Doctrine of a Plurality of Worlds," *Annals of Science* 1 (1936), passim, esp. p. 398.
[70]Nouvelle édition (Paris, 1771), 3:453; see 1:215.

that we know and the causes that we assign to them, or the ends for which we believe them to exist. (3:454)

Huygens does not discuss this matter. For if he did, he would also have to reopen the whole Pandora's box of theological and dogmatic questions that, like other enlightened philosophers of plurality, he had so decidedly closed and pushed aside (pp. 7–8). So this dilemma demonstrates once again how, as thinkers turned away from dogmatic issues, they were confronted by new, philosophical problems they were not fully able to cope with. For all the differences in their treatment of the subject (which through their works became something of a fashion in philosophy), Huygens and Fontenelle have this at any rate in common.

3. In Newton's Train: Bentley, Bolingbroke, Leibniz, Wolff, Wright, Kant, Herder, Bonnet, Swedenborg, Bernardin de Saint-Pierre, Bode, and Others

The problematic conclusions drawn by Fontenelle and Huygens were philosophical inferences drawn from the Copernican-Galilean world model; both of them adhered (Huygens more critically than Fontenelle) to the new formulation of that system by the corpuscular physics of Descartes, according to which tiny particles of matter throughout the universe form "vortices," at the centers of which are the stars. Neither Fontenelle (in later editions) nor Huygens took as his basis the cosmologic system that made the vortex theory untenable, Newton's theory of gravitation. This theory, however, provided the Copernican conception of the universe with a more durable foundation and thereby gave a new impetus to the plurality idea, especially in literature.

The *Philosophiae Naturalis Principia Mathematica* (1687) played a lasting role—not so much directly as through popularizations—in molding the general consciousness. The reason for this was that Newton succeeded in identifying with mathematical precision the physical laws that guaranteed the ordered motion of the heavenly bodies in the universe—first in our own solar system but then, analogously, in all other solar systems. For although he disliked the formation of hypotheses, Newton had in the famous "Scholium Generale" of the second edition of the *Principia* (1713) admitted at least the possibility that the fixed stars were suns, though he showed not the slightest interest in the matter.[71] He had even, in the *Optice* (1706), somewhat mysteriously conceded that it must be admitted that God might have created worlds of another kind ("diversa specie") elsewhere in the universe (p. 347). This suggestion that the Newtonian model might possibly be multiplied, without necessarily leading to absolutely identical systems, was the aspect of Newton's world view that continually fascinated poets, particularly in the eighteenth century. It is no coincidence that the apotheosis of Newton, which began in his own lifetime, took place above all in lyric poetry. Again and again poems about Newton himself and also poems taking their themes from the *Principia* speak of the "worlds beyond worlds" and the "millions of suns with planets circling round," everywhere governed by the same laws, of which the "Prince of the new philosophy" had made people aware.[72]

Newton himself had avoided expressing even a suspicion that other planets, in our solar system or in others, might be inhabited. When, in the "Scholium Generale," he let himself be lured away from his ascetic distaste for any but purely scientific thought in order to defend himself against the charge of atheistic mechanism, he did speak of a continuously active, indeed "ruling," God as the originator of the world-machine; he maintained too that it was possible to know of his Being at least so much: that in his creation of the world he pursued certain aims and purposes, "the wisest and best structures of things and final causes" (*sapientissimas & optimas rerum structuras & causas finales*, 2:763). Newton thus drew a dividing line between himself and Descartes's neo-Atomism, which was suspected of being atheistic. But not only did Newton say nothing about the content of this teleology—about what *were* God's purposes; he also did not suggest by so much as a word that they might also relate to the other planets (and their possible humanlike inhabitants, about whom he was careful to say nothing in print). This conclusion was, however, eagerly drawn by his many followers and pupils. A scientific expert like Alexandre Koyré was of the opinion that "doubtlessly" Newton himself shared his disciples' belief in plurality.[73] This must, however,

[71]*Principia*, ed. Alexandre Koyré and I. Bernard Cohen (Cambridge, Mass., 1972), 2:760.

[72]See William Powell Jones, "Newton Further Demands the Muse," *Studies in English Literature* 3 (1963), 287–306, quotations pp. 291–93; Meadows, *High Firmament*, chap. 6; Crum, *Scientific Thought in Poetry*, chap. 4; Bernhard Fabian, "Newton und die Dichter im achtzehnten Jahrhundert: Popularisierung durch schöne Literatur," *Medizinhistorisches Journal* 12 (1977), 309–24; Fritz Wagner, *Isaac Newton im Zwielicht zwischen Mythos und Forschung* (Munich, 1976), pp. 106–24; Dick, *Plurality of Worlds*, chap. 6.

[73]*From the Closed World to the Infinite Universe* (New York: Harper Torchbooks, 1958), p. 189.

remain mere conjecture, since Newton makes no such statement either here or in his theological writings.

The first and certainly one of the most influential of Newton's popularizers, who saw in his physical conception of the universe the basis of a contemporary gospel of the plurality of inhabited worlds, was the classical philologist Richard Bentley. His philosophical position can be gauged from the fact that he was chosen in 1692 to give the first of the Boyle Lectures, which, under the will of the physicist Robert Boyle, were intended to defend Christianity against all forms of unbelief, including deism. In the sixth, seventh, and eighth of these lectures, titled *The Folly and Unreasonableness of Atheism* (1693), Bentley demonstrates his thesis on the basis of Newton's cosmology: a universe ordered by gravity can only have sprung from the wisdom of a personal God. And Bentley sought confirmation of his ideas on the philosophical meaning of Newtonian physics—if not approval of his own actual text—from the master himself. Newton wrote him four letters in which he elucidated the religious convictions that, though unexpressed, underlay his system,[74] much as he did later in the "Scholium Generale." The motions of the heavenly bodies, he said, could not be explained by natural—mechanical or accidental—causes, but only by "the counsel and contrivance of a voluntary Agent"—a sovereign, "supernatural" Creator-God as postulated by theism (p. 204; see also pp. 207, 210). The plurality of worlds is touched on only briefly and inconclusively: there may perhaps have been other world systems before the present one, Newton says, but it seems to him absurd to suggest that new systems could emerge from old ones without "mediation of a divine power." He does not, however, make it clear whether such a process would be more plausible if one *did* assume divine intervention (p. 211).

Bentley himself is more confident. Seldom has the anti-anthropocentric view of God's purpose been stated more decisively. For him, it is nonsense to believe that the whole visible universe was created "for us," as many educated people in the seventeenth century, and some physicotheologians until far into the eighteenth century, took for granted. With even more stringent logic than John Ray in his highly successful, indeed popular, book, *The Wisdom of God Manifested in the Works of the Creation* (1691),[75] Bentley argues in *The Folly and Unreasonableness of Atheism*:

[74]Letters published in *The Works of Richard Bentley*, ed. Alexander Dyce, vol. 3 (London, 1838), pp. 200–215. This volume, to which all page references relate, also contains Bentley's *The Folly and Unreasonableness of Atheism*.

[75]Fourth ed. (London, 1704), pp. 17–21, 202–16. His illogicality lies in the fact

Who will deny but that there are great multitudes of lucid stars even beyond the reach of the best telescopes; and that every visible star may have opaque planets revolve about them, which we cannot discover? Now, if they were not created for our sakes, it is certain and evident that they were not made for their own. For matter hath no life nor perception, is not conscious of its own existence, nor capable of happiness, nor gives the sacrifice of praise and worship to the Author of its being. It remains, therefore, that all bodies were formed for the sake of intelligent minds: and as the earth was principally designed for the being and service and contemplation of men, why may not all other planets be created for the like uses, each for their own inhabitants which have life and understanding? (P. 175)

And this declared defender of Christianity now adds his voice to the chorus of those who consider questions of theological dogma to be obsolete, when he dismisses the questions about descent from Adam and redemption by Christ as "frivolous disputes" (p. 175). But his reason for saying this is highly significant; for it presents the problem in a light liable to upset even (or particularly) his enlightened hearers. For Bentley, the dogmatic problems do not arise because it is not certain that the rational creatures on other planets are also human beings. On the contrary, man is only one instance of a rational form of life. It is conceivable that God, in the infinite fecundity and freedom of his creative power, has not only encased "human" minds in nonhuman bodies but also created minds with a lower or higher degree of reason than that of humans and provided them with a human body (just the sort of suggestion that filled Huygens with horror!):

Now, God Almighty, by the inexhausted fecundity of his creative power, may have made innumerable orders and classes of rational minds; some in their natural perfections higher than human souls, others inferior. But a mind of superior or meaner capacities than human would constitute a different species, though united to a human body in the same laws of connexion; and a mind of human capacities would make another species, if united to a different body in different laws of connexion. . . . God, therefore, may have joined immaterial souls, even of the same class and capacities in their separate state, to other kinds of bodies, and in other laws of union; and

that despite all his polemic against theological anthropocentrism, he (like Addison, incidentally; see the Fontenelle discussion, sec. 2 above) takes comfort from the Christian paradox expressed in Psalm 8, that God is nevertheless mindful of man. *Wisdom of God* went through more editions; the 10th came out in 1735.

from those different laws of union there will arise quite different affections, and natures, and species of the compound beings. So that we ought not upon any account to conclude, that if there be rational inhabitants in the moon or Mars, or any unknown planets of other systems, they must therefore have human nature, or be involved in the circumstances of our world. (P. 176)

For a professional theologian (Bentley was ordained in 1690 as a deacon of the Church of England and became a doctor of theology in 1696), this is indeed an astonishing statement. What constitutes a human being is not the human mind by itself, nor the human body, but the combination of both. That on the basis of this definition there may be human minds in nonhuman bodies is not so disturbing as the idea, obviously regarded as a genuine possibility, that there may be higher "intelligences" in either human or nonhuman bodies on other planets. For Bentley is clearly thinking not of the traditional hierarchies of angels but of the rational beings whose existence is inferred by analogy from the similarity of the planets in the heliocentric world systems. The nightmare of enlightened reason, that rational man may be only a quite inferior being, is given support by this enlightened churchman—who makes not the slightest effort to allay the fear induced by this thought. His praise of the Newtonian God as the "intelligent and good Being that formed it [the world] in that particular way out of choice and design" (p. 177) has, indeed, that overtone of exaltation that drowns out fears. The question What is man?—which Bentley himself asks more or less rhetorically and answers only in formal terms: a mind with a certain degree of reason in a certain kind of body—takes on an unexpectedly sinister ambiguity.

Whereas in Bentley, this ambiguity appeared unexpectedly and was perhaps unintended, soon afterward the same disturbing inference is drawn deliberately and pointedly by Lord Bolingbroke, the statesman and philosopher who belonged to the circle of Swift and Pope. The whole argument of his "Fragments and Minutes of Essays" (not later than 1744) is directed against the "chorus" of "theistical philosophers" and "divines" for whom human reason, as the gift of God which distinguishes man, represents the highest level of understanding.[76] Bolingbroke concedes that man is the most intelligent creature on his own Earth, but does it follow that the all-wise Creator made our Newtonian planetary system for man and for his happiness?

---

[76]*Works* (London, 1809), 8:168. Page references in the text are to this edition. On the dating, see 7:278.

Not at all! Bolingbroke looks at it the other way round: the fact that we enjoy congenial conditions on Earth does not mean that the Earth (let alone the system of the planets and stars) was made for us, but that we are made to suit the Earth. This immediately prompts the question What sort of beings are made to suit the other planets? Bolingbroke states quite bluntly that this way of looking at things—a Copernican revolution in itself—arises from "modern discoveries in astronomy"; in what sense, he makes perfectly clear:

> We cannot doubt that numberless worlds and systems of worlds compose this amazing whole, the universe; and as little, I think, that the planets, which roll about our sun, or those which roll about a multitude of others, are inhabited by living creatures, fit to be the inhabitants of them. When we have this view before our eyes, can we be stupid or impertinent and vain enough to imagine, that we stand alone or foremost among rational created beings? We who must be conscious, unless we are mad, and have lost the use of our reason, of the imperfection of our reason? Shall we not be persuaded rather, that as there is a gradation of sense and intelligence here from animal beings imperceptible to us for their minuteness, without the help of microscopes and even with them, up to man, in whom, though this be their highest stage, sense and intelligence stop short and remain very imperfect; so there is a gradation from man, through various forms of sense, intelligence, and reason, up to beings who cannot be known by us, because of their distance from us, and whose rank in the intellectual system is even above our conceptions? This system, as well as the corporeal, that is the whole physical or natural system, for such the two properly are, must have been alike present to the Divine Mind before he made them to exist. (Pp. 173–74)

Such is the boldness and vigor with which Bolingbroke subdues the pride of mankind, as Fontenelle would have said. If man were to meet the extraterrestrials envisaged by Bolingbroke, he would, in the familiar analogy, be the Indian and no longer the European. But in sharp contrast to the urbane and confident Fontenelle, who used that analogy, this angry opponent of church Christianity has no intention of surreptitiously restoring *Homo sapiens* to preeminence. What is more, he also makes the connection with the question of theodicy, which so preoccupied his age: it is, he says, a contradiction for his theological opponents to say in the same breath that man and his world are in a bad way and yet that the whole universe was created "for the sake of mankind" (pp. 174–75). There is no suggestion that the hierarchy or "gradation" of beings could itself offer some metaphysical consolation (be it that the imperfection of the Earth could be justified, as it later was by Kant, as one necessary level in that gradation [see sec. 1 above, at n. 33] or, as the upholders of palingenesis thought, that our imperfection could be made acceptable by viewing it as a transitional phase on the way to higher forms of being). On the contrary, Bolingbroke seems, astonishingly, already to sense the fundamentally problematic nature of the "enlightened" theodicy. Ehrenberg was at about the same time expressing the popular rationalist consensus view that our reason exactly corresponds to the divine reason,[77] or rather that it *must* correspond to it, since otherwise all philosophy would be doomed to failure from the outset. Bolingbroke radically challenges this basic tenet, which is that of all physicotheology but also of atheism, when he says that both the theologian and the atheist judge by moral attributes of God which they assume a priori. This, he says, is an absurdity, "since their ideas of these attributes are very human ideas, applied arbitrarily to the divine nature, not founded in any knowledge of it that they have really" (p. 175). This is, in a nutshell, Goethe's reminder to any theology that human and divine reason are "two very different things."[78] That being so, the question of theodicy remains unanswerable.

This, then, is Bolingbroke's contribution. He expresses what Bentley probably could not admit to himself.

It was a notable philosophical achievement on Bentley's part to have led up to such questions. For it was clearly more usual for Newton's followers to maintain the existence of a plurality of worlds in the planetary systems held together by the force of gravity, and perhaps to praise the majesty of God made manifest by this plurality, but then to evade the question of how man might compare with his extraterrestrial counterparts. This is true for instance of Nehemiah Grew, William Whiston, John Keill, Pierre Louis Moreau de Maupertuis, and especially William Derham, in his *Astro-Theology: or, A Demonstration of the Being and Attributes of God, from a Survey of the Heavens* (1715), which went through many editions and became a kind of home reference work for the educated and half-educated. The book is

---

[77]Ehrenberg, *Curiöse und wohlgegründete Gedancken*, p. 160.
[78]Remark to Eckermann, 15 October 1825; in Johann Wolfgang von Goethe's *Conversations with Eckermann*.

introduced by a special discourse, "Plurality of Worlds," with a Newtonian basis, as a sort of key to what follows. We learn from this that for Derham too the whole universe is not there for the sake of the needs and convenience of man:[79] he too extends teleology to include the planetary worlds, which, because they are habitable, must be inhabited.[80] But Derham, an Anglican clergyman, dismisses curiosity as to the kind, status, and nature of the "proper Inhabitants" of the planets both within and beyond our solar system, as a useless exercise of the imagination. He briefly refers to Huygens's reflections—as merely amusing speculations—before he too lifts his voice in the obligatory hymn praising the Creator (p. 41):

> But now the next Question commonly put is, What Creatures are they inhabited with? But this is a difficulty not to be resolved without a Revelation, or far better Instruments than the World hath hitherto been acquainted with. But if the Reader should have a mind to amuse himself with probable Guesses about the Furniture of the Planets of our Solar Systeme, what Countries 'tis probable are there, what Vegetables are produced, what Minerals and Metals are afforded, what Animals live there, what Parts, Faculties and Endowments they have, with much more to the same purpose; he may find a pleasant entertainment enough in the great Mr. *Christian Huygens's Cosmotheoros* and some other Authors that have written on the Subject. To which I shall refer him, rather than give either him or my self any farther trouble about these matters, which are merely conjectural. (Pp. liv–lv)

A distant echo of this restraint in pursuing the teleological line of thought can still be heard in James Ferguson's simply written textbook *Astronomy Explained Upon Sir Isaac Newton's Principles* (1756): "From what we know of our own System, it may be reasonably concluded that all the rest are with equal wisdom contrived, situated and provided with accommodations for rational inhabitants."[81] But as to the nature of the inhabitants of our neighboring planets and those of the satellites of other suns, he is content merely to conjecture that they acknowledge and venerate God as we do: "These similarities leave us no room to doubt, but that all the Planets and Moons in the System

---

[79] I find inexplicable John Dillenberger's opinion that "for Derham, the whole universe was still geared to the necessities and comfort of man" (*Protestant Thought and Natural Science* [Garden City, N.Y., 1960], p. 150).

[80] Third ed. (London, 1719), pp. xlvii, liv. Page references are to this edition. The ninth ed. came out in 1750.

[81] Sixth ed. (London, 1778), p. 4. Subsequent page reference is to this edition.

[and in other systems] are designed as commodious habitations for creatures endowed with capacities of knowing and adoring their beneficent Creator" (p. 6). Even in the early nineteenth century, William Paley, a major figure in the popularization of physicotheology, shows this same way of thinking (confident in making general assertions but hesitant when it comes to detail), which bears only an apparent resemblance to the genuinely religious use of the topos of inscrutability by, say, Gassendi. In the many editions of his *Natural Theology* (1802), Paley expresses the view that, altogether, astronomy is not the best way to prove the existence of a rational Creator.[82] In particular the idea of a plurality of inhabited worlds does not help at all, in Paley's view; but it is interesting to note that a mere subsidiary clause refers to the inhabitants of the planets, so completely does he take them for granted: "Our ignorance, moreover, of the *sensitive* natures, by which other planets are inhabited, necessarily keeps us from the knowledge of numberless utilities, relations, and subserviences, which we perceive upon our globe" (p. 411). This is all he has to say on this topic. Physiology, botany, and zoology provide more conclusive arguments.

It is in this vague form that the Newtonian-inspired belief in the plurality of worlds was received on the Continent. In particular, it was introduced into the German discussion of the subject by Leibniz—a curious fact, given that Leibniz's own rationalistic conception of the world, untouched by empirical science, already made him a "pioneer of the pluralistic view of the world."[83] Of course, in his famous reference to the "best of all possible worlds," the plural refers to the *un*created worlds present only in the mind of God, which he could have created but, on the principle of choosing the best, did not create. The "monde" that was in fact created is a world in the sense not of Terra—Earth—but of a cosmos or universe.[84] But Leibniz, in so far as he thinks about the universe in astronomical terms, is most definitely a Copernican, and he has no qualms about taking the heliocentric system as a model for the planetary systems of the fixed stars, so that his philosophy can certainly admit a plurality of worlds (in the sense which, from Copernicus onward and still more as a result of Newton's work, was the only one taken seriously until well into the nineteenth century: a plurality of planetary worlds in solar systems). A hint of this idea is already present in the *Nouveaux essais* (written in 1704);

---

[82]Third ed. (London, 1803), p. 409. Subsequent page reference is to this edition.
[83]Zöckler, *Geschichte*, 2:61.
[84]*Essais de théodicée* (Amsterdam, 1710), "Essais sur la bonté de Dieu . . .," § 8. On the following sentence see § 19, quoted below. Quotations are taken from Leibniz, *Theodicy*, trans. E. M. Huggard (London, 1951).

interestingly, it occurs there in association with the old conception of gradations of being, though this is now applied not to the old vertical hierarchy but to the diffuse order of the new science (the problems inherent in this maneuver were outlined in sec. 1 above, at n. 37). The law of "Continuité," we are told in the *Nouveaux essais*, requires that the harmonious gradation of life-forms should reach from the simplest forms to man and continue beyond him, not necessarily on our Earth but on other planets in our solar system or on planets belonging to other solar systems. These forms of life would thus be more perfect than those we know and would not be merely "souls," spirits," or "intelligences" but beings with bodies that, in accordance with their greater spiritual or intellectual perfection, would also be more perfect than those familiar to us (bk. 3, chap. 6, § 12).

In his *Théodicée* (1710), Leibniz discusses somewhat more fully the idea of such a plurality of worlds within God's Creation of the "best" universe, and here his language is wholly that of contemporary astronomical science. Speaking of the "true greatness of the divine State," in the "Essays on the Goodness of God, the Freedom of Man and the Origin of Evil," he writes:

It seemed to the ancients that there was only one earth inhabited, and even of that men held the antipodes in dread: the remainder of the world was, according to them, a few shining globes and a few crystalline spheres. To-day, whatever bounds are given or not given to the universe, it must be acknowledged that there is an infinite number of globes, as great as and greater than ours, which have as much right as it to hold rational inhabitants, though it follows not at all that they are human. It is only one planet, that is to say one of the six principal satellites of our sun; and as all fixed stars are suns also, we see how small a thing our earth is in relation to visible things, since it is only an appendix of one amongst them. It may be that all suns are peopled only by blessed creatures, and nothing constrains us to think that many are damned, for few instances or few samples suffice to show the advantage which good extracts from evil. Moreover, since there is no reason for the belief that there are stars everywhere, is it not possible that there may be a great space beyond the region of the stars? Whether it be the Empyrean Heaven, or not, this immense space encircling all this region may in any case be filled with happiness and glory. It can be imagined as like the Ocean, whither flow the rivers of all blessed creatures, when they shall have reached their perfection in the system of the stars. What will become of the consideration of our globe and its inhabitants? Will it not be something incomparably less than a physical point, since our earth is as a point in comparison with the distance of some fixed stars? Thus since the proportion of that part of the universe which we know is almost lost in nothingness compared with that which is unknown, and which we yet have cause to assume, and since all the evils that may be raised in objection before us are in this near nothingness, haply it may be that all evils are almost nothingness in comparison with the good things which are in the universe. (§ 19)

One significant feature of this fairly detailed passage is that Leibniz, speaking the language of contemporary science, stirs up philosophical problems but does not even begin to discuss them. He says that the extraterrestrials need not be human beings (whether they were or not was one of the bones of contention in the debate on plurality). They are not only, as in the *Nouveaux essais*, more perfect; they are possibly also happier than we are. This pointedly raises the problem of theodicy, but it is not discussed: even if the universe is the best of all possible ones, the world on our planet is obviously not the best of all imaginable ones. But the theodicy debate of the day was, after all, concerned with this world; and the idea at the end of the quotation, that as our planet is infinitely reduced in size by the cosmic perspective, so the evil in this world is reduced to insignificance, is a mere playing with words rather than an argument, and hardly provides a solution. Finally, what of man's rank in the hierarchy of beings postulated by Leibniz? He can clearly no longer, with the confidence characteristic of the Enlightenment, see himself as the embodiment of the highest or divine reason. This will be stated explicitly by Bolingbroke. Leibniz, like most of the Newtonians of the day, says nothing.

The eighteenth-century authors who make Leibniz's ideas the common property of the German reading public still show the same combination of conviction that extraterrestrial beings must logically be assumed to exist and of vagueness as to their nature. Christian Wolff enlists the support of the by-now-standard arguments of analogy and of teleology (*nihil frustra*, "nothing happens to no purpose") to prove on the basis of the Copernican-Newtonian model of the universe that on the planets, but also on the moon and even the comets, there are plants and animals, rational beings, even humans.[85] But having inferred the existence of extraterrestrials, he does not enter into a consideration of their specific characteristics beyond calculating,

[85]Wolff, in his many writings, often returns to the subject. The main passages are paragraphs 488 and 526 in *Elementa Astronomiae* in the third volume of the *Elementa Matheseos Universae*, editio nova (Halle, 1735). The belief that comets are inhabited is particularly stressed in Wolff's *Natürliche Gottesgelahrtheit*, vol. 1 (Halle, 1742), § 446.

in the *Elementa Astronomiae*, the bodily size of the inhabitants of Jupiter. It is a curious passage (§ 527), which attracted some ridicule not only from Voltaire (in a letter to Maupertuis, 10 August 1741) but also from the *Encyclopédie* (12:705). For Wolff calculates, with odd precision, that because of the limited amount of light reaching Jupiter, the Jovians must have larger pupils, and consequently larger eyes, . . . and larger bodies. Wolff's philosophy of plurality does not go beyond this speculative beginning. It is noteworthy, however, that such extraterrestrial anthropology should be included, in all seriousness, in a textbook or manual intended for use by academic institutions, just as in his *Mathematisches Lexicon* (1716) Wolff includes "Selenites" among the technical terms listed.

Johann Christoph Gottsched, in his popularization of the ideas of Leibniz and Wolff in the *Erste Gründe der gesammten Weltweisheit* (1733–34 [9th ed., 1778]) goes a step further in the philosophical exploration of the diverse life-forms on the planets. Coincidentally he too starts with the inhabitants of Jupiter, whose existence he cannot help but assume for teleological reasons, against the general background of Newtonian cosmology. For God will not have created Jupiter's moons "for nothing," "for the benefit of a barren land," but in the interests of the rational beings of that planet, to illumine the night for them. Like Fontenelle, he thinks it probable that these rational beings are different from us; unlike Fontenelle, but like Leibniz and Bolingbroke, he also thinks it probable that they may be superior. Yet, whereas Leibniz steered clear of the consequent philosophical problems, for instance the questions of God's justice and of man's dignity, Gottsched extricates himself somewhat precipitately by means of the idea of the transmigration of souls, which he imagines as a process of perfection taking place on those other planets.[86] Leibniz had not ventured to determine whether there is any relationship between us and those more perfect creatures in other worlds; at most he had hinted that there was a progression toward perfection and that such cosmic dwelling places of the soul might be the stations of that progress.[87] Wolff, for his part, was skeptical of the idea of transmigration.[88] It is therefore not surprising that Gottsched too does not discuss the idea in more

[86]*Erste Gründe, Erster, Theoretischer Theil* (Leipzig, 1733), § § 782–86; *[Zweiter,] Praktischer Theil*, 7th ed. (Leipzig, 1762), pp. 497–503 (not in earlier editions). See Schatzberg, "Gottsched as a Popularizer."

[87]See the passage from *Théodicée* (§ 19) quoted above, and also Leibniz's letter to the Electress Sophie of 4 November 1696, in *Correspondance de Leibniz avec l'électrice Sophie de Brunswick-Lunebourg*, ed. Onno Klopp, vol. 2 (Hanover, n.d.), pp. 14–18.

[88]*Vernünfftige Gedancken von Gott, der Welt und der Seele des Menschen, auch allen Dingen überhaupt*, 7th ed. (Frankfurt and Leipzig, 1738), § 950.

detail. But by mentioning it he introduces a motif that was later to figure prominently in discussion, especially among the followers of Newton, and with good reason. As suggested earlier, the assumption of more perfect beings challenges man to develop a new conception of himself in relation to these other forms of being. The hypothesis of the transmigration of souls offered a solution that was extremely popular in the Age of Enlightenment. It allayed the anxiety caused by the thought of higher life-forms (since the human soul would itself reach this level in future stages of its development on other planets); and, what is more, it made possible a new anthropocentrism (the whole universe with its myriad worlds was intended for this soul).

The hypothesis of the transmigration of souls is most closely linked with Newtonianism by Thomas Wright, one of the most imaginative amateurs among eighteenth-century astronomers. In this respect he presents an interesting contrast to the other English Newtonians, who are not prepared to become involved in specific questions of philosophical anthropology. A cursory glance through his volume of fictional letters, *An Original Theory or New Hypothesis of the Universe* (1750),[89] reveals many quotations from lyric poetry interspersed among the cosmologic diagrams and mathematical calculations, giving the strong impression of a scientist who may be eccentric but is certainly gifted with imagination and who, unlike his revered master, Newton (p. 6), has not set his mind against all philosophical speculation. In the philosophical domain, he is interested in what the universe signifies for man. Of course for Wright, too, the day of naïve anthropocentrism is past; in his teleology the purpose of the heavenly bodies relates not to us but to their own inhabitants (pp. 32–33, 35, 84). Even so, his whole aim is to interpret the structure of the inhabited universe in such a way that it holds a special meaning for man on Earth. In other words, the idea that the other worlds were created for the extraterrestrials is nevertheless made to form part of an overall teleology that is anthropocentric; and this is achieved by means of the transmigration hypothesis.

As Copernicus, and after him, Newton, used mathematical calculation to discover the order of the solar system, so it is Wright's ambition to find an analogous order underlying the whole universe. Where Giordano Bruno and his followers saw an expanse with no

---

[89]Reprint with an introduction by Michael A. Hoskin (London and New York, 1971). Page references are to this edition. See also Hoskin, "The Cosmology of Thomas Wright of Durham," *Journal of the History of Astronomy* 1 (1970), 44–52.

central point, Wright, making physical calculations to satisfy a religious need, sees a spherical structure of gigantic size with a shining or dark body as its absolute center, around which all the stellar masses of the universe gravitate, with their satellites (p. 79). This central point of Creation can only be discovered mathematically, since from our peripheral position, observation gives a distorted impression of the true structural relationships. The center is primarily of religious significance, as the "throne of God": "Should it be granted, that the Creation may be circular or orbicular, I would next suppose, in the general Center of the Whole an intelligent Principle, from whence proceeds that mystick and paternal Power, productive of all Life, Light, and the Infinity of Things" (p. 78). It is the center "where the divine Presence, or some corporeal Agent, full of all Virtues and Perfections, more immediately presides over his own Creation. And here this primary Agent of the omnipotent and eternal Being, may sit enthroned, as in the *Primum Mobile* of Nature, acting in Concert with the eternal Will. . . . Thus in the *Focus*, or Center of Creation, I would willingly introduce a primitive Fountain, perpetually overflowing with divine Grace, from whence all the Laws of Nature have their Origin" (p. 79).

The religious line of thought remains firmly attached to the science of natural laws, and the same is true of Wright's discussion of the "worlds" he assumes exist around this center of the universe. He calculates that there are about 170 million such worlds "of a terrestrial or terraqueous nature," not counting the comets (p. 76). The birth and the perishing of such worlds, with their inhabitants, are daily events like our births and deaths. Since Wright, as an astronomer, believes that all the worlds, without exception, move in orbits around a common central point, he infers, as a religious mystic, that the various species of men on these bodies must be seen as forming a gradation of many levels of perfection—perfection in a narrowly moral sense that, despite the mysticism of the whole conception, verges on the banal. Not merely the orbits of the planets but the levels of perfection are arranged in a predetermined order: the most virtuous rational beings are closest to the hidden center; the most wicked are furthest from it. This is, however, a dynamic, mobile order: the universe with its 170 million abodes is the setting in which souls undergo probation, are perfected, and strive toward "final perfection" (pp. viii, 33), toward the purity of the Divine in the absolute center (pp. 81–84). In this scheme, the individual planets are the setting for punishments and rewards exactly appropriate to every degree of vice and virtue; the soul on its journey of perfection must pass through each one until it reaches the central star (pp. 33, 35). This is the unifying idea in the philosophy of man which parallels the idea of a single center in physics: all human creatures in the universe are incarnations of fundamentally similar souls at various stages of their (moral) development, striving toward their unchanging goal, the throne of God the Father. "To this common Center of Gravitation, which may be supposed to attract all Vertues, and repel all Vice, all Beings as to Perfection may tend" (p. 79). The originally neo-Platonic idea of the ascent of the soul to the source of being has changed into the moralistic conception of a process in which the virtuous are progressively rewarded (p. 81).

What is our place in this cosmic hierarchy of being? Like Leibniz and also Locke,[90] and later also Bonnet and Herder and many others, Wright sees man as being far removed from perfection (p. 50). But unlike, say, Bolingbroke, he sees this not as a reason for bitter scorn and disparagement of man, but as a signal for hope of future perfection to be achieved by interplanetary palingenesis. Through the migration of souls, God in his justice has provided the soul with the possibility of at last coming before his presence and partaking of joy (pp. 33, 35, 84). "Here we may not irrationally suppose the Vertues of the meritorious are at last rewarded and received into the full Possession of every Happiness, and to perfect Joy. The final and immortal State ordain'd for such human Beings, as have passed this Vortex of Probation thro' all the Degrees of human Nature with the supream Applause" (p. 81). The Christian yearning for eternity and the Enlightenment's philosophy of happiness are blended to form a new whole, with science as an additional element. For Wright does not shrink from describing the final goal ("paradise" in theological terms) also in physical, even climatological, terms: the "primogenial Globe," we are told, suffers neither extremely high nor extremely low temperatures (pp. 80–81)! And when Wright, the mystical calculator, moralist, and observer, gives detailed descriptions of the other "worlds" ("of all Formations, suited no doubt to all Natures, Tastes, and Tempers, and every Class of Beings"), he is, with his combination of science and theology, venturing into the domain of much later science fiction:

> Here a Groupe of Worlds, all Vallies, Lakes, and Rivers, adorn'd with Mountains, Woods, and Lawns, Cascades and natural Fountains; there Worlds all fertile Islands, cover'd with Woods, perhaps upon a common Sea, and fill'd with Grottoes and romantick Caves. This Way, Worlds all Earth, with vast extensive Lawns and Vistoes,

[90] *An Essay Concerning Human Understanding*, bk. 3, chap. 6, § 12; bk. 2, chap. 2, § 3.

bounded with perpetual Greens, and interspersed with Groves and Wildernesses, full of all Varieties of Fruits and Flowers. That World subsisting perhaps by soft Rains, this by daily Dews, and Vapours; and a third by a central, subtle Moisture, arising like an Effluvia, through the Pores and Veins of the Earth, and exhaling or absorbing as the Season varies to answer Nature's Calls. Round some perhaps, so dense an Atmosphere, that the Inhabitants may fly from Place to Place, or be drawn through the Air in winged Chariots, and even sleep upon the Waves with Safety; round others possibly, so thin a fluid, that the Arts of Navigation may be totally unknown to it, and look'd upon as impracticable and absurd, as Chariot flying may be here with us; and some where not improbably, superior Beings to the human, may reside, and Man may be of a very inferior Class; the second, third, or fourth perhaps, and scarce allow'd to be a rational Creature. Worlds, with various Moons we know of already; Worlds, with Stars and Comets only, we equally can prove is very probable; and that there may be Worlds with various Suns, is not impossible. And hence it is obvious, that there may not be a Scene of Joy, which Poetry can paint, or Religion promise; but somewhere in the Universe it is prepared for the meritorious Part of Mankind. (p. 81).

It is surprising to find an astronomer, however amateurish, giving such free rein to his imagination. It is even more thought-provoking to see that Kant in his *Allgemeine Naturgeschichte und Theorie des Himmels* (General natural history and theory of the heavens, 1755), written in the period before his *Critiques,* pursues ideas of this kind with all his customary seriousness and, while making occasional references to Wright, arrives at conclusions diametrically opposed to Wright's. In these early years—following the motto "Who will show us the boundary where well-founded probability ends and arbitrary inventions begin?"[91]—he made an intensive study of the question of the plurality of worlds. The third part of his *Allgemeine Naturgeschichte* is wholly devoted to an "attempt to compare, on the basis of the analogies of Nature, the inhabitants of various planets." Unlike Wright, he limits himself to the planets of our own solar system, a system that he describes, as his subtitle states, "on Newtonian principles." But even

[91]*Werke,* ed. Ernst Cassirer, vol. 1 (Berlin, 1912), p. 367. Page references are to this edition. Kant probably knew only a review of Wright's work; see Hoskin in his edition of Wright, *Original Theory,* p. xxvii. In his "Critical" period, Kant returns to plurality occasionally, if rather more tentatively; see *Kritik der reinen Vernunft,* "Methodenlehre" 2:3; *Kritik der Urteilskraft,* § 91; "Idee zu einer allg. Geschichte in weltbürgerl. Absicht," 6th Statement; see also *Logik,* "Allg. Elementarlehre," § 84.

on this much smaller stage he sees a graduated hierarchy of "thinking beings" (*denkende Naturen*, p. 353).

First, Kant too takes it for granted that the presence of more or less rational beings on the planets, or at any rate on most of them—and perhaps, in due time, life will emerge on those at present uninhabited!—forms part of "Nature's purpose" or the intention of the Creator-God (pp. 354, 358, 360). And unlike Huygens, but like Fontenelle and the followers of Newton and of Leibniz, Kant adheres to the principle of *diversity* rather than unity in nature. From this diversity among rational beings on the planets he infers a gradation that he believes is so convincing as to be practically a certainty but which is the exact opposite of the hierarchy envisaged by Wright, and earlier, in relation only to our own planetary system, by Gottsched.[92] Whereas Fontenelle's theory presented terrestrial man as the golden mean, Kant, like Gottsched and Wright, sees the degrees of perfection among the planet dwellers as forming a logical sequence corresponding to the distances of the planets from the center, so that our position in the middle is, rather, a sign of mediocrity. But Kant reverses the direction of the progression: perfection is to be found not close to the center (of the solar system according to Gottsched, of the whole universe according to Wright) but at the periphery. Leibniz had already considered something of the kind to be possible, though in relation to the universe as a whole (see § 19 of his *Théodicée* already quoted).

The reasoning by which Wright had justified *his* hierarchy had been inspired essentially by religious ideas; Kant's thoughts are presented in the guise of scientific and philosophical arguments. He is convinced that "the distances of the heavenly bodies from the sun lead to certain conditions that exert a profound influence on the various characteristics of the thinking beings that are present upon them" (pp. 353–54). The further a planet is from the sun, the more subtly organized is the material, the "physis" or "animal economy" (*animalische Ökonomie*) of its inhabitants: "The material of which the inhabitants of different planets, and even the animals and plants on them, are formed must be altogether of a lighter and finer kind, and the elasticity of the fibers together with the advantageous arrangement of their structure must be the more perfect, the greater the distance from the sun" (p. 360). But this state of affairs in the "material of the machine," in man's physical being, affects his "mental capability."

[92]*Erste Gründe der gesammten Weltweisheit, Erster, Theoretischer Theil* (Leipzig, 1733), § 653.

Kant thinks he may legitimately conclude "that the excellence of the thinking beings, their speed in forming conceptions, the clarity and liveliness of the ideas they receive from external impressions, together with the ability to combine them, and finally their rapidity in putting them into practice, in short, the whole extent of their perfection is subject to a certain rule according to which they become more excellent and more perfect in proportion as their habitations are more distant from the sun" (p. 361).

This perfection encompasses not only liveliness, the use of reason, and morality, but also happiness. Accordingly these qualities are present in the highest degree on Saturn, whose inhabitants were characterized for Fontenelle by lethargy (pp. 361–63). Man's planet is midway between Mercury and Saturn, and thus he occupies the "middle rung" (p. 361). He can feel flattered or humiliated according to whether he compares himself with the Mercurians and Venusians or with the Jovians and Saturnians. "Looking from one side we saw thinking beings among whom a Greenlander or Hottentot would be a Newton, and on the other side others marveling at him as at an ape"—as Pope had suggested (pp. 361–62).

Kant makes the comparison in both directions; unlike Bolingbroke he does not gain philosophical satisfaction from humbling the pride of enlightened man. Even so, his definition of man's intermediate position—no longer Fontenelle's golden mean but an "intermediate stage, both physically and morally" (p. 368)—represents a challenge to man's view of himself in the age of philosophy: man is "infinitely distant" from the "highest grade of beings" (p. 355). Moreover, as it seems that for Kant, this "intermediate state" in which man finds himself, "between wisdom and unreason," is the only one in which it is possible to sin (p. 367), Kant cannot bypass the question of God's justice, particularly as he has been at pains to interpret the universe as the result of "the wisdom of God" (p. 365). Wright had solved this problem by relating the differing degrees of happiness on the various planets to the differing degrees of virtue of the souls inhabiting them and by promising to every soul, on the basis of the theory of palingenesis, the possibility of attaining moral perfection. And even Kant, who does not think in such straightforwardly moralistic terms, rather hastily adds to the third part of his *Allgemeine Naturgeschichte und Theorie des Himmels* a separate "Conclusion" in which he too argues for the hypothesis of transmigration: "Can it be that the immortal soul . . . will never be enabled to look more closely at the other wonders of Creation?" (pp. 368–69). "Perhaps" after death the soul will develop to a higher level and accordingly occupy "new habitations" on other globes, where ultimately it will be granted supreme happiness. This assumption would vindicate God: here, then, is another way in which Kant at least implicitly links the plurality of worlds with theodicy (see sec. 1 above, at n. 33).

And he does this in a work whose subtitle promises to give an account of the "mechanical origin of the whole universe, on Newtonian principles"! The difficulty of attempting to combine metaphysics and mechanics, teleology and naturalism, could hardly be thrown more starkly into relief. In Fontenelle and Huygens, it was on the basis of Cartesianism; in Kant, of Newtonianism: on the one hand he speaks of the plan, the wisdom and "intention" of the Creator-God (p. 365), while on the other—specifically in his account of the origin and development of life on a plurality of planets, as determined by natural laws—he speaks in evolutionary, mechanistic terms. Kant fully acknowledges this troublesome philosophical legacy Newton left to his successors, and it is not satisfactorily explained away by his rhetorical question "Must not the mechanics of all natural motions have a pronounced tendency to bring about only consequences that, given the whole range of interconnections, are in accord with the plan of the highest Reason?" (p. 365). This is dubious even stylistically. Here and elsewhere the reference to the Creator has the ring merely of a pious phrase. It is no wonder, therefore, that even in Kant's lifetime other Newtonians rejected this compromise between mechanics and the teleological solution involving the wisdom of the Creator and instead opted for either one or the other.

On one side is Kant's friend Johann Heinrich Lambert with his *Cosmologische Briefe über die Einrichtung des Weltbaues* (Cosmologic letters on the arrangement of the universe, 1761). These "letters" were written in a pleasant, conversational tone like Fontenelle's *Entretiens* (indeed, the preface states that the author would like them to be regarded as a sequel to the *Entretiens*), and they were far more successful than Kant's academic treatise. For Lambert, too, "the Newtonian law of gravity," which he sees as operating in the "whole world," including the myriad systems based on the fixed stars and the "host of Milky Ways,"[93] provides a justification for setting up "habitability and diversity in every [!] solar system" as a contemporary article of faith (p. 96). The deciding factor, however, is not the automatic operation of evolution but metaphysical teleological argument, to which, as a typical disciple of Newton, he declares his adherence in the preface.

[93]First edition (Augsburg, 1761), p. 131 and preface, p. xxvi. Page references are to this edition.

Indeed, he takes the argument of "God's intention" (p. 108) further, invoking the principle of plentitude, which has nothing whatever to do with scientific thought but had been used to support the idea of the plurality of worlds in neo-Platonic–Christian cosmology: "Could the world be an effect of the infinitely effective Creator without life and activity, thoughts and impulses being present in the creatures in every part of it?" (p. 62). "The Creator, the eternal source of all life, is far too active not to have implanted life, powers, and activity in every grain of dust," in other words, not to have made "the whole universe inhabited" (p. 108). Thus, at the height of enlightened thinking, the scientific theory of the plurality of worlds paradoxically also returns to its prescientific theological basis.

But this was only one solution to the problem inherited from Newton and apparent in Kant. As the century goes on, the Newtonians in particular ask ever more insistently whether the subject cannot be adequately dealt with without recourse to the Creator. The first writer explicitly to exclude this highly problematic metaphysical aspect of the doctrine of plurality was William Herschel, the distant follower of Newton from Hanover, in a famous essay published in the *Philosophical Transactions* of the Royal Society in 1795. What is left is an immanent interpretation of the mechanics of natural processes. That the planets both within and beyond our solar system are inhabited, that even the moon and sun are inhabited, Herschel says he deduces (and emphasizes that he does so) solely from "astronomical principles," namely by analogical reasoning based on the laws of physics, and not from theological or poetic "fancy."[94]

At the stage of the discussion represented by Kant, the problem of mechanics and metaphysics is not the only dilemma to become apparent. Another arises both from Kant's own particular theory (set up in opposition to Wright's) of the hierarchy of beings and from the philosophical conviction, linked with it, of the migration of souls. Looking back at the various attempts, from Fontenelle and Leibniz onward, to discern some sort of hierarchy among the manlike inhabitants of the celestial bodies, one can hardly help finding them astonishingly arbitrary, despite the absolute certainty or scientific plausibility each thinker claims for the particular hierarchy that for him is manifestly the true one. Teleology, it seems, speaks in many tongues, and so does the theodicy associated with it. There are yet more contributors

---

[94]"On the Nature and Construction of the Sun and Fixed Stars," *Philosophical Transactions of the Royal Society* 85 (1795), 63, 65.

to this intense discussion of metempsychosis on an interplanetary basis (in other words, presupposing a plurality of inhabited worlds); as we listen to a few more, the confusion of voices becomes a veritable Babel.[95]

Johann Gottfried Herder, who was a pupil of Kant's at Königsberg, had a strong interest in astronomy from his student days onward; as early as 1765, when he was twenty-one years old, he came up against the problem of the plurality of worlds in his (unpublished) "Anfangsgründe der Sternkunde" (First principles of astronomy).[96] He was from an early stage sympathetic to the neo-Platonic idea, revived by Leibniz, of the hierarchy of beings in the universe. But when he defines *Homo sapiens*, not very originally, as an "intermediate species between rational being [*Geist*] and animal" (32:355), yet elsewhere as an "intermediate creature between angel and animal" (31:216), this variation in terminology points to the vacillation between Christian and latitudinarian modes of thinking that is generally characteristic of Herder. His thoughts on transmigration, by which he tries to resolve this complex of ideas, are also confused (return to this planet? migration to other planets? ascent from animal to angel?).[97] Some clarification is provided by the three dialogues "Über die Seelenwanderung" (On the transmigration of souls) that appeared in 1782 in Christoph Martin Wieland's journal *Teutscher Merkur* (German Mercury) and in 1785, reworked, in Herder's *Zerstreute Blätter* (Scattered pages).

In these dialogues, Theages and Charikles discuss the various historical conceptions of transmigration. As they step out into the "holy stillness" of the clear, starry night with its "suns and worlds," the doctrine of metempsychosis, the reincarnation of the soul on this Earth, seems reduced to pitiful provinciality. Cosmic palingenesis now takes its place. The stars are inhabited, not by creatures who could in some way threaten man's view of himself as the being made in God's image, but, once again, by the human soul itself at various stages of perfection. For Theages-Herder, too, the ascent of the soul from world to world is a return to God. Remarkably for a Lutheran clergyman, Herder identifies God with the astronomical center (as did Wright,

---

[95]See also Rudolf Unger, *Herder, Novalis und Kleist* (Frankfurt, 1922), pp. 2–3; Friedhelm Pamp, " 'Palingenesie' bei Charles Bonnet (1720–1793), Herder und Jean Paul: Zur Entwicklung des Palingenesiegedankens in der Schweiz, in Deutschland und in Frankreich" (Ph.D. diss., Münster, 1955).

[96]See H. B. Nisbet, *Herder and the Philosophy and History of Science* (Cambridge, 1970), p. 237. Page references in the text are to Herder's *Sämmtliche Werke*, ed. Bernhard Suphan (Berlin, 1877–1913).

[97]See Pamp, " 'Palingenesie' bei Charles Bonnet," pp. 31–90; Nisbet, *Herder*, pp. 232–39.

with whose work Herder was familiar).[98] So Herder's progression is in the opposite direction to Kant's, not centrifugal but centripetal. The setting for this palingenesis is, however (as in Kant but not Wright), not the whole universe but only our solar system, as calculated by Newton. Newton's system, says Herder, is the "edifice" of human immortality, of the "everlasting progression and upward flight" to God, who is symbolized by the sun (15:274–75). The significance of the soul's progression from planet to planet is that it becomes progressively more perfect; its goal, determined by God, is happiness:

> Imagine for a moment that our starry edifice may be joined together according to the *moral state* of its inhabitants in the same way as it is unquestionably joined together by its *physical* state, and that there is only one sisterly chorus praising its Creator in diverse tones and proportions but in the harmony of *one* strength. Suppose that from the furthest planet right up to the sun there are gradations of creatures as there are gradations of light, of distance, of mass, of powers (and nothing is more probable); now let the sun be the great *meeting point* of all creatures of the system that it rules, just as it is the queen of all light and all warmth, all beauty and truth, which it distributes to creatures everywhere according to degree; see the great ladder that everything climbs, and the long way that we still have to go before we reach the center and fatherland of that which, in our stellar system alone, we call *truth, light, love.* (15:276)

So we Earth dwellers occupy a middle position between the Saturnians, whom Herder, like Fontenelle, assumes to be lethargic, and the lively Mercurians; but for Herder, as for Kant, this position represents not Fontenelle's golden mean, where man is fortunate enough to be able to stay, but merely a transitional stage:

> Ch[arikles]: So, the greater the distance from our sun, the darker, the coarser; the nearer to the sun, the brighter, lighter, warmer, faster? The creatures on Mercury, which is always hidden in the rays of the sun, must of course be of a different kind from those sluggish inhabitants of Saturn, the dark Patagonian giants, who barely circle once round the sun in thirty years, and for whom five moons hardly illumine their night. Our Earth, then, would be somewhere in the middle—
> Th[eages]: And perhaps that is why we are such *intermediate creatures*, between the dark Saturnian kind and the light sunlight, the

[98]Nisbet, *Herder*, p. 238.

source of all truth and beauty. Our reason here is really only just dawning; and our free will and moral energy are nothing much, either; so it is a good thing that we do not have to remain forever on Planet Earth, where we would probably not make anything much of ourselves. (15:276–77)

In this hierarchy, as in Kant's, it is easy to see how the hierarchical Chain of Being has been transferred to the Newtonian universe. But as soon as one thinks of Newton's solar system as "our solar system," that is, just one of infinitely many, then here too the image becomes so confused as to be meaningless.

Let us look back again. Identical human life on all planets (Huygens) or very diverse planet dwellers with man in the middle as the unchanging ideal (Fontenelle); the soul undergoing a process of perfection by progressing from planet to planet toward the center (Gottsched, Wright, Herder) or toward the periphery (Kant); the whole universe as the scene of this development (Wright) or only our solar system (Kant, Gottsched, Herder). The confusion would be comic but for the fact that these philosophical efforts are the expression of a human consciousness that has more or less freed itself from the habit of being guided by Christian ideas about salvation but now has to strive to work out its own image of man in the context of the new cosmology. Kant supplied a motto for this endeavor by his use of the quotation from Pope that he placed (in German) at the head of the third part of his *Allgemeine Naturgeschichte und Theorie des Himmels*, which attempts to "compare the inhabitants of various planets":

He, who through vast immensity can pierce,
See worlds on worlds compose one universe,
Observe how system into system runs,
What other planets circle other suns,
What varied Being peoples every star,
May tell why Heaven has made us as we are.[99]

Philosophical anthropology has become cosmology; cosmology has become philosophical anthropology. This is Newton's legacy to an age that finds it difficult to turn to Christianity for a solution, as he did, but itself fails to reach a new consensus.

A further variation on these dream creations of the philosophical imagination is offered by the mystic and scientist who in the latter half

[99]Kant, *Werke* 1:351; Pope, *Essay on Man* 1, lines 23–28.

of the eighteenth century was one of the most influential advocates of the idea of the plurality of inhabited worlds and who also influenced Herder's formulation of his ideas: the Lausanne botanist, zoologist, and physiologist Charles Bonnet.[100] He places interplanetary palingenesis into a context so strongly theological that, however "modern" the astronomy, there emerges a conception of the universe that seems to have a strong affinity with the closed Aristotelian-Ptolemaic cosmos.

The reasons for this cosmologic aberration, or at least confusion, may be that on the one hand Bonnet came to the idea of the plurality of worlds, and the migration of souls from one to another, not from astronomy but from botany and zoology, and that on the other hand the outcome of his thought is determined in advance by Christian eschatology. Both in his *Contemplation de la nature* (1764) and in his *Palingénésie philosophique* (1769), which substantially recapitulates it, the dominant concept that guides the powerful momentum of his arguments is once again the mystical–neo-Platonic idea of the Chain of Being leading from inorganic matter via plants, animals, and humans up to the highest "intelligences" in the aura of the Creator-God. But Bonnet's conception of this chain is a dynamic one: all living things are engaged from the start in a continuous ascent to higher forms. Embryology reveals the earlier stages of development, and philosophy those of the future: perhaps one day elephants and apes will bring forth the Newtons and Leibnizes; by that time mankind will no longer exist on this Earth but will, metamorphosed into a still higher form, be peopling other planets.[101] The details of Bonnet's speculation are often lost in obscurity. In the *Palingénésie,* however, where he concentrates mainly on man, he does attain at least the clarity of internal logic. Strictly speaking, he focuses on the *soul* of man (*l'âme*) and its ascent. After death the soul will not float away as a pure spirit to higher regions but will instead be resurrected in a new body, a "corps spirituel" (*Palingénésie* 2:438). This body will develop from the indestructible organic "seed," preformed from the beginning of Creation, in which the soul resides and which Bonnet assumes to be located somewhere in the brain, perhaps in the extended spinal cord, perhaps in the corpus callosum. The future body of the soul consists of ether

---

[100]On Herder see Nisbet, *Herder,* p. 238. On Bonnet's influence apart from that on Herder see Pamp, " 'Palingenesie' bei Charles Bonnet"; also Jacques Marx, *Charles Bonnet contre les Lumières, 1738–1850* (Oxford, 1976); Raymond Savioz, *La philosophie de Charles Bonnet de Genève* (Paris, 1948), pp. 340–372.

[101]*La palingénésie philosophique* (Geneva, 1769), 1:203–4. All references in the text are to this, the first edition. References to *Contemplation de la nature* are to the first edition (Amsterdam, 1764).

or light or analogous "substances"; it will retain unimpaired memory and will be furnished with more refined and also more numerous senses and organs, capable of incomparably different perceptions and feelings.[102]

In various such "resurrections" (Bonnet freely mixes the vocabulary of physiology and of Christian revelation) the soul is gradually perfected, spiritually and morally (*Palingénésie* 2:442). It is the "sole purpose" (*fin*) of the myriad "worlds" (*mondes*) of the innumerable solar systems to provide the settings for the stages of this development (*Contemplation* 1:14–15). In this sense, Bonnet is one of those who, unlike Huygens, assume the planets to contain an unimaginable *diversity* of life-forms. If there are scarcely two wholly identical leaves or cabbages, caterpillars or human beings, he says, then all the more certainly there are no two identical worlds. Compared to our world, some of the others are "imperfect," some "on the contrary so perfect that they have only beings belonging to the higher classes [*des Etres propres aux Classes supérieures*]. In these latter worlds, the rocks are organic, the plants feel, the animals reason, the humans are angels" (*Contemplation* 1:25).

This is the kind of bold fantasy one finds in today's science fiction, even if Bonnet does not give a detailed picture of those worlds (unlike those "seers of ghosts," Swedenborg and Jacques Henri Bernardin de Saint-Pierre, who are akin to him and are discussed shortly). But at any rate, Bonnet indicates his belief that each planetary world has its own, uninterchangeable inhabitants, products, and laws; for duplication would mean sterility (*Palingénésie* 2:425). He does also hint that the living creatures on other worlds may compare to us as men compare to an ape, or to a "globule of air" (*Contemplation* 1:83). For Bonnet too, of course, the demotion of man that this implies is made acceptable by the idea of the progressive upward development of the soul.

This is the significance of Bonnet's predilection, already noted, for using Christian terminology when describing his grandiose conception of the history of the soul. It is not merely a matter of style. On the contrary, the Christian terminology helps to define more closely Bonnet's understanding of the upward progression of the soul from planet to planet and above all the final goal of this progression. Thus in the passage already quoted, about the angels as inhabitants of the most perfect worlds, he continues: "How excellent, then, is the HEAVENLY JERUSALEM where the ANGEL is the least of the INTELLIGENT BEINGS!" (*Contemplation* 1:25). And his vision of the universal

---

[102]*Palingénésie* 2:131–56, 416–21; *Contemplation* 1:85–90.

history of life culminates, at the end of the *Palingénésie,* in a hymnic celebration of Christian eschatology which uses the vocabulary of Revelation. At the end of Time, the soul will come into the presence of the Lord of Hosts, and the mysteries of his governance of the world, including the purpose of evil, will be revealed to the understanding (here too the plurality of worlds is linked with theodicy). The mysticism of the love of God will find its fulfillment in supreme happiness; virtue will become perfect (*Palingénésie* 2:428–46). "REVELATION," Bonnet concludes, "also adds its *sanction* to these *philosophical* principles" (*Palingénésie* 2:445).

The fact that scientific philosophy thus receives biblical sanction means, however, that in Bonnet's thought, the cosmic framework of this grandiose vision of interplanetary migration becomes blurred. All of Bonnet's predecessors whom I have mentioned thought unmistakably in terms of a Cartesian or Newtonian universe. Bonnet does indeed mention Newton as the representative of the highest intelligence possible on our planet (*Palingénésie* 2:426–27), but his use of the technical term *tourbillon* (i.e., vortex) places him in the Cartesian camp (*Contemplation* 2:15, 84). Either way, however, he would appear to be a Copernican, and what he says about other suns (the fixed stars) encircled by planets also bears this out. It is therefore all the more surprising to find that the key passage of the eschatological vision at the conclusion of the *Palingénésie* suggests images more reminiscent of the structure of the Aristotelian-Ptolemaic model of the universe. For Bonnet does not place the goal of the soul's ascent at the center, like Wright, whose conception of palingenesis in the framework of the whole universe perhaps most closely resembles his, but at the periphery—not merely of our solar system, like Kant, but of all systems, rather as Leibniz had imagined the place of "happiness" to be beyond the stars. But for Bonnet, this destination of the soul is a Beyond that is essentially similar to the empyrean of the medieval theologians. Just as the empyrean was assumed to be beyond the spheres that enclosed the Earth, so Bonnet's God, toward whom all souls strive, is enthroned beyond the "spheres" of innumerable Copernican systems that, taken together, seem to form a closed universe:

> But man's reason penetrates beyond all the planetary worlds; it rises up to Heaven where God resides; it contemplates the august throne of the Ancient of Days; it sees all the spheres turning beneath his feet, obeying the impulse that his mighty hand has imposed on them; it hears the acclamations of all intelligent beings (*Intelligences*), and, mingling its adoration and praise with the majestic singing of those

hierarchies, it cries out, with a deep sense of its own nothingness:
Holy, holy, holy is he who is! The Eternal One is the sole Good! Glory
be to God in the Highest; goodwill toward man! (*Palingénésie* 2:427)

Here the scientist in Bonnet plainly gives way to the religious
enthusiast. Scientific and Christian ideas about the plurality of worlds
merge and become blurred.

Perhaps it is precisely this cosmologic vagueness that accounts
for the widespread influence of Bonnet's ideas about transmigration.
Once again, as with Lambert and Herschel, two examples may show,
this time in the narrower context of transmigration as imagined by
Bonnet, how the idea of a plurality of inhabited worlds offers a meeting
ground for diametrically opposed philosophical viewpoints. On the
one hand, the interplanetary migration of souls plays a certain role in
the philosophical framework of Louis Sébastien Mercier's futuristic
novel *L'an 2440* (Amsterdam 1771), which sets out to be a utopian
work of the kind typical of the Enlightenment. In the twenty-fifth
century, the long-standing contest for precedence between the Book
of Nature and the Book of Books has long since been settled in favor of
the former. It is the *Encyclopédie*, not the Bible, that contains everything
worth knowing on all subjects; and the greatness of God, a deistically
conceived Creator, can be directly experienced not by a mystical illumination but by looking through a telescope. The worlds whose existence
the telescope either reveals or at least gives indications of are all
without exception intended for man, places where his soul develops
to a higher level in a series of stages that come to an end only in the
presence of the "Sovereign Being." Mercier is not usually regarded as
a follower of Bonnet,[103] but one feels that he must have had the
Palingénésie (2:443–45) open before him when he wrote, in the nineteenth chapter of *L'an 2240:*

> We think, further, that all the stars and all the planets are inhabited
> but that nothing one sees or feels in one is found in another. This
> unbounded magnificence, the endless chain of these different worlds,
> this radiant circle, was a necessary part of the vast plan of Creation.
> Well, these suns, these worlds that are so beautiful, so large, so
> diverse, seem to us to be habitations that have all been prepared for
> man; they intersect with one another; they correspond and are all
> subordinated to one another. The human soul ascends to each of
> these worlds as on a shining, graduated ladder that at every step

[103]Marx (*Charles Bonnet contre les Lumières*, p. 429) quotes a favorable remark by
Mercier about Bonnet.

brings it closer to the greatest perfection. On this journey it does not lose the memory of what it has seen and learned; it preserves the storehouse of its ideas, which is its most cherished treasure; it carries it everywhere with it. If it soars up to some sublime discovery, it passes over worlds peopled by inhabitants who have remained inferior to it; it rises by reason of the knowledge and virtues that it has acquired. (Pp. 123–24)

On the other hand, there is the Christian enthusiast Johann Kaspar Lavater. Bonnet was his favorite philosopher, and he translated the *Palingénésie* into German. In his *Aussichten in die Ewigkeit* (Prospects of eternity, 1768–78), in which he frequently quotes Bonnet, the planets of the solar systems are specifically, and apparently exclusively, intended for *Christians*, as the places where their souls wait for their ultimate resurrection. Lavater's eschatology does not make it wholly clear whether the souls have to perfect themselves by passing through many such stages. But there is, among others, a "planet of the damned" (22d letter). On one point, however, Lavater's cosmology is more explicit than Bonnet's: the celestial bodies (of all systems) circle around a common center, the "heaven of heavens," the seat of God, which is the soul's goal and its "eternal home" (9th letter). So God resides not beyond the periphery, as in Bonnet's quasi-medieval view of the universe, but once again at the absolute center, as in Wright (though to localize God at one particular point in a cosmologic model is in principle alien to Protestantism, as the Zurich pastor should have known).

This arbitrary mingling of the assumption of an infinite number of scientifically plausible inhabited worlds with ideas that are theologically desirable and symbolically meaningful but scientifically unprovable is taken a step further by the two fantasts among the scientists of the eighteenth century, Emanuel Swedenborg and Jacques Henri Bernardin de Saint-Pierre. Both had serious scientific ambitions and were certainly more than well informed laymen. Swedenborg, who is seen as a precursor of Kant and Laplace in connection with the nebular theory, was in 1718 offered a professorship of astronomy at the University of Uppsala.[104] Bernardin de Saint-Pierre was for a time, during the French Revolution, director of the Jardin des Plantes. But as soon as they start to speak on their favorite topic, the plurality of worlds, all scientific restraint deserts them and poetical imagination takes over.

---

[104]See Inge Jonsson, *Emanuel Swedenborg* (New York, 1971), p. 33 (and chap. 2, passim); J. G. Dufty, *Swedenborg the Scientist* (London, 1938), p. 27.

Their cosmic visions, however different from one another, are both most appropriately classed as "literature."

After a successful scientific career, Swedenborg moved across to this field after his experiences of religious illumination during his stay in Holland and England in 1743–45. On his return to Sweden he laid aside his scientific studies and devoted himself wholly to his idiosyncratic exegesis of the Bible based, as he states, on the communications of spirits with which he believed himself to have been in daily contact since his religious awakening. Whether or not Swedenborg, the theosophist, was mentally ill, his revelations from the spirit world—or hallucinations—had considerable influence in his day, even among serious-minded readers. "The writings of so strange a man, who has caused such a stir, deserve somewhat more than the attention given to ordinary fantasies," writes a reviewer in 1780 in the *Allgemeine deutsche Bibliothek*.[105]

Those of the visions from the spirit world that relate to the plurality of worlds are gathered together and systematically set out (reported, incidentally, in quite a matter-of-fact style) in the book *De Telluribus in Mundo Nostro Solari Quae Vocantur Planetae* (London 1758). Kant, in 1766, dismissed these abstruse accounts as "Träume eines Geistersehers" (the dreams of a seer of ghosts). But no less level-headed a man than Lessing studied this book at the very end of his life, and in his conversations with Friedrich Heinrich Jacobi declared his sympathy, if not with Swedenborg's views, at least with Bonnet's hypothesis of the migration of souls, which is essentially comparable. For while Bonnet occasionally suggests (*Contemplation* 2:88–89; *Palingénésie* 2:421, 439) that after death the soul will move freely about the cosmos, Swedenborg makes this idea the very cornerstone of his thought. But in the case of this Stockholm inspector of mines, the Christian God has deemed his soul worthy to experience such cosmic adventures in *this* life: not only do spirits from all planets come to Swedenborg; his soul also travels out to the far reaches of the universe.

The structure of his universe is surprising for a Cartesian who in 1710–13 studied Newtonian physics in England, perhaps under Newton himself, but less surprising for the mystic par excellence as portrayed by Ralph Waldo Emerson in *Representative Men* (1850).

---

[105]Appendix to volumes 25–36: vol. 2 (1780), 1016. On Swedenborg's influence see Jonsson, *Emanuel Swedenborg*, chap. 9. The library books found in Lessing's house after his death included the German version of Swedenborg's *De Telluribus* (information given by Wolfgang Milde, Wolfenbüttel). Swedenborg also had a significant influence on Lavater; see Ernst Benz, "Swedenborg und Lavater," *Zeitschrift für Kirchengeschichte* 57 (1938), 153–216.

Swedenborg's universe has the form of a man, *maximus homo*, and each of the myriad worlds is related to particular parts and faculties of this cosmic body (*De Telluribus*, § 9). The worlds are, without exception, peopled by more or less identical "human beings" (§ 2–4). Huygens's anthropocentric "imperialism" is here resurrected in more theological guise. For when Swedenborg claims that "the Lord" created the anthropomorphic universe solely for man, he means not only man on our planet (nor does he mean it in the sense suggested by the theory of transmigration, which assumes future incarnations of terrestrial man on other planets): he means the countless humankinds, wholly analogous to us and contemporaneous with us, of the "worlds" of the universe. Thus, at a time when the popularity of the plurality idea is at its height, man on Earth, who had for a long time been deposed from his position as the supreme purpose of Creation, now once again at least shares that status, in a way that is both anthropocentric and antigeocentric and that, according to Swedenborg, is directly evident to "enlightened reason" (*ratio aliquantum illustrata*, §§ 112, 126). (Swedenborg likes to present his inspirations as rational insights.) Accordingly, the humans of all the planet-worlds revere the one God in human form (*Deum sub Humana forma*, § 121). This is of course the Christian God; Swedenborg's assumption of many humankinds in the universe seems to him in no way to conflict with his strictly biblical Christian belief. It is no coincidence that *De Telluribus* is a compilation of material taken from his visionary exegesis of the Bible in eight volumes, *Arcana Coelestia* (1749–56); and a special chapter (§ 113–22) explains why Christ, the Lord in human form, appeared on our Earth and on no other planet. A gloss on "In the beginning was the Word" is intended to settle once and for all this long-standing point of contention: only on this Earth did man develop the art of writing, on tree bark, parchment, or paper, and therefore the gospel of God's incarnation could be recorded for posterity only here, and disseminated only from here, and so on.

This dissemination is effected by those spirits who are the very element of Swedenborg's universe and of his philosophy. For every planetary world, inhabited by humans, is surrounded by the spirits of the departed, which are in constant contact with each other and thus can also transmit Swedenborg's brand of Christian teaching "from mouth to mouth" (*ore tenus*, § 120) and from planet to planet. In Godwin's *Man on the Moone*, the fact that the moon dwellers fall on their knees at the mention of the name of Jesus Christ is an unexplained miracle; in Swedenborg, the missionary work of the spirits provides an explanation.

It is the spirits of the dead of other planets who, in *De Telluribus*, give Swedenborg information about the inhabitants of the planets of our solar system. (The planets of other solar systems are visited by his soul itself.) They give an account, as the title page of the 1770 German translation states, "of their way of thinking and acting, their form of government, administration, divine worship, marriage, and their habitations and way of life in general." And this is no empty boast. In the description of this universe teeming with life, scientific matters play a very subordinate role: at most, outlandish features of life on a planet are sometimes related to the physical conditions there. There is no hierarchical relationship between the worlds, since Swedenborg does not speak of a progression of the soul through stages of perfection on a succession of planets. So he is able to concentrate on giving numerous descriptions, in no particular order, of the planetary worlds and their human inhabitants. There is much that is comme chez nous, and Swedenborg only very seldom exploits the more exotic elements for utopian or satiric purposes, and then very marginally; as one who consorts with spirits, he is wholly uninterested in such earthly matters. The ubiquitous references to Swedenborg's brand of Christianity are the one element that could perhaps be seen as binding together this loose collection of science fiction portrayals of other worlds. The narrator never tires of stressing that the cosmic humankinds everywhere revere as their Savior the same God in human form (§§ 7–8)—hardly a century and a half after Giordano Bruno suffered martyrdom at the stake for the idea of the plurality of worlds. This and the fact that Swedenborg sees man on whatever planet as (theologically speaking) the same creature, made by God, means that the situation where terrestrial man meets his counterpart in the universe and is defined by the encounter hardly arises. He is already defined, and the extraterrestrials are almost identical reflections of that definition. Their strangeness is hardly different from the strangeness of other peoples on Earth as described in the travel books proliferating just then. This can be seen in the following examples, which, however, also give an impression of the bizarre, erratic fantasy—and banality—that are the hallmarks of Swedenborg's visionary thinking on plurality.

There is a "correspondence" between the Mercurians (§§ 9–45) and the *memory* of the cosmic *maximus homo*. A desire for knowledge is their chief characteristic, but it is directed only toward intellectual and spiritual matters: for them even the philosophy of Christian Wolff is too much concerned with the senses and material things. In physique the Mercurians hardly differ from us, though they are of slimmer build, with smaller facial features. The men wear clothes without pleats or frills. Their oxen are like ours, but smaller. The inhabitants of Jupiter (§§ 46–84) are peace loving, contented, honest, and interested above all in the upbringing of children. To keep their faces clean and protected from the sun is an almost religious practice among them. They do not really walk upright but sometimes support themselves on their hands, without thereby becoming four-legged creatures. They sleep with their faces turned not toward the wall but into the room, that is, toward God. Their meals, which they eat sitting on fig leaves, last a long time, because they enjoy conversation. They want to have nothing to do with science. They revere God not as their king (which would be too worldly), but as their Savior. The walls of their dwellings are hung with tree bark, which by its color and the way it is decorated makes the walls look like the starry sky. The Martians (§§ 85–96) live not in states but in small groups, from which troublemakers are expelled. The lower half of the Martian face, up to the ears, is black, the upper part yellowish. For lighting their dwellings they use liquid fire. The Saturnians (§§ 97–104) live mainly on fruit and legumes; they dress in a thick animal skin or a coat; they do not bury their dead but cover them with twigs. The Venusians (§§ 105–10) are divided into two races, a friendly and humane race in one hemisphere and a wild and brutish one in the other. The whole life of the latter consists of plundering and enjoying the spoils; they will nevertheless be saved, after suitable punishment. Finally, the moon (§§ 111–12) is inhabited by dwarflike men with long heads; their speech comes not from the lungs but from the abdomen, because—and this is one of Swedenborg's few concessions to physical factors in his depiction of his fantasy worlds—the moon has a different kind of atmosphere.

The descriptions of the various humankinds become shorter with each planet discussed. When describing the entirely human dwellers on planets outside our solar system, Swedenborg's imagination gradually runs dry—which is surprising since, as he stresses in recommendation of these descriptions, he is here speaking of things he has found on his own travels and not merely of what the spirits have told him (§ 126). In one of the worlds belonging to another star (§§ 127–37), the peasants dress like European peasants; the men walk with a confident stride; the women walk humbly, but without loving their husbands any the less. There is, by the way, a hell there too. Another such far-off world (§§ 138–47) has meadows, gardens, fruit trees, lakes abounding in fish, blue birds with golden wings; the people live in groves, it is usual to have between ten and fifteen children, and whores go to hell. In the most distant world visited by that indefatigable tourist, Swedenborg's soul (§§ 168–77), the tribes gather every thirty days for an open-air religious service. This world is, incidentally, blessed with milch cows that are also woolly like sheep, and as if this were not paradisiacal enough, it enjoys perpetual spring and summer. Social mores are less luxurious: by divine decree, the people are monogamous. With this revelation, *De Telluribus* ends, with typical abruptness.

In a similar and yet unmistakably different way, mysticism and absurdity, the wildly fantastic and the all-too-earthly form a curious blend in Bernardin de Saint-Pierre's *Harmonies de la nature* (1815), the compendium of mystic-poetical natural philosophy at which the author of *Paul et Virginie* labored from 1792 until his death in 1814. Bernardin's theory of the reciprocal harmony and correspondence among all phenomena of animate and inanimate nature need not be discussed in detail here,[106] since his theory of the plurality of worlds is for the most part derived independently from the by-now-familiar combination of analogical and teleological thinking. With Bernardin as with Swedenborg, a particular brand of Christian belief, in this case Catholic, underlies the belief in plurality. But apart from a reference to the omnipotent Creator, the Frenchman, unlike the Swede, has no explanation to offer for this link between the two. Instead, in his account of the planetary worlds, he emphasizes the scientific grounds for believing in a plurality of worlds, which Swedenborg had largely ignored in his more intuitive account. Thus Bernardin begins the final book of his *Harmonies*, which deals with the "Harmony of the heavens, or the worlds," with a section on cosmology that shows him to be well versed in the theories and observations of Newton, Leibniz, and Herschel, the "Columbus of astronomy" (p. 290); and he opens each of the subsequent chapters on the individual planets of the solar system with an account of the astronomical data and topographical and climatic conditions according to current scientific knowledge.

Although the planets are very different from one another, Bernardin adheres strictly to the principle of basic analogy in describing their inhabitants: the highest beings on all planets, at least in our solar system, are humans like us (p. 304). Unlike earlier extrapolations going back to Kepler and Hill, Bernardin even assumes, with Huygens, that the humankinds on all globes are of the same size. But against this background of uniformity he stresses the differences between them. These are the differences between peoples and cultures familiar to us.

[106]See Albert Duchêne, *Les rêveries de Bernardin de Saint-Pierre* (Paris, 1935); Basil Guy, "Bernardin de Saint-Pierre and the Idea of 'Harmony,' " *Stanford French Review* 2 (1978), 209–22. Page references in the text relate to *Oeuvres complètes*, ed. Louis Aimé-Martin, vol. 10 (Paris, 1826).

Indeed, he makes each planet in its entirety reproduce, in heightened form, the topographical and climatic conditions and the flora and fauna of a specific region (or several regions) of our world, and he forms a picture of its human inhabitants based on the corresponding nations and peoples on Earth. This makes the "harmony" simple but no less abstruse and, despite the astronomical foundation, highly imaginative.

Mercury with its luxuriant and brightly colored vegetation corresponds to that land of Cockaigne, India, and accordingly it is the Indian people that the Mercurians most resemble. Venus too is a paradise, pictured with poetical extravagance as a combination of the landscapes of Switzerland, France, Polynesia, and the Moluccas. Its inhabitants, true to the traditional associations of Venus, devote themselves mainly to love. "Some, grazing their flocks on the mountain tops, lead the life of shepherds; the others, on the shores of their fertile islands, give themselves over to dancing, to feasts, divert themselves with songs, or compete for prizes in swimming, like the happy islanders of Tahiti" (p. 315). Mars, as pictured by Bernardin's precise imagination, is a copy of the polar regions, complete with polar bears and penguins, sea lions and whales. The inhabitants engage in hunting, as the occupation most nearly resembling warfare; surprisingly, they resemble not the Eskimo but the Tartars, Poles, and North Germans. The inhabitants of Jupiter are the sailors and deep-sea fishermen among the planet dwellers, thus resembling in "character" the Danes, Dutch, and English. "Provided with light by constant auroras, which mingle with the soft brightness of the moons, when they milk their large herds in their vast grasslands, or spread out their nets, richly filled with fish, on their sandy beaches, they bless Providence and cannot imagine any more beautiful days or happier nights" (p. 337). The Saturnians are a shepherd people living on endless plains where cedars and broom grow. Finally, the planet Uranus (discovered by Herschel) is inhabited by the extraterrestrial Laplanders, who, Bernardin says, surely have their reindeer and faithful dogs and use whale-oil lamps to light the long nights. Like the Lapps, the Uranians have neither libraries, theaters, nor monuments to warriors.

Of the planets that he thinks probably exist around the other suns in the firmament (p. 370) and of their inhabitants, Bernardin says nothing (p. 377) and thus avoids the inevitable repetition to which Swedenborg had fallen prey. Thinking, as he did, in terms of harmony, he would no doubt have postulated life-forms there analogous to those of our solar system.

One is astonished to find Bernardin stressing that his detailed descriptions of landscapes and cultures are not "produced by my imagination exalted by the sense of an all-powerful Providence"! He is able to give only a sketch, but one "drawn with astronomical precision" (pp. 348–49). This certainty derives ultimately less from the Providence that he frequently invokes or the conclusive evidence provided by astronomical research than from a view of the universe that is at bottom geocentric. In each case, the population of a planet in the solar system corresponds to a population or culture known on Earth, or a combination of two or three. Only the Earth, which alone has the whole range of climatic zones and types of landscape, possesses the diversity of all peoples and cultures. Our Earth is the meeting point of all life in the universe, a sampler of the whole animate world. It has thus ideologically recaptured the central position of which Copernicus deprived it. Man need not feel lost in an alien universe of countless inhabited worlds; on the contrary, Providence (which Bernardin constantly brings into the discussion) has singled us out for especial attention.

One may perhaps see this maneuver—like those, at once similar and different, of Fontenelle, Huygens, Swedenborg, and others—as a surreptitious attempt to restore terrestrial man to his rightful position as the measure of all things, which initially he had forfeited as he became aware of the plurality of worlds.[107] Certainly the attempt to see the Earth and its people as being "singled out" reveals a desire somehow to make good the "loss" inflicted by the idea that "we are not alone." More than anything else, it is this aspect of Bernardin's abstruse theory of plurality that makes it significant, or at least thought provoking, in the context of the history of ideas.

It is noteworthy, too, that this endeavor to give some special distinction to the Earth is most marked in those advocates of plurality in whose writings what starts as scientifically based extrapolation becomes, to put it mildly, uncomfortably eccentric. It is a relief, therefore, to turn to the sober observations of a professional scientist who more or less puts the finishing touch on this development in Enlightenment thinking and the manifold speculations about a hierarchy of life-forms in the universe. These observations are contained in *Allgemeine Betrachtungen über das Weltgebäude* (General observations on the universe, 1801 [3d ed., 1808]) by Johann Elert Bode, the Berlin astronomer who was a member of the most respected European academies of the day. His interest in the plurality of worlds is perhaps most

[107]On Bernardin de Saint-Pierre as an Enlightenment figure, see Guy, "Bernardin de Saint-Pierre," p. 220.

apparent in his 1780 translation with commentary, already mentioned, of Fontenelle's *Entretiens sur la pluralité des mondes*. Through his *Allgemeine Betrachtungen* he made many more people aware of the subject, however; for it is in fact a separate printing of the final section of his phenomenally successful astronomical textbook, *Anleitung zur Kenntniss des gestirnten Himmels* (A guide to an understanding of the firmament). This final section first appeared in the second edition of 1772 and was last published in Bode's lifetime, after several revisions, in the textbook's ninth edition of 1823. So Bode's ideas, which take account of just about everything said on the subject of the plurality of worlds, were being published and read for half a century. It is thus appropriate to discuss them here, at the end of this survey of the Enlightenment philosophy of the plurality of worlds.

Like so many supporters of the idea in the Age of Enlightenment, Bode is a Newtonian, and like so many other Newtonians he is convinced, on the basis of both analogical and teleological thinking, that God created all the planets in our solar system and beyond for "rational and sentient beings." In "those immeasurable domains" these beings cause "songs of gratitude and rejoicing to rise up to the throne of the Ruler of the whole universe, who is eternal love" and who created worlds and living beings in order to "make them happy." "What innumerable kingdoms of God!"[108] There were worlds before us; there will be worlds after us: Creation is not finished (p. 652). Coming into being and perishing are both subject to God's "wise plans" (p. 656). These do not exclude more perfect or less perfect humankinds on other planets (p. 643). Such inequalities do not, however, cast doubt on God's justice; for Bode too (in an astronomical textbook filled with tables and calculations!) admits the possibility of the transmigration of the human soul, a migration "guided by the fatherly hand of the All-Benevolent, step by step to ever happier regions and higher perfections" (p. 659). But Bode says only that this development will take place in "more perfect worlds" (p. 658), without saying which planets these are and in what order they stand to one another. Thus he says nothing specific about the structure of the universe with reference to the plurality of worlds.

It is precisely this that differentiates him from all the other theoreticians of plurality so far mentioned. He gives a purely physical reason, among other reasons, for this abstention from speculation, namely the speed of light. Given that star systems, indeed whole galaxies, are

---

[108] *Anleitung zur Kenntniss des gestirnten Himmels*, 9th ed. (Berlin, 1823), pp. 640–42, 632. Page references are to this edition. See Pascal, chap. 3, sec. 1 above.

still coming into being (or being created) and that such new phenomena, which form part of the structure of the universe as a whole, can only be seen from the Earth after a considerable lapse of time, any statement about the structure of the universe is by definition inaccurate (p. 654). The implication for the theory of the plurality of worlds is that any assumption of a hierarchical arrangement (in physical terms) of the worlds in the universe, for instance around a common center, must be dismissed as irresponsible speculation. In this context Bode mentions Kant's gradational theory and himself extends it to apply to the whole universe, placing the creatures with the most refined organization at the outer edge of the overall, spherical structure of the countless solar systems that revolve around a "common central sun" (pp. 643–44)—the precise opposite of Wright's cosmologic model of a plurality of worlds. But Camille Flammarion, the nineteenth century's popular authority on plurality, is not justified in concluding that Bode's suggestion represents his own conviction.[109] It is simply an idea he mentions as a possibility without committing himself to it, after he has given a résumé of the opinions of others; he speaks in the conditional mode. At most he admits, elsewhere, that there may "perhaps" be such a central sun (p. 650). In any case he closes his reflections, as he looks back over a century of the most bewildering speculations on plurality, with a remark that strikes the historian almost as deliberately polemical: "But what inhabitant of the Earth will be so bold as to inquire into these mysteries?" (p. 645). He sees all the hierarchical theories as arrogant, "as if the Infinite One, in designing the whole, had taken the dot on which we live as a guideline or measure" (p. 642). By the same token, we humans are not the measure of all things. Bode therefore advocates "humility," not only in theories about the structure of the universe, but also in conjectures about the "physical build," the "mental faculties," and the hierarchy of the "rational inhabitants" of the other globes (p. 642)—though he is, of course, convinced in principle that "we are not alone" in the universe. The reason why he is prepared to discuss the *detailed* speculations, so suspect to him, about the plurality of worlds and the hierarchy of their inhabitants at all is that the thought of the "myriad worlds full of rational beings" is "too important for me not to entertain it. It brings with it a wealth of consequences" (p. 650).

What consequences does Bode himself draw from the plurality of worlds (which he conceives of only in the most general terms)? His book ends with an expression of belief in the interplanetary migration

---

[109]*Les mondes imaginaires et les mondes réels*, 12th ed. (Paris, 1874), pp. 533–34.

of the soul, which again he does not define in detail. All earthly things are thereby placed in the context of eternity, and he thus achieves that reassurance which since the seventeenth century had often been expressed but had been in conflict with the many causes for anxiety: sometimes, indeed, anxiety had shown through between the lines of actual expressions of reassurance. In Bode's "reassuring thoughts," the conflict between science and theology sparked off by Copernicus is resolved by the combination of an unfailingly down-to-earth scientific approach with an undogmatic but also vague piety:

> From these considerations I learn rightly to estimate the true worth of all earthly things around me, and to look upon and judge from a quite different point of view from the usual the physical changes and events on the small globe which I inhabit, and the moral and political destinies of its peoples and of my individual fellow citizens; and here I attain a truer conception of a self-sufficient Providence constantly watching over us and of the plan, often incomprehensible to us but always wise, by which it governs the world. What incomparable, what reassuring thoughts this allows me! How many great and sublime things I discover while still here below! But what knowledge awaits me beyond the grave! How much will still remain for me to discover throughout the boundless length of my existence in the hereafter! (P. 662)

4. Space Voyages in the Novel and Epic: Morris, Kindermann, Wilson, Klopstock, Wieland, Swift, Voltaire, and Others

If one is aware of the scientific and philosophical background, described above, of eighteenth-century literature, then the literary treatment of the subject of the plurality of worlds acquires a wealth of references and perspectives that would hardly be apparent without such a survey of the intellectual context of the time. In other words, this overview enables one to see what possible interpretations of the theme of the "thousand worlds" are being rejected or ignored when one particular interpretation is used in literature and perhaps presented in an original variation. The spectrum of possible interpretations is broad; it extends, with many a variation, from the neo-Platonic-Christian doctrine of plenitude, via the teleological thinking of deism and analogical speculation based on empirically obtained data, to mechanistic thinking. And it is all the more necessary to determine the position of individual literary treatments of the subject within this spectrum since in the "century of philosophy" the plurality of worlds is a more frequent literary theme than in the preceding age. At this time, German-speaking countries too begin to produce literary works that can be seen as part of the European development of the theme (not until the 1720s did German literature begin to take cognizance of the Copernican revolution and post-Copernican cosmology,[110] which are the indispensable basis for the "worlds upon worlds"). In eighteenth-century literature, then, in Germany as well as in England and France, names such as Copernicus, Kepler, Galileo, Huygens, Fontenelle, and above all Newton occur in connection with the "thousand worlds." The use of these names does not in itself, however, reveal how terrestrial man is seen in the light of those worlds and their inhabitants.

The genre most directly concerned with this topic is of course that of the planetary novel or epic. Leibniz himself may have given the cue. When stating the basic idea of his *Théodicée*, that our world, for all its evils, is the one that the Creator judged to be the best, he adds at one point that one *could* imagine "possible worlds" without sin or misery and make some sort of novels out of them.[111] It almost seems as if the novelists had accepted this challenge. The genre had a large readership at the time; in some cases only the titles now survive, though other works have recently been reprinted. Of those books of which copies can still be found, a disproportionate number are utopian or satiric, sometimes both. These are not concerned with the encounter between *Homo sapiens* and his counterpart in space. Instead, the space voyage is merely one possible vehicle for more or less specific social criticism; what takes place on a planet could equally well have been set on a remote island or in a neighboring country. These novels need therefore be only briefly mentioned in our context: some of them by their very titles reveal themselves to be something like a foil to space novels of more philosophical weight, such as those of Swift and Voltaire. Nevertheless, as we shall see, these more significant works still have points in common with those others, showing that there is a more than accidental connection between them.

It is significant that in many such satiric and utopian space novels, either the action is set on the moon or a moon dweller visits the Earth.[112] Once it was known that the moon lacked the conditions

[110]Confirmed again by Karl Richter, *Literatur und Naturwissenschaft*, p. 139. On Gryphius as an exception see chap. 3, sec. 1 above, at n. 23.

[111]*Théodicée*, "Essais sur la bonté de Dieu . . . ," § 10.

[112]Many such works are discussed by Marjorie Nicolson in *Voyages to the Moon* (New

necessary for any kind of life, it could become the setting for imaginative creations that, unlike Kepler's moon novel, deliberately chose to convey a kind of truth different from that which science made plausible; one could be certain from the outset that what was depicted was unthinkable. Hence the moon was also a popular choice as the setting for comic poems like Samuel Colville's "mock poem" *Whiggs Supplication* (1695) and Saverio Bettinelli's "poema eroico-comico" *Il mondo della luna* (1754) and also for comic operas and operettas like Haydn's *Il mondo della luna* (1777, libretto by Carlo Goldoni); Maximilian Blaimhofer's Singspiel *Die Luftschiffer oder der Strafplanet der Erde* (The space travelers; or, Earth's penal colony on a planet, 1786); Louis Abel Beffroy de Reigny's *Nicodème dans la lune* (1791); Friederike Helene Unger's *Der Mondkaiser* (The emperor of the moon, "a farce translated from the French" 1790); and others.

The moon novels of this period that satirize society, some of which are clearly influenced by *Gulliver's Travels*, include for instance Defoe's *The Consolidator* (1705); "Murtagh McDermot's" *A Trip to the Moon: Containing Some Observations and Reflections Made by Him during His Stay in that Planet, upon the Manners of the Inhabitants* (1728), which is a barely disguised journey to the England of coffeehouses and parliamentary politics; *A New Journey to the World in the Moon* (2d ed., 1741); the three anonymous *Relations du voyage fait dans la lune* (1751), which promulgate philosophical truths supposed to have been brought from the moon; William Thomson's *The Man in the Moon, or Travels in the Lunar Regions* (1783); the novel of metempsychosis attributed to Carl Friedrich Bahrdt, *Zamor oder der Mann aus dem Monde* (Zamor; or, The man from the moon, 1787); the anonymous *Reisen in den Mond, von einem Bewohner des Blocksberges* (Voyages to the moon, by an inhabitant of the Blocksberg, 1789); *A Voyage to the Moon, Strongly Recommended to All Lovers of Real Freedom* by "Aratus" (1793); and Pierre Gallet's *Voyage d'un habitant de la lune à Paris à la fin du XVIIIe siècle* (1803). In this genre the book that achieved the greatest international success was "Captain Samuel Brunt's" *A Voyage to Cacklogallinia* (1727), in which the ostensible author, "Captain Samuel Brunt" describes how, with the help of rationally endowed giant chickens, he flies to the Selenites and becomes acquainted with their mores and customs.

York, 1960); see also Roger Lancelyn Green, *Into Other Worlds: Space-Flight in Fiction, from Lucian to Lewis* (London, 1958), chap. 5: "A Lunatick Century." *"The Man in the Moone" and Other Lunar Prophecies*, ed. Faith K. Pizor and T. Allan Comp (New York, 1971), is an anthology of such lunar novels containing social criticism. Peter Leighton, *Moon Travellers* (London, 1960), gives summaries of contents. Philip Babcock Gove, *The Imaginary Voyage in Prose Fiction* (New York, 1941), is an important bibliographical source.

In most of these satires a utopian element is at least marginally present. It is more prominent or indeed given central importance in works such as *Besondere Begebenheit eines Bürgers aus dem Mond, wie derselbe, ohne sein Wissen, ohngefähr auf unsere Erde gekommen, und mit einem Erdbürger verschiedene merkwürdige Unterredungen geführet* (Adventure of a citizen of the moon, an account of how, inadvertently and by chance, he came to our Earth . . . , n.d.); *Reise eines Europäers in den Mond, nebst einer Reise-Beschreibung eines Monden-Bürgers von seiner Reise auf unsere Erd-Kugel, worinne absonderlich die Sitten, Gebräuche und Gewohnheiten derer Innwohner des Mondes und derer Europäischen Völcker nachdencklich beschrieben werden* (A European's voyage to the moon . . . , 1745); Daniel de Villeneuve's *Le voyageur philosophe dans un pais inconnu aux habitans de la terre* (1761); Francis Gentleman's *A Trip to the Moon: Containing an Account of the Island of Noibla, Its Inhabitants, Religious and Political Customs, etc.* (1764); *Robinsons Luftreise nach dem Mond, oder Beitrag zur Geschichte der Seelenwanderung* (Robinson's aerial voyage to the moon . . .) by one Sommer (Vienna 1785). Lunar novels that tend more toward didactic allegory, dealing with human virtues and vices, include *Philophili Reise in den Mond* (Philophilus's journey to the moon, 1707), which is recommended as being "exceedingly useful and necessary" for the "art of investigating human dispositions," and the voyage to the moon in *Le char volant* by Cornélie Wouters, Baroness of Wasse (1783). Pure fantasy and playfulness are the keynote of Johann Krook's *Tanckar om jordens skapnad* (Thoughts on the creation of the Earth, 1741); the *Histoire intéressante d'un nouveau voyage à la lune* (1784) attributed to Antoine François Momoro; and the lunar episode in Gottfried August Bürger's popular classic, *Wunderbare Reisen zu Wasser und zu Lande . . . des Freiherrn von Münchhausen* (Baron Münchhausen's wondrous journeys by water and land . . . , 1786). David Russen's *Iter Lunare* (1703), finally, is essentially a mere paraphrase, with commentary, of Cyrano's lunar novel.

So much for the novels about the moon. To the literary historian, it is perfectly logical that the other celestial bodies that contemporary astronomy was far more willing to credit with being habitable than the airless, waterless moon cannot compete at all with the moon as a setting for satiric, utopian, or didactic-allegorical space voyages at this period. The sun, which as late as 1795, William Herschel considered to be habitable, is the scene for the allegorical action of *Aletophili Reise in die Sonne, auff welcher er die Tugenden und Laster der Menschen zu untersuchen und zu erkennen gelernet* (Aletophilus's journey to the sun, 1708). The planet Uranus, discovered by Herschel in 1781, is the setting for *A Journey Lately Performed through the Air, in an Aerostatic*

Globe . . . to the Newly Discovered Planet, Georgium Sidus (1784), by "Vivenair." Finally, Carl Ignaz Geiger locates three bad regimes and an ideal state on Mars in his *Reise eines Erdbewohners in den Mars* (An Earth dweller's journey to Mars, 1790), to which Johann Christoph Röhling wrote a similarly satiric-cum-utopian response, *Reise eines Marsbewohners auf die Erde* (A Martian's journey to Earth, 1791).

All this is far removed from the depiction of an "encounter." But the impression conveyed by surveys of eighteenth-century space literature (see n. 112), that the narrative literature of the day does not give serious consideration to the theme of the imagined meeting between men and aliens, is misleading. True, there are not many instances of such treatment, and the theme is more often merely touched upon than fully worked out. But the eighteenth-century space novel by no means ignores the subject of the "encounter." We will look briefly at a few borderline cases before examining those novels that are intellectually on a higher level and have closer links with science.

Not many intellectual demands are made for instance by a few accounts of space journeys at this period, where even the problem of transport is solved without the aid of science. In the *Relation du monde de Mercure* (1750), the Chevalier de Béthune recounts how his soul separates itself from his body and is miraculously transported to Mercury, where small, flying human beings who never sleep lead a fairy-tale existence. The narrator of *Träume* (Dreams, Halle 1754) travels in his sleep to all the planets of our solar system, as well as to the sun itself and the moon (18th–22d dream). The author is Johann Gottlob Krüger, who, like Gottsched, made a name for himself by his popularization of science, which included spreading the idea of the plurality of worlds as an implication that can be extrapolated from Newton's work.[113] In his "Dreams" there are more or less rational "creatures" everywhere in the universe. The book has reminiscences of Fontenelle and of *Gulliver* and also allusions to mythology, as when the inhabitants of Venus are totally dedicated to love, those of Mars to warfare. Such literary and mythological allusions, as well as barbs of social criticism and ironic references to contemporary literature and philosophy, mean of course that the subject of the plurality of worlds is not seriously thought through as a philosophical problem. One philosophical theme that does make an appearance is that of anti-anthropocentrism: the Selenites are of the opinion that the Earth "must have been created solely for the sake of the moon," that is, for the sake of the

[113]On Krüger see Schatzberg, *Scientific Themes*, pp. 57–61.

moon dwellers (pp. 104–5); the inhabitants of Jupiter, who are far advanced in science, see the tiny Earth as an utterly insignificant celestial body and are surprised to find that the earthling is not wholly lacking in intelligence (pp. 112–14); on Saturn, heretics are burned for maintaining "that there is a star which they call the Earth" (p. 117), and so on. But this anti-anthropocentrism is no longer remarkable, particularly in this banal form, coming in the wake of Cyrano and of Voltaire's *Micromégas*.

The same applies to "The History of an Inhabitant of the Air, Written by Himself, with Some Account of the Planetary Worlds," a lengthy insertion in *A Voyage to the World in the Centre of the Earth* by an anonymous author whose identity has still not been established (London 1755). Certainly the aerial spirit's story is placed in a slightly more scientific context by the fact that it opens with a long discourse on the likelihood that both the visible celestial bodies and the unseen planets of other suns are inhabited. The author's way of thinking, which is both analogical and teleological, demands that God create nothing "in vain"; the worlds in the heavens must therefore be intended for the benefit of extraterrestrials, some of whom are undoubtedly as far superior to us as we are to the smallest and least sentient reptile (p. 107). For man, or more precisely the Christian, to think that everything was created for him is therefore ludicrously parochial (p. 108); and the remarkable story of the aerial spirit which follows is, so to speak, a practical demonstration of the theory. The spirit comes from Jupiter; in various incarnations, his soul has experienced life on the planets of the solar system and on a comet. As on Cyrano's sun, the rational inhabitants of Jupiter are birds and are far superior in "perfection" to the human beings of our Earth (p. 122), even if they are not immune to the temptations of parricide and political trickery. On the comet, the most extreme tyranny reigns; the Saturnians are "the vilest and most abject Slaves in Nature" (p. 139); Mars is the scene of the Golden Age, while the Earth is the home of all kinds of social and moral evils. So here, once again, the subject becomes a vehicle for topical satire, though the lively and eventful narrative of the aerial spirit's many-faceted life is some compensation. Lastly, the author of *Voyages de Milord Céton dans les sept planettes* (1765), by Marie-Anne de Roumier, Madame Robert, almost completely neglects the opportunity to treat the subject from a philosophical angle. In the days of Cromwell, the young English lord and his sister fly, with the help of an aerial spirit called Zachiel, from planet to planet and also to the sun. What they find there is simply the reality of astrological and mythological tales: a world of love on Venus, of warfare on Mars, of avarice on

Mercury, of rational lucidity on the sun, and so on. Although the description of the world of the sun does at least touch on science (the sun is the abode of terrestrial astronomers, among them Copernicus, Galileo, Descartes, and Newton), the inventive author is decidedly more interested in weaving a colorful and eventful story than in the philosophy of the plurality of worlds.

Philosophy makes a hesitant entry into the novel of entertainment in Ralph Morris's *A Narrative of the Life and Astonishing Adventures of John Daniel* (London 1751), which was apparently much read in the eighteenth century and which has recently been rediscovered as an early example of science fiction.[114] The—unintended—journey to the moon by means of a sort of sailplane is only one short episode (chap. 15) in this novel, which is a pure adventure story and deals mainly with the Robinson Crusoe–like existence of an English family on an island in the Indian Ocean, and also with a journey by air to another island inhabited by half-human, half-amphibianlike "monsters." But in the chapter about the moon, the author is not concerned, like the writers of the novels mentioned so far, with reflecting European civilization in a distorted or contrasting lunar world; only once, in a different context, does he briefly express a hatred of European civilization (p. 186). What interests Morris, and his airborne travelers John Daniel and his son, is the strange, marvelous world of the "most romantick country I had ever beheld" (pp. 169–70). It is a fabulous land whose nights and days each last two weeks, with crystal-clear water, mountains, fields, trees, and exotic animals, also inhabited by small men ("as I call them"), long haired and light footed, with a coppery sheen, who talk with the animals and live on plants. Only briefly do the new arrivals feel afraid of these moon dwellers (p. 174). For in fact the natives approach them with childlike friendliness and "in the humblest manner," that is, with the demeanor of colonial peoples toward the conqueror in the wishful thinking of the official version of the "white man's burden." In keeping with this benevolently imperialistic mode of thinking, the humans are most keenly interested in the moon dwellers' religion (a question Kepler had not yet ventured to broach and which Melanchthon had feared). "I was very desirous," John Daniel relates, "of informing myself, whether they had any religion amongst them." Their religion is a sun cult: at the monthly setting of the sun, the moon dwellers form ceremonial

---

[114]Second ed., 1770; 3d ed., 1801; newly issued as vol. 1 of Library of Impostors (London, 1926); 1st ed. reprinted New York, 1975. Extracts in Pizor and Comp, "*The Man in the Moone.*" Page references are to the first edition.

processions praying for the life-giving star to return, and its return is greeted by a corresponding procession. John Daniel reacts to this with the sense of superiority of the English Christian colonialist: "I pitied their ignorance, and attempted to shew them, that the sun which they had been making supplications to, was so far from being the giver of life itself, that it was but a creature (as we were) of the great Giver of life, and Maker of the world; and would have demonstrated to them, that he was only to be beheld by the understanding, not by the bodily eye; but I fear, that all my endeavours for their information was [sic] abortive, for I could not discern them a whit the wiser for them" (pp. 177–78).

Here one can clearly see an attempt being made by an enlightened Christian or Christian deist of the eighteenth century to fit the "new world" on the moon into the current religious (if no longer into a dogmatic) way of thinking, an attempt familiar from the encounter with the exotic peoples, in need of protection and enlightenment, of the worldwide British Empire. But Morris makes only the obligatory gesture in this direction; he spares himself any more intellectually demanding discussion by letting the visitors understand the moon people's language only "in trivial matters." The attractions of exoticism and adventure override philosophical curiosity as to the nature of the rational moon dwellers and their status in the context of the plurality of worlds.

What happens when interest in a lively sequence of events is subordinated to the Christian conqueror mentality is shown, in a most tedious manner, by a novel from what was then the least imperialistic nation: *Jonas Lostwaters, eines holländischen Schiffsbarbiers, Reise nach Mikroskopeuropien, einem neuerer Zeit entdeckten Weltkörper* (The voyage of Jonas Lostwater, a Dutch ship's barber, to Microscopeuropia, a recently discovered planet) by Johann Jakob Hertel (Glückstadt 1758). The first-person narrative begins excitingly—after all kinds of adventures at sea, an explosion that blows the ship up into the air propels the ship's barber to an unknown planet—but the main part of this seaman's yarn is heavily didactic. No sooner has Lostwater explained the incredible event by a scientific account of interplanetary gravitation than it emerges that the people on that planet are extremely similar to ourselves. Not only do they speak a language as close to Dutch as German is; Lostwater's host is working on plans for social reform that are indistinguishable from those of contemporary enlightened states. In the one instance where Hertel even begins to come to grips with the philosophical implications of the theme of the plurality of worlds, it is apparent that, while he is emancipated from the dogmatic approach of the seventeenth century, he still has chiefly theological interests at heart. What the ship's barber has to offer the extraterrestrials (who, be it noted, have not come to their planet from Earth but were "separately created" [*besonders erschaffen*, p. 174]) is "rational" Christianity (p. 160). And indeed the planet dwellers, represented by Lostwater's wise host, are immediately convinced of the superiority of this doctrine (pp. 40, 43); they are already predisposed to it (pp. 158–59), and the host is at once filled with a truly missionary zeal. After his death, the death of an extraterrestrial *anima naturaliter Christiana*, the ship's barber is, to his astonishment, brought back to Earth in a manner worthy of Münchhausen.

This kind of Christian imperialism extended to other parts of the universe sinks to the level of a religious tract in the small anonymous publication, *A Journey to the Moon, and Interesting Conversations with the Inhabitants, Respecting the Condition of Man*, "by the Author of 'Worlds Displayed' " (London, J. Evans, n.d.). This slim volume is "interesting," according to its author (the pastor John Campbell?) not because it gives insight into the moon's inhabitants, which might in turn shed some light on the human condition, but solely because it presents the fundamental tenets of Christianity in easily comprehensible form—for the instruction of the moon people. The narrator travels to them in a mysterious way: one night he is observing the moon when he is lifted into the air, together with his telescope, and several days later descends onto the moon's surface. The "inhabitants," of whom there is no description whatever except that there is no death among them, greet him ceremoniously and, after establishing some cosmic personalia, ask him questions solely concerned with the religious ideas of the Earth's inhabitants. The narrator, who for his part has no questions to ask of the moon dwellers, explains to them that the death to which earthly man is destined is a punishment for the sin committed in Paradise and goes on from there to expound, with well-meant pedantry, the doctrine of salvation by the crucifixion of Jesus Christ. He ends with a lurid description of the misery of those who do not obey God's commands, in comparison with which his picture of the bliss awaiting those who acknowledge Christ is brief and colorless. (So much, at least, this innocent author has in common with Dante and Milton.) The moon dwellers learn all this with astonishment, but it all seems to be irrelevant to them, since as immortal human beings who have not succumbed to original sin, they are clearly not in need of conversion. So the author, in his concluding sentences, turns to his *readers*, urging them to weigh the joys of faith against the deceptive pleasures of sin. In the course of this appeal, he not only neglects to describe how he returned to Earth but also omits to explain how someone holding his beliefs can come to terms with the assumption of a sinless human race on the moon—and why he introduces it at all, since for his purposes it seems unnecessary and indeed pointless.

Various such attempts were made by unsophisticated thinkers, not to reject out of hand the idea of the plurality of humankinds but to come to terms with it intellectually,[115] starting out from the familiar framework of religious beliefs. Among these we may look first at one group of authors whose treatment of the subject of inhabited worlds is still expressly aimed at finding an accommodation with the Christian world view. In the history of thought, they represent a transitional phenomenon between theological-dogmatic and philosophical-secular habits of thought. In their effort to "place" extraterrestrial life and define it more closely, they look, as it were, for the blank areas on the map of dogma and there let loose their imagination: the stellar humankinds are beings created by God who are damned for their sins or rewarded with blessedness for their piety, or who have not suffered a Fall. This is one possible form of the belief in plurality, one that theological thinkers from the seventeenth century onward had already considered and that was occasionally put forward in the eighteenth century in the context of Newtonian cosmology.[116] As theology it is somewhat anachronistic in the Age of Enlightenment, but in the realm of literature such ideas open the way for a species of imaginary voyage, at once scientific and imaginative, that often has great charm.

C. S. Lewis is thus not the first writer whose scientifically trained literary imagination has been fired by such ideas. Two hundred years before his trilogy of novels, the idea of sinless human beings on other planets entered science fiction in a brief story of a space voyage that has remained more or less unknown, although it was republished in 1923 as the seventeenth volume in the Erasmus series. It is entitled *Die geschwinde Reise auf dem Lufft-Schiff nach der obern Welt, welche jüngs-*

---

[115]Such rejection does, by the way, occur even in literary works whose authors one would unhesitatingly class as representatives of the Enlightenment: as late as 1756 the popular exponent of Enlightenment Michael Richey of Hamburg issues a vigorous rebuke to the "strong spirit" and empirical scientist who "has sent inhabitants to every wandering star": "You make everything clear *[du klärest alles auf],*" he accuses; "we can see what you intend, but you have not yet made clear the truth" (*Deutsche Gedichte* [Hamburg, 1764], 2:228). Already in 1720 Daniel Defoe had proclaimed the "truth" in the appendix to the third volume of his *Robinson Crusoe* ("Vision of the Angelick World"): the climatic conditions on the various planets are unsuited to human life; the Creator has placed only spirits, devils, and angels in outer space.

[116]See Robert Jenkin, *The Reasonableness and Certainty of the Christian Religion: Book II* (London, 1700), p. 222. There is also a suggestion of this idea in § 19 of the "Essays on the Goodness of God . . ." in Leibniz's *Théodicée* quoted in sec. 3 above.

thin fünff Personen angestellet, um zu erfahren, ob es eine Wahrheit sey, daβ der Planet Mars den 10. Jul. dieses Jahrs das erste mahl, so lange die Welt stehet, mit einem Trabanten oder Mond erschienen? (The swift journey by airship to the upper world, recently undertaken by five persons in order to discover whether it is true that on 10 July of this year the planet Mars appeared, for the first time since the world began, with a satellite or moon, 1744; 2d ed., 1784). The unnamed author is the astronomer Eberhard Christian Kindermann, who had already, five years earlier, published an astronomical textbook entitled *Reise in Gedancken durch die eröffneten allgemeinen Himmels-Kugeln* (Journey in thought to all the discovered celestial globes), and in his preface to the *Geschwinde Reise* he particularly refers to the new edition of the *Reise in Gedancken*, now called *Vollständige Astronomie* (Complete astronomy, 1744), in order to show that even in this space fantasy he writes as a serious scientist.

But in his astronomical guide, Kindermann had already gone far beyond a purely scientific approach, with its exact indications of distances, orbiting speeds, intensity of light, temperatures, and surface conditions on the various celestial bodies. What he is concerned with most of all in this book, alongside the topography, flora, and fauna of the "worlds," is their "rational" inhabitants, whose existence he assumes on analogical and teleological grounds (*Reise in Gedancken*, 1739, pp. 28–29, 46). What he has to say of them is beautifully simple: the more "splendid" a planet appears, the more noble, happy, refined, and rational in the knowledge and veneration of God are its inhabitants. With the exception of the "Marsites," all the planetary human-kinds of the solar system (including the Lunarians), and incidentally those of Sirius's system too, are far superior to us in this respect. But, this astronomer and "Christian artist" (as he calls himself on the title page) defines the superiority of the extraterrestrials and the inferiority of man, both quite frequently to be found in the Enlightenment, in a purely theological manner, expressing them in Christian terms: on those planets there has been no Fall; the cosmic paradises are still out there with their Adam-like human beings who know nothing of death and sin, of work and the hardships of life (and to whom God speaks in Hebrew). That the inhabitants of Mars are sinful is conclusively proved by the inhospitality of their planet, though they too are to be saved somehow, if not by Christ; still, the sinful inhabitants of Earth are better off in that God sent his son to them in human form (p. 97). If, even so, there is still more than one religion on Earth, this is an effect of the Fall, says Kindermann with all the pious parochialism of his noncolonizing nation (p. 55).

In the *Geschwinde Reise*, Kindermann transfers these theological-cum-scientific extrapolations to the genre of the space novel. The five passengers on the airship are, oddly enough, allegorical figures representing the five senses—Auditus, Visus, Odor, Gustus and Tactus—and their adventures are related by Fama. Yet this is not simply an allegorical dream-vision. As in Cyrano de Bergerac there is a precise description, on physical principles, of the airship and its equipment—which includes a telescope—in order to give the journey a modicum of scientific credibility. The universe is of course that of post-Copernican astronomy. The travelers land on the newly discovered "Moon of Mars," a fairy-tale world with brightly colored vegetation, strange fabulous animals, and "creatures" that "in their external appearance resembled the people of the world below" (1744, p. 15), even if they are organized less "crudely" and more ethereally (p. 17). As in *John Daniel* and in Krüger's Eighteenth Dream, the astronauts' first reaction, despite the affinity, is one of fear: the shock of what is quite unknown, quite different. So they display neither the superiority a reader of Fontenelle would anticipate nor the recognition of familiar traits common to all humanity that a follower of Huygens might expect. Here too, however, that fear is soon shown to be groundless: the lunar beings live in their natural paradise in "love and harmony" (*Liebe und Verträglichkeit*), completely without "wants or infirmities." "Fear God" and "Love one another" are the only commandments by which they live (p. 19). They are "a thousand times more intelligent" than their human visitors (p. 24) and, naturally, happier; indeed, God reveals himself to them, "speaks" to them (p. 22).

The reason for this greater happiness provides the theological framework: in contrast to "humans, who are fallen, utterly and completely, indeed, to the very depths," the inhabitants of the moon of Mars and their ancestors have not fallen but are still party to the "mysteries" of God (pp. 22, 24). It is, however, most revealing that Kindermann is not content only to downgrade man theologically: he also speaks as a representative of the Enlightenment. For his moon dwellers, who have not developed technologically at all, take the construction of the airship as a sign that by God's will "your reason, completely darkened by the Fall" is growing "light again." A typical Enlightenment metaphor, it is, however, at once interpreted as the fulfillment of the eschatological biblical prophesy that says that "in the last days, when your world is close to transformation, many mysteries lost through the Fall will once again become clear." But one of these Christian mysteries proves to be the scientific view that "each of the twinkling stars is a world inhabited by creatures, a certain number being placed around each fixed star" (p. 24). Thus Giordano Bruno, burned as a heretic, would be vindicated by both scientific and Christian modes of thought! The violence of the effort necessary to give theological sanction to scientific beliefs is all too plain: by the middle of the eighteenth century, such attempts are obsolescent.

Similar ways of thinking are applied to the whole inhabited universe in one of the most curious space novels of the eighteenth century, *The History of Israel Jobson, the Wandering Jew* (1757). The author is Miles Wilson, vicar of a remote Yorkshire village, who has no other claim to fame.[117] His titular hero, Jobson, the Wandering Jew, begs God to alleviate his punishment, with the result that the Lord sends him his guardian angel, who travels with him for a year, not around the Earth but to the worlds of the heavens. He travels by means of a flying chariot, like Elias on his ascent to heaven. Despite this biblical vehicle, however, we are decidedly in the world of the new science. The very title page announces that the author will adhere to the "Principles of Natural Philosophy, concording with the latest Discoveries of the most able Astronomers." Newton is the chief source of the information about the distances and sizes of planets, orbiting times, and gravitational conditions, which the angel tirelessly imparts to his charge in their conversations. We are worlds away from the singularist orthodoxy of Kircher's guiding angel, but this scientifically well-informed modern angel has also retained his Christian way of thinking. This is most apparent in his commentary on the more or less humanoid beings that populate all the celestial bodies, since God (here we have teleology straight from the angel's mouth) created nothing "in vain" (p. 54).

The cosmic humankinds the angel shows Jobson on his celestial voyage each have their theological place in a frame of reference that is extrapolated from specifically Christian ideas, however secular and exotic Wilson's portrayal of the extraterrestrial worlds is in other ways. The moon's inhabitants (whose existence is by this time hardly supported any more by the astronomical textbooks) are only three feet tall, but otherwise quite human, except that they are made of clanking metal; their surgeons are plumbers, who have their hands full, given the moon dwellers' love of warfare. Their moral status, it seems, is probably the same as ours (p. 24). On Mars, in contrast, live nine million happy "Rational Beings" of uncertain form—sexless, scarlet in color, immobile as trees; their only occupation is praising the Creator,

[117]See George K. Anderson, "The History of Israel Jobson," *Philological Quarterly* 35 (1946), 303–20. Page references are to the 1757 edition.

but they have received no revelation and are not destined for eternal bliss. The inhabitants of Jupiter, however, gigantic in stature and intellectually far superior to humans, are undergoing probation in their land of Cockaigne, from where they will proceed either to hell or to paradise depending on what powers of resistance they have shown to the temptations of sensuality. The Saturnians, on the contrary, are irrevocably destined for eternal bliss. They have one eye in the forehead and another in the back of the head and live in a state of innocence; no Fall has ever taken place in their world. The same is true of the inhabitants of a distant star in the Milky Way, of which Wilson paints a ravishing picture in the manner of the prevailing fashion for the exotic:

> But how shall I describe this Fair, this Fragrant, this charming Land of Love! the Delectable Vales and Flowery Lawns, the Myrtle Shades and Rosy Bowers, the Bright Cascades and Chrystial Rivulets, rolling over Orient Pearls and Sands of Gold, which here spread their silent Waves into broad Transparent Lakes, smooth as the Face of Heaven, and there break with rapid Force thro' Arching Rocks of Diamond and Purple Amethists! Plants of immortal Verdure creep up the Sparkling Cliffs, and adorn the Prospect with unspeakable Variety: Whatever can raise Desire, whatever can give Delight, whatever can satisfy the Soul, in all the boundless Capacities of Joy is to be found here: The Inhabitants my Conductor told me, who indeed were charming Creatures, were exempt from all Evil, blest to the Height of their Faculties and Conceptions, and were priviledged with Immortality. (Pp. 78–79)

A similar world, somewhere in our galaxy, belongs to those who have passed their test, and paradise itself is also like this. Jobson is able to look at it briefly before the heavenly chariot takes him back to Yorkshire, where the angel promises him ultimate redemption. Thus each of these worlds—except perhaps for the grotesque moon-world, where the clerical story-spinner freely indulges his sense of humor—has a specific theological meaning and place value, and this gives Wilson's conception of the universe, for all its highly imaginative and also banal elements, a certain seriousness of intellectual effort.[118] Wilson's ambition is both theological and scientific. His main aim may have been to spread astronomical knowledge, but what makes his folksy book interesting in the context of the history of thought is his

---

[118]See also Anderson, ibid., p. 306: "The pamphlet was meant to be taken seriously on the whole."

attempt to find a common denominator for astronomical facts, the plurality theory of "natural philosophy," and the Christian system of thought.

There is no doubt, however, that the most influential exponent of this kind of thinking, a writer whose style is in total contrast to Wilson's, is Friedrich Gottlieb Klopstock. In his *Messias* (Messiah, 1748–73) he places the biblical subject of the Fall and the redemption in the universe of countless worlds vouched for by science, thereby risking conflict with orthodoxy. But he solves the conflict between the plurality of humankinds and the doctrine of salvation by assuming that the inhabitants—the "human beings" or "people" (*Menschen*) of other worlds such as Hesperus or Jupiter (canto 1, line 460)—have not fallen prey to sin. Not descended from Adam and Eve, they are not comparable with us from the theological point of view: "People, like us in form, but full of innocence, not mortal people" (5, line 154, quoted below). As in Wilson and Kindermann, this assumption could be regarded as dogmatically safe to the extent that there was no dogma unequivocally supported by the Bible that made a judgment on this matter: it was, as I have said, a blank area on the map of dogma. Later, in Canto 17, Klopstock puts forward another assumption that he probably regards as defensible for the same reason. This is the assumption that the Lord chose a certain planet, not precisely localized, on which to place the sinners who drowned in the Flood and are not yet saved (17, lines 85 ff.).[119]

Klopstock pictures these cosmic "worlds" in a wholly "earthly" way, with hills and fields, trees and flowers; and his extraterrestrial people are wholly modeled on terrestrial ones. The following patriarchal family scene taking place on an unnamed star peopled by unfallen, immortal beings suggests some Swiss idyll:

Gott ging nah an einem Gestirne, wo Menschen waren;
Menschen, wie wir von Gestalt, doch voll Unschuld, nicht
    sterbliche Menschen.
Und ihr erster Vater stand voll männlicher Jugend,
Ob in dem Rücken des Jünglinges gleich Jahrhunderte waren,
Unter seien unausgearteten Kindern. Das Auge
War ihm nicht dunkel geworden, die seligen Enkel zu schauen;
Noch zu der Freudenträne versiegt. Sein hörendes Ohr war
Nicht verschlossen, die Stimme des Schöpfers, der Seraphim Stimme,

---

[119]See Hans Wöhlert, *Das Weltbild in Klopstocks "Messias"* (Halle, 1915), esp. pp. 15–17, 36–37.

Und aus dem Munde der Enkel dich, Vaternamen, zu hören.
An der Rechte des Liebenden stand die Mutter der Menschen,
Seiner Kinder, so schön, als ob itzt der bildende Schöpfer
Ihres Mannes Umarmungen erst die Unsterbliche brächte;
Unter ihren blühenden Töchtern der Männinnen Schönste.
An der linken Seite stand ihm sein erstgeborner,
Würdiger Sohn, nach des Vaters Bilde, voll himmlischer Unschuld.
Ausgebreitet zu seinen Füßen, auf lachenden Hügeln,
Leichtumkränzt mit Blumen ihr Haar, das lockichter wurde,
Und mit klopfendem Herzen, der Tugend des Vaters zu folgen,
Saßen die jüngsten Enkel. Die Mütter brachten sie, eines
Frühlinges alt, der ersten Umarmung des segnenden Vaters.

(5, lines 153–72)

God walked close to a star where there were people; people like us
in form, but full of innocence, not mortal people. And their first
father stood full of manly youthfulness (even though this youth had
centuries behind him), amid his undegenerate children. His eye had
not grown too dim to look upon his happy grandchildren, nor too
dry to shed a tear of joy. His hearing ear was not closed to the voice
of the Creator and the voice of the seraphim, nor to you, O name of
"Father," from the mouths of the grandchildren. On the right of the
loving one stood the mother of that human race, of his children, as
beautiful as if the Creator who formed her had only just brought her,
the immortal, to the embraces of her husband; the most beautiful of
the women, among her blooming daughters. To his left stood his
first-born, worthy son, in his father's image, full of heavenly innocence. Spread out at his feet on smiling hills, their hair, growing more
curly, lightly wreathed with flowers, sat the youngest descendants,
their hearts pounding with the desire to emulate the father's virtue.
Their mothers brought them, after they had lived through one spring-
time, for the first embrace and blessing from the father.

Klopstock's Christianized conception of the plurality of worlds is
shared, as one would expect, by Johann Jakob Bodmer in his *Noah*
(1752). But he goes further: with bold anachronism he links the world
of the Old Testament far more directly with the new science than
Klopstock would ever have thought of doing. Thus the telescope has
already been invented in biblical times, and as Wieland says in his
"authorized" discussion of the poem, *Abhandlung von den Schönheiten
des epischen Gedichts 'Der Noah'* (Treatise on the beauties of the epic
poem "Noah," 1753), Sipha uses it to make "all the discoveries that
Galileo, Scheiner, Cassini, Hevel, and Lahire have made in modern

times."[120] Significantly, Wieland in his commentary attempts to reconcile the scientific plausibility of inhabited worlds with Christian doctrine in the spirit of Bodmer: he goes on to say that only "people whose brains have frozen solid" could possibly still imagine themselves to be at the center of a universe created for them; "there is therefore nothing more credible than that the Patriarchs, so rational and so free from all prejudices, should have had no difficulty in admitting the supposition that the heavenly bodies were separate worlds, especially as they could obtain confirmation of that supposition through their intercourse with the angels." Especially!

The extent to which the young Wieland, in his own literary work too, was committed to the course Klopstock and Bodmer set with regard to the plurality of worlds is most clearly shown by the first work produced in his Zurich days, while he was staying at Bodmer's house: *Briefe von Verstorbenen an hinterlassene Freunde* (Letters from deceased people to friends left behind, 1753). His own paraphrase of the ninth of these letters shows him thinking along the same lines as his models: "Theotima describes one of the planets that exist among the innumerable stars of the Milky Way, a planet that is inhabited by innocent human beings (*Menschen*), and tells the story of the Creation, the temptation, and the victory of the first parents of the race of these happy creatures."[121] And in the fourth letter he draws up, as a sort of framework for the whole, the cosmology of a plurality of inhabited worlds, which, significantly, uses traditional Christian concepts: spheres, cherubim, seraphim, angels, "movers of the stars," and God's throne.

So long as the scientifically based subject of the plurality of worlds is allowed to occupy only those areas left unoccupied by the tenets of Christian dogma, it barely undermines man's view of himself, whether by arousing fear of alien life-forms (which, when it arose in Kindermann and others, was quickly dispelled) or by causing the soul to be overcome by giddiness ("Seelen-Schwindel" as Brockes calls it, anticipating the language of Jean Paul).[122] Yet, as soon as authors no

[120]Christoph Martin Wieland, *Werke* (Berlin: Hempel), pt. 40, p. 465.

[121]*Gesammelte Schriften*, sec. 1, vol. 2, ed. Fritz Homeyer (Berlin, 1909), p. 81. But, in two didactic poems written in the year before he moved to Zurich, *Die Natur der Dinge* (bk. 4) and *Zwölf moralische Briefe* (12th letter), Wieland had already treated the idea of the plurality of worlds in his own way, that is to say unrestrained by Christian dogma and adhering particularly closely to science. Thus the perfection of the beings on other celestial bodies was not that of beings untouched by the Fall, but that of the "virtue" and "love of humankind" (*Menschenliebe*) of the Golden Age (sec. 1, 1:306–7); similarly, the fact that man on our planet is given the flattering poetic title of "Nature's favorite" does not betoken a very Christian attitude (sec. 1, 1:85).

[122]*Irdisches Vergnügen in Gott* 4:8. Page references after quotations from Brockes are all to the Peter Lang reprint, which follows the last edition of each part (see n. 44).

longer feel bound to adopt this comforting frame of reference, the question of man's nature and status thrusts itself forward far more urgently and disturbingly.

The most noteworthy instances of this in eighteenth-century space novels are two works in which the topic is reached by way of satire (toward which the planetary novel in this century has a general tendency). These are the voyage to the flying island Laputa in the third part of Swift's *Gulliver's Travels* (1727, chaps. 1–4) and Voltaire's *Micromégas* (1752). At first sight it would appear that in both works the subject that interests us is wholly buried beneath the satire; a closer look, however, reveals unmistakably that this subject in fact forms the very substance of these philosophical novels.

The journey to Laputa, in particular, is thickly encrusted with satire. Here Gulliver relates how, after pirates have captured his ship, he drifts ashore on an island in the South China Sea which later turns out to be the country, or continent, called Balnibarbi. Hardly has Gulliver's boat drifted ashore on this landmass when a "flying island" approaches—round, four and a half miles in diameter, and with buildings visible on it. It consists mainly of a metal ("adamant") that is both magnetic and diamondlike. The humanlike beings ("people," "men") who live on this "island" pull Gulliver "on board" by means of a chain, and here he comes to know a "race of mortals" that is the least familiar of all those he has met on his sea voyages—a truly extraterrestrial race. They hold their heads bent either to the right or to the left; one eye looks inward, the other toward the zenith; their clothing is decorated with suns, moons, and stars; and their intellect, or at least the intellect of the men (whose wives run away) is continuously engaged—making them totally absent-minded—in mathematical and astronomical speculations, from which they have to be roused from time to time by means of a rattle. Even the king in his palace, surrounded by his court, is occupied in this way. Every aspect of life on Laputa is governed by mathematics: meat is served cut up into geometric shapes, beauty is judged on the basis of the same geometric relationships, the tailor takes measurements with the aid of a quadrant and compasses, and so on. That abstract science is being ridiculed for its uselessness and irrelevance to real life is obvious, and it has been shown that this mockery has specific targets; altogether, in this third part of the *Travels*, Swift makes mock of contemporary scientific research as practiced by members of the Royal Society in particular.[123]

[123]For further detail see Marjorie Nicolson, "The Scientific Background of Swift's Voyage to Laputa," in *Science and Imagination* (Hamden, Conn., 1976), pp. 110–54; Colm Kiernan, "Swift and Science," *Historical Journal* 14 (1971), 709–22.

Yet the science pursued on Laputa also has a disturbing side that cannot simply be laughed away by common sense and that is a pointer to the science fiction theme concealed in this apparently absurd story. This disturbing aspect has to do with the nature of the flying island and its movement and thus with the basic scientific facts underlying this fiction, which Swift sets out at some length in the third chapter. Mounted inside the flying island is a six-yard-long magnet with one attracting and one repelling pole. As the interior of the Earth beneath Balnibarbi has a particularly strong magnetic field, it is possible by turning the magnet to set the flying island in motion either horizontally or vertically and to steer it with precision (though only within the range of the Earth's magnetism, which Swift estimates at four miles). It has been shown that Swift derives this idea from William Gilbert's experiments with magnetism. His use of them in a literary fantasy, however, introduces, strictly speaking, not only one but two motifs associated with the theme of the plurality of worlds.

For Swift deliberately leaves us in doubt as to whether this flying island, which has the "usual" geologic layers but also landing steps and observation "galleries," is to be thought of as a natural celestial body, like a planet, or rather, as the product of an advanced technology. In the former case, the mutual attraction and repulsion of Laputa and Balnibarbi—operating across the space between them—would be a somewhat idiosyncratic model of Newtonian universal mechanics. Those who have argued for this interpretation have stressed that Gilbert regarded the Earth as a magnet and, in his experiments with the variable influence of two magnets on each other, called the smaller one "terrella."[124] Swift would then be presenting a model, limited to two planets, of the plurality of inhabited worlds. The difficulty of course is that here the determining force is not gravitation but magnetism and that Swift's "terrella" can be manipulated by the Laputian engineers, which greatly diminishes the analogy with a planet. It therefore seems more logical to see the flying island as the gigantic spaceship or space colony of a technologically more advanced manlike "race." True, Swift may be using this idea too for satiric purposes, making fun of the various seventeenth- and eighteenth-century attempts to construct airships, at least one of which (by Athanasius Kircher) was based on the same magnetic principle.[125] That does not

[124]Marjorie Nicolson, "Swift's 'Flying Island' in the *Voyage to Laputa*," *Annals of Science* 2 (1937), 413–19; for the following sentence, see the same article.

[125]Nicolson, "Swift's 'Flying Island,' " pp. 419–30; Paul J. Korshin, "The Intellectual Context of Swift's Flying Island," *Philological Quarterly* 50 (1971), 630–46 (p. 641 on Kircher).

rule out, however, that Swift is playing seriously with the science fiction idea of a higher extraterrestrial civilization making contact with us (instead of the other way round, as was more usual in eighteenth-century space voyages).

That the scientists on Laputa are superior to the English astronomers of the day (Newton's day!) Swift stresses repeatedly in the second, third, and fourth chapters; as the representative of a world that is backward in science, particularly in mathematics and astronomy, Gulliver is an object of contempt, in so far as he receives any attention at all. No one on Laputa has the slightest interest in the laws and systems of government, history, religion, and customs of the countries of Earth, except for one courtier who is regarded as "the most ignorant and stupid person among them." But not only is man, seen through Swift's misanthropic eyes, downgraded intellectually; the Laputians are also superior in power. The narrator interprets the name of the flying island etymologically as "high governor"—reluctantly; but Swift takes such unmistakable pleasure in elaborating on Laputa's oppressive superiority over Balnibarbi that we may legitimately see this emphasis on power as an early instance of one particular motif of science fiction. And, as we might expect of Swift, this superior power is not that of benevolent, guardianlike despotism such as we find later for instance in Kurd Lasswitz (*Auf zwei Planeten*), A. C. Clarke (*Childhood's End*), and Steven Spielberg (*Close Encounters of the Third Kind*), but the alien, brutal superiority in strength of an ethically indifferent superrace, as in H. G. Wells (*The War of the Worlds*) and countless authors of light fiction. Laputa has subjugated the island or continent above which it hovers, like a flying saucer or a space station, and exacts tribute from it. Any opposition is punished simply and effectively: by altering the position of Laputa at will, its engineers can deprive the people below of sunshine and rain; they can bombard their towns with missiles and even wipe them out completely by simply letting the flying island drop down on to Balnibarbi. This is not only, if at all, a satire on the inhumanity of science (as has been claimed even recently, with a sideways glance at the threats posed by science in our own day);[126] the Laputa episode can more plausibly be read as a scientific-technological nightmare of a cosmic disaster hanging over the Earth, a catastrophic encounter with an incalculably strange and superior extraterrestrial life-form, in which man is doomed to annihilation. This is the interpretation that suggests itself if one considers the work in the light of the later development of the genre and not in that

---

[126]Kiernan, "Swift and Science."

of the then-usual themes of the space novel. The two perspectives are not mutually exclusive; but by stressing the link with the later development, one is able to identify in this part of *Gulliver's Travels*, which is too often interpreted as merely contemporary polemic, a theme that is still productive and still relevant today. Even if Swift's vision, apocalyptic in a secular sense, was inspired by a misanthropic schadenfreude, there is clearly more to it than that. We sense in it an early manifestation of the cosmic fear that characterizes a large proportion of serious modern space novels.

The story of an "invasion" of our planet from outer space, a staple of contemporary science fiction, is also to be found in Voltaire's *Micromégas* (1752). Unlike Swift's Laputians, the two visitors here— one from Saturn, the other from one of the planets of Sirius—are not at all hostile to man but are animated by a benevolent curiosity, so that we have here the other one of the two typical possibilities available to modern treatments of the "cosmic invasion" theme. Yet although this is the friendly type of encounter, Voltaire goes much further than Swift in calling man's identity into question and undermining his sense of security, and the reason for this is that from the outset Voltaire takes a more philosophical approach to the subject. What the two extraterrestrials voyaging through space and visiting the Earth are engaged upon is "a short philosophical tour" (pp. 88/127);[127] to be more precise, it is concerned with "natural philosophy," the *philosophia naturalis* whose principles Newton had established for his age. Voltaire was one of the leading Newtonians among the writers of his century. He studied the cosmology of Newton and his followers intensively for years and popularized it in his *Éléments de la philosophie de Newton* (1738). The very title of this work shows the nature of this encounter between the philosophical thinker and the rigorous scientist: Voltaire is interested not in mathematical data but in the philosophical conclusions to be drawn from the new conception of the universe. These in turn provide inspiration for his imagination as a literary writer, and the result is *Micromégas*, which was probably begun in the late 1730s.

Certainly the scientific substratum of the novel derives from Voltaire's study of the new science.[128] The nature of the philosophical

---

[127]For quotations from *Micromégas*, two page references are given. The first is to Voltaire, *Candide and Other Tales*, with introduction by H. N. Brailsford, trans. Tobias Smollett, rev. J. C. Thornton (New York: E. P. Dutton, Everyman's Library, 1937; repr. 1971), from which the English quotations are taken; the second is to the French critical edition by Ira O. Wade in his book *Voltaire's "Micromégas": A Study in the Fusion of Science, Myth, and Art* (Princeton, N.J., 1950).

[128]See Colm Kiernan, "Voltaire and Science, a Religious Interpretation," *Journal of Religious History* 4 (1966), 14–27.

conclusions he draws in *Micromégas* from Newton's confirmation and completion of the Copernican revolution does not, however, emerge with total clarity. The fundamental conclusion *is* clear: for Voltaire too, at least in this work, the Copernican-Newtonian system, or rather systems, make it necessary to assume the plurality of inhabited worlds on analogical grounds. That he should have used this idea purely as a vehicle for satire and devastating mockery of all kinds of abuses prevalent in the civilized world (or even mockery of the genre of the literary space voyage itself)[129] does not seem likely when one learns from other, reliable, sources that Voltaire, like so many Newtonians of his generation, was firmly convinced of the plurality of worlds both before and after he wrote *Micromégas*.[130] In depicting our world, therefore, as seen through the eyes not of a South Sea islander but of a Saturnian and an inhabitant of a planet of Sirius, he is not simply adopting a literary convention or, in other words, merely employing a thematically irrelevant artistic device. In Voltaire's description of the planet-trotting extraterrestrials, there is certainly an incidental element of polemic—just as he made fun (as noted in sec. 3 above) of Christian Wolff's speculation on the physical size of the Jovians and of the assumption, made for instance by Nicholas Hill, of a relationship between the size of a globe and that of its inhabitants; but this does not mean that the theme is *primarily* a vehicle for polemic.

True, the opening of the story is largely shaped by this polemical interest. Micromégas, the young man from a planet close to Sirius, is eight leagues high, in accordance with the greater diameter of his planet of origin (pp. 83/120), and he knows of inhabitants of other planets who are incomparably bigger, "a single foot of whom is larger than this whole globe on which I have alighted" (pp. 98/139). The Saturnian accompanying him on his educational journey (who is incidentally, in allusion to Fontenelle, the secretary of the Saturnian Academy) is, in accordance with the dimensions of his planet, a mere dwarf six thousand feet high (pp. 85/122). The two of them make use of the laws of gravitation to travel on sunbeams and comets and finally reach the Earth, landing on the north coast of the Baltic Sea, and walk right round the Earth in thirty-six hours. They are at first convinced that this world is uninhabited, because they are unable to see any living

---

[129]This is suggested in particular by Nicolson, *Voyages to the Moon*, pp. 214–16. See Richter, *Literatur und Naturwissenschaft*, p. 176.

[130]Wade, *Voltaire's "Micromégas,"* gives a few quotations relating to this conviction, pp. 42–45. See also the angel's speech at the end of *Zadig*; A. Owen Aldridge, *Voltaire and the Century of Light* (Princeton, N.J., 1975), p. 113; Colm Kiernan, *The Enlightenment and Science in Eighteenth Century France*, 2d ed. (Banbury, 1973), p. 52.

creature, however small. But the author so arranges it that, close to the Baltic, a diamond necklace worn by the giant from the Sirian system breaks; gathering up the diamonds, he notices that they act as magnifying glasses, and with their aid, he discovers first a whale and then a ship on which the philosophers and naturalists of Maupertuis's expedition to Lapland are returning home. Once the extraterrestrials have realized that the ship, complete with passengers, is not an animal, they designate the teeming human beings as mites and are amazed that it should be possible for such tiny creatures to exist. The idea that these "atoms" could possess reason, a soul, and language seems to them absurd—until Micromégas lifts the ship onto his thumbnail and is to some extent able, by means of an ear trumpet made from a fingernail, to hear and understand the confused buzz of voices: it is an inexplicable "jest of Nature" (pp. 96/137).

As Voltaire juggles in this way with the astonishing effects produced by the scientifically calculated relative sizes, his aim obviously goes far beyond esoteric polemics. He in fact succeeds in creating a disturbing variation of the Pascalian theme of the "disproportion of man" (see chap. 3, sec. 1 above)—the merely relative view of man, placed as he is midway between the infinitely great and the infinitely small—which was disconcertingly topical in an age when microscope and telescope vied with each other to reveal undreamed-of worlds. Voltaire underlines this topicality by speaking directly and in scientific terms of contemporary discoveries in the field of the infinitely small: "What wonderful skill must have been inherent in our Sirian philosopher that enabled him to perceive those atoms of which we have been speaking. When Leuwenhoek and Hartsoeker observed the first rudiments of which we are formed, they did not make such an astonishing discovery" (pp. 94/135). As the relative sizes are of course used symbolically in a number of ways, the effect of this demonstration of man's position in between two abysses is, in every case, to demote the pinnacle of Creation in a disturbing manner. Or, to put it in terms more appropriate to enlightened secularism, the measure of all things is declared invalid.

Man, the "mite," "so near akin to annihilation" (pp. 97/138), is barely visible on the Sirian's thumbnail, even with the aid of a powerful magnifying glass, and is politely pitied for his small size by Micromégas and asked whether he has a soul; and yet he philosophizes about heaven and Earth, confidently claiming validity for his conclusions—which, incidentally, differ greatly from one thinker to another (chap. 7). True, the visitors from space are modest and well mannered enough not to see the power of man's intellect as being necessarily in direct proportion to his bodily size; the Creator could have made beings even smaller than man yet with reason greater than that of the giants whom Micromégas has seen on his grand tour "in Heaven" (pp. 98/139). In the cosmic perspective, any logical argument based on a gradation is thus declared absurd. Nevertheless, the arrogant self-assurance with which humans, with their worm's-eye perspective, express the most contradictory views about the nature of the soul and of matter (the ship holds representatives of all schools of philosophy of the time) causes Micromégas to smile at the unrealized ignorance of the scholars. Both visitors break into Homeric laughter when one of the little creatures (*animalcules*), a theologian at the Sorbonne, declares to them "that their persons, their [worlds], their suns and their stars were created solely for the use of man" (pp. 102/145).

This mockery of anthropocentric "pride," particularly in this theo-logical-dogmatic form, is usually seen as the point of this small "his-toire philosophique" (the subtitle added in later editions).[131] Certainly "that inextinguishable laughter" forms the climax; but in the middle of the eighteenth century—a century after Cyrano—mere anti-anthro-pocentrism is hardly a sensational outcome for a discussion of the plurality of worlds. It is in the closing sentences that Voltaire seems to point to the true nub of the matter. No sooner has Micromégas recovered from his laughter (which is good natured but also tinged with annoyance) at the terrestrials' anthropocentric teleology, than he promises the disputing theologians a philosophical book from which they will discover "the purpose of existence" (pp. 103/146; "*le bon [le bout] des choses*; see p. 179). He keeps his promise; but the book he gives them contains nothing but blank pages.

The ultimate purpose of existence is unknown, undiscoverable, or nonexistent. Looking back from this point, we now notice that in *Micromégas* and elsewhere, Voltaire justifies the plurality of worlds only on the grounds of analogy and never supports it, as was so common at the time, with the teleological argument (i.e., there is a "design" for the extraterrestrials). If, then, there is no question for Voltaire of such a guiding purpose on the part of God, what can possibly be the purpose or meaning of a plurality of humankinds in the universe? Here we approach the hidden core of his little space fantasy. One thing is perfectly clear: the inhabitants of the various other planetary worlds are essentially neither better nor worse off than we are. Some humankinds may be more intelligent and also physically

[131]Thus for instance W. H. Barber as recently as 1976: "Man is put in his cosmic place, so to speak" ("Voltaire's Astronauts," *French Studies* 30, p. 36).

larger, able to live longer, and endowed with more senses (on Saturn they have seventy-two, on the planets of Sirius about a thousand, and far more elsewhere). But everywhere the same complaints are made about the brevity of life, the inconstancy of love (an erotic idyll on Saturn lasts barely one hundred years), ignorance about the "last things," inferiority compared to the human beings on other planets. Everywhere there are all-too-human failings such as unjustified arrogance and both political and religious intolerance (thus Micromégas has to leave his native planet because his scientific book offends the censor, who has not read it). Of course, in thus ironically pinpointing the similarities between the inhabitants of all planets, Voltaire is obviously satirizing the imperfect, ever-discontented human race that pursues its existence on *this* "little mud-heap," as he calls it (pp. 84–85/ 121, echoing writers from Pico della Mirandola onward). Indeed, the strength of his impulse to satirize terrestrial man causes him to depart from his general principle of finding analogy throughout the universe: the two extraterrestrials express wonder at the disparity between terrestrial man's advanced position in science and mathematics and his moral backwardness. This disparity is of course a favorite motif in later science fiction, particularly in the visions of the ending of our world through atomic self-destruction.

This satirizing of earthly man is, however, only incidental. Voltaire lays more stress on the sameness of the human condition from the philosophical point of view, the uniform unhappiness or irrationality of the beings in the many "worlds" who regard themselves as rational. He seems thereby to be pointing to a fundamental problem in the philosophy of plurality, a problem that had figured in the debate at least since Saint Thomas Aquinas made his contribution to it. (At the end of the novel the doctor from the Sorbonne refers to Thomas in connection with his own anthropocentrism.) The argument is that many *diverse* worlds cannot all be perfect, and thereby call into question the justice of the Creator, while many *identical* worlds are an absurdity (see the end of chap. 1 above). This insoluble dilemma had already emerged in consequence of Huygens's theory of plurality, which postulated complete uniformity (see above, at the end of sec. 2). For his part, Voltaire fails to fulfill his apparent promise to reveal, at the end of *Micromégas*, an all-embracing purpose, "le bon des choses," an intention on the part of the "author of nature" (pp. 87/ 125). Rather like Cyrano de Bergerac, Voltaire asks why those many worlds should exist in which, in a sense going far beyond satire of specific aspects of society, everything is essentially the same as here. What is the point of a multiplicity of worlds, if the only individuality they show is that inhumanity manifests itself in one as brutal censorship and in another as brutal warfare, if one cannot discern any true superiority of one humankind over another? This being so, is the Homeric laughter at the end really the laughter of wisdom?[132] Does the book with the blank pages really express, in disguised form, the hope of receiving revelation?[133] Or are we not faced here with the nightmare of reason: that the universe may indeed be absurd and lack any meaning—the fearful vision that confronted reason at a specific stage in its grappling with this subject, when it was still convinced that the analogical argument was sound but no longer trusted the teleological inference, usually linked with it, of the wisdom of the Creator of the plurality of worlds?

In these questions implicitly raised by Voltaire, and in Swift's secularly apocalyptic vision, the space literature of the eighteenth century reaches the peak of its achievement as "philosophical literature." A glance back at the satiric or utopian space novels shows how far *Micromégas* and the Laputa episode have advanced beyond the range of ideas of that genre, from which they nevertheless developed.

5. *Variations in Lyric Poetry from Brockes to Schiller*

Whereas eighteenth-century novelists are capable of covering the whole range of questions relating to the idea of the plurality of worlds, the lyric poetry of the period generally proves less suitable for the expression of all aspects and variations of the theme. Of course the subject occurs in the lyric, very frequently in fact. It was, after all, on everybody's lips, and—an unmistakable sign that it was well known to the general public—even the comic muse appropriated the subject. We find it in comic epics, farces, and operettas and also in anacreontic poems and songs like Johann Wilhelm Ludwig Gleim's "Der Sternseher" (The stargazer: "O könnt' ich doch im Monde / Mit jenen Mädchen spielen!"—Could I but play with those maidens on the moon!) and Lessing's "Die Planetenbewohner" (The planet dwellers):

Mit süßen Grillen sich ergötzen,
Einwohner in Planeten setzen,
Eh man aus sichern Gründen schließt,
Daß Wein in den Planeten ist:

---

[132]Thus Wade, *Voltaire's "Micromégas,"* p. 115.
[133]As suggested by Barber, "Voltaire's Astronauts," p. 37.

Das heißt zu früh bevölkern.

To enjoy sweet fancies and place inhabitants on planets, before one has concluded on reliable grounds that there is wine on the planets, is to populate them too soon.

But only very seldom does the more serious reflective poetry of the Enlightenment attempt the sort of profound philosophical exploration of the theme that might arrive at independent conclusions and thus make its own contribution to the history of ideas. More often in the lyric, the idea is introduced only as a subsidiary theme in some broader context.[134] Thus in John Gay's poem, "A Contemplation on Night" (1713), it appears as a mere possibility, as an almost absurd example to illustrate the unimaginable omnipotence of God:

> Whether those stars that twinkling lustre send
> Are suns, and rolling worlds those suns attend,
> Man may conjecture, and new schemes declare,
> Yet all his systems but conjectures are;
> But this we know, that Heaven's eternal King,
> Who bade this universe from nothing spring,
> Can at his word bid numerous worlds appear,
> And rising worlds the all-powerful word shall hear.
> (Lines 27–34)

For Matthew Prior, too, the possible existence of myriad inhabited worlds in the firmament is merely one of many examples used to show that man's ability to know is limited and always will be (*Solomon*, 1718, bk. 1, lines 512–62). This use of the plurality idea simply as a means to an end is far removed from the theme of the "encounter"; and Thomas Gray is no closer to it in his "Luna Habitabilis" (1737), in which he imagines English trade and colonial rule extended in the near future to the continents of the moon, so that Britannia rules the air as well as the waves.

Of course hymnic nature poetry takes up the motif of the many worlds that are the legacy of the Copernican revolution in thought. In German literature such treatment of the topic is well represented by Klopstock's odes (written between the late 1750s and the turn of the century). The "countless worlds" and "Earths" are constantly apostrophized by the poet, who is filled with solemn emotion by the

---

[134]I am indebted to Schatzberg, *Scientific Themes*, for valuable source references to treatments of scientific motifs in German literature.

unlimited creative power of the "Eternal" and "Infinite" that is revealed when one gazes up at the starry sky.[135] True, in Klopstock's
"Frühlingsfeier" (A celebration of spring), the "I" of the poem, overwhelmed by his extreme smallness—almost nothingness—in the universe, tries to define himself in relation to this "ocean of worlds" in
which his own Earth is no more than a "drop":

> Wer sind die tausendmal tausend, wer die Myriaden alle,
> Welche den Tropfen bewohnen, und bewohnten? und wer bin ich?
>
> (Lines 17–18)

> Who are the thousand thousand, who are all the myriads who inhabit
> the drop, and have inhabited it? and who am I?

But all this hymnic eloquence hardly ever confronts the beings
with whom, of course, Klopstock, too, peoples his "worlds." At most
there is, in "Die Verwandelten" (The transformed), a reminder that
the "lot" of the inhabitants of Saturn is "more joyful bliss than we
know"; or in "Die höheren Stufen" (The higher rungs), a dream in
which the immortal "spirits" of Jupiter, who seem to speak by altering
their form, are "happy," indeed are growing ever happier as they
ascend toward the sun. Usually there is a lyric mood of generalized
religious enthusiasm, in which more clearly defined ideas would strike
a discordant note.

The same is true of the enthusiastic, emotional nature poetry being
written in England, for instance the poems of the Newtonians Henry
Baker (*The Universe*, 1734), Richard Blackmore (*The Creation*, 1712),
David Mallet (*The Excursion*, 1728), and Henry Brooke (*Universal
Beauty*, 1735).[136] Already stereotyped, the phrases recur again and
again: the millions of worlds, the new worlds, the other worlds, the
worlds upon worlds, the suns upon suns, the innumerable "systems"
with their inhabitants who may be nobler than we or remain veiled in
mystery. Again and again the poems culminate in praise of the divine

---

[135]See for instance "Die Allgegenwart Gottes" (The omnipresence of God), "Die
Frühlingsfeier" (A celebration of spring), "Die Gestirne" (The stars), "Die Zukunft"
(The future), "Dem Unendlichen" (To the Infinite One), "Die Welten" (The worlds),
"Die Verwandelten" (The transformed), "An die nachkommenden Freunde" (To the
friends who come after), "Die höheren Stufen" (The higher rungs).

[136]I need only give a brief outline here, as Jones, *Rhetoric of Science*, gives a full
discussion (with quotations) of these and similar poems by John Hughes, Soame Jenyns,
Aaron Hill, Richard Gambol, Bezaleel Morrice, and others. For French literature there
seems to be no comparable study touching on our topic.

Architect who with his almighty hand and supreme wisdom created "this mighty system, which contains / So many worlds":

> Homage to thee let all obedient pay.
> Let glittering stars, that dance their destin'd ring
> Sublime in sky, with vocal planets sing
> Confederate praise to thee, O Great Creator King![137]

If the rational creatures of these other worlds are mentioned at all in more than a merely perfunctory way, it is as beneficiaries of God's wise provision. Thus in Henry Baker's *The Universe* we are told, with reference to the climate on Saturn, which is quite different from ours, that

> Who there inhabit, must have other Pow'rs,
> Juices, and Veins, and Sense, and Life than Ours.[138]

But the conclusion dispels any idea that such differences among living creatures in the universe might betoken injustice on the part of the Creator:

> Eternal Goodness certainly design'd,
> That ev'ry one, according to its kind,
> Should Happiness enjoy [.]
> (Pp. 34–35)

For the most part, then, lyric poetry offers strong but not clearly defined expressions of feeling in response to the new worlds that were Newton's unintended gift to the imagination. The more lyric the tone, one might say, the vaguer the ideas.

It is nevertheless instructive to look more closely at the lyric in the German language, which until the beginning of the eighteenth century had played no significant part in formulating the themes associated with the plurality of worlds. This is all the more justified because German literature—which, as I have said, only now begins to share the Western European awareness of science—does, at times in the eighteenth century, treat the philosophical problems of the plurality theme more circumspectly, more profoundly, and with greater imaginative flexibility than contemporary English or French poetry. In these

---

[137]Richard Blackmore, *The Creation*, in *The Works of the English Poets*, ed. Alexander Chalmers, vol. 10 (London, 1810), p. 380. The previous quotation ibid., p. 349.

[138]Henry Baker, *The Universe* (London, n.d. [c. 1760]), p. 12.

latter countries there was by now a rich tradition of philosophical discussion of the subject, so that there was perhaps less incentive for poetry to develop philosophical aspirations of its own.

The point when such philosophical aspirations enter German literature is marked by the young Barthold Hinrich Brockes, who as the son of a Hamburg patrician family had a good knowledge of intellectual developments abroad and in particular of the English literature, philosophy, and science of the late seventeenth and early eighteenth centuries. Various poems in the monumental physicotheological book that was his life's work, *Irdisches Vergnügen in Gott* (Earthly delight in God, 1721–48, 9 vols.),[139] rehearse the philosophical motifs associated with the idea, relatively new in Germany, that "we are not alone", and they do so with some originality and a considerable awareness of the question of man's status in an infinitely animate universe. And as one of the most influential and popular writers in German in the early years of the Enlightenment, Brockes, more than any other writer, familiarized the reading public, which was growing by leaps and bounds, with the idea of the plurality of worlds—and with its problems.

For Brockes too, of course, the dogmatic problems have ceased to be relevant. He himself captured this change, a turning point in the history of thought, in his poem for New Year's Day, 1722, in which a Christian and a philosopher discuss, somewhat pedantically, the plurality of worlds (which in France had long been a topic of conversation in elegant society). One by one the Christian believer's dogmatic objections to the new heresy of "so many thousand Earths" are refuted (1:430–31). While Klopstock and others try to see whether the structure of Christian dogma can provide a niche somewhere for the poetically attractive idea of countless worlds, Brockes prefers to reject out of hand such dogmatic quibbling about paradise, the Fall, and redemption; to distance himself from questions such as "Can Christ have died for only one world? or did the first Adams on all of them fall?" (1:435). "Religion," we are assured, is "not at all endangered" by the assumption of inhabited worlds (1:436); on the contrary, "the greater the impression we form of God, in all respects" (1:430), the more convincingly God's omnipotence and glory are made manifest (1:437). And the "greatest object known to our senses" is of course that "space, so immensely great" with its innumerable inhabited worlds, in which God is "constantly creating, eternally active" (1:430–33). To doubt this is—we recall the designation of Galileo as the second Columbus—as

[139]Page references are to the reprint cited in n. 122.

anachronistic as to doubt the existence of the antipodeans (1:435) whom Columbus was thought to have discovered in the New World. To believe in the plurality of worlds, conversely, "does not diminish, it heightens God's glory" (1:436), once the anthropomorphic conception of God is abandoned: "God is not an old man . . . he is an eternal, omnipresent All" (1:436-37). A significant feature of this bold departure from dogmatism and reinterpretation of the "most grievous heresy" as piety is that the belief that the whole universe is inhabited is still justified by the neo-Platonic principle that Christian theology had long ago assimilated, the principle of plenitude: if God worked so inexhaustibly on our "grain of sand" (1:432), it is a denial of his omnipotence to deny that he did "the same" elsewhere with the "innumerable host of suns, stars, Earths" (1:436-37). Similarly, Brockes writes in a later poem, "Das Sonnenreich" (The realm of the sun):

> Denn wer glaubt nicht, GOtt zur Ehr,
> Daß der Raum ohn alle Gräntzen nicht von Creaturen leer;
> Sondern ebenfals von Wundern seiner Macht und Weisheit voll?
> Wer dem wiedersprechen wollte, denckt fürwahr nicht wie er soll.
> Heischet es der Menschen Pflicht, von der GOttheit stets das Gröste,
> Herrlichst', Allerwürdigste, das Vollkommenste, das Beste
> Zu gedencken und zu glauben; so wird man ja dieß nicht fassen,
> Daß der Schöpfer solches Raums tieffe Tieffen leer gelassen;
> Leer von allen Gegenwürffen seiner Weisheit, seiner Liebe,
> Die ihn doch allein die Wunder, die er schuf, zu schaffen triebe[.]
>
> (5:17)

For who does not believe, to the greater glory of God, that boundless space is not empty of creatures, but equally full of the wonders of his power and wisdom? Anyone who would wish to deny this does not, indeed, think as he should. As it is man's duty always to think and believe of the Divinity whatever is greatest, most splendid, most worthy, most perfect, and best, one cannot conceive that the Creator left empty the deep depths of this space—empty of all objects of his wisdom, of his love; it was only this, after all, that led him to create the wonders that he did create.

In this way, Brockes creates a space within which the subject of the plurality of worlds can be discussed without regard to whether or not it can be defended in dogmatic terms. No sooner is this done, however, than the spirit thus liberated is assailed by fear. It is all too easy to imagine that the inhabitants of at least some of the other bodies possess greater "perfection" than man. True, it is possible for this idea to lead to still more enthusiastic praise of God's power, wisdom, and goodness, as it does for instance in Brockes's poem for New Year's Day, 1730. But already in this poem the "almost too bold flight of questing thoughts" is quickly restrained, and the "excessive thinking" (*übertriebne[s] dencken*) is forcibly steered toward pious contemplation (4:485). But when the poet's thinking in his "philosophical poetry" does not allow itself to be channeled in this way, it suffers the shock of finding man reduced to a merely relative status, and the painfulness of this experience cannot be simply converted into praise of the Lord of Hosts. We see this in the poem "Nothwendigkeit der Betrachtung" (Necessity of contemplation) in the same volume of *Irdisches Vergnügen*. The idea of the plurality of worlds leads with irrefutable logic to the assumption that "all this splendor" which surrounds us is "no more made [for man] than for the animals," but the corollary is that "the wonders of nature, in their wonderful splendor, were created and made for quite other [!] creatures, and not for you"; and this leads to a particularly disturbing variation of the question of man's nature in the light of the cosmic humankinds; the pluralistic teleology, logical as it is, delivers a crushing blow to the individual:

> Mir schauert fast die Haut, wann ich dieß überlege,
> Daß unser Geist hiedurch fast aus der Menschen Orden
> Herausgerissen sey, und gantz zum Vieh-Geist worden.
> (4:236)

> I almost shudder when I consider that by this our spirit is, as it were, wrenched from the order of mankind, to become no more than the spirit of a beast.

Here it is hard to see any way of overcoming the cosmic shudder and the crisis of human identity. Suspiciously, the poem does not so much come to an end as break off, with the exhortation that one should take more delight in God's Creation.

This anxiety caused by the demotion of man from his position as the pinnacle of Creation (or, in deistic terms, as the most rational creature of a rational Creator) shows itself again and again in Brockes's work in his fascination with the idea of space travel. For the motif of the space voyage involves the confrontation of *Homo sapiens* with other creatures, a confrontation that defines him, and in a new way. This happens for instance in the poem "Traum-Gesicht" (Dream vision). In it, the dreamer is transported to an exotically beautiful imaginary world in a solar system (we are not told which). Here feathered creatures float in the ether; they have only one sense, that of smell, but this makes them so happy, giving them something approaching heavenly bliss, that they incessantly thank their Creator (4:195–96). Another planet on which the dreamer finds himself has creatures that can only hear, and they too, in their "perfect joy," sing to the Creator "a constant song of praise" (4:197–98). Unlike Wieland, who takes over these two bizarre notions in the fourth of his *Briefe von Verstorbenen*, Brockes makes them the cue for some reflections on the status and nature of man. Those inhabitants of distant planets with their single sense may be happier than men, but in an unmistakable attempt to raise the status of man, that denizen of a "grain of sand" somewhere in this disorienting new universe, Brockes reminds the reader that man enjoys the privilege of having five senses. In this attempt, it is most revealing that he does not pursue the speculation, familiar at least from Fontenelle, if not before, that on other planets there might be creatures with *more* than five senses, in other words, creatures even more privileged than man. He is more concerned to admonish man to recognize his good fortune in having five senses, instead of being discontented (*unvergnügt*) and "trying only to make money here [on Earth]" (4:198).

He thus broaches the satiric and utopian possibilities of the imaginary space journey, which all through the eighteenth century were exploited chiefly in the novel. Brockes himself approaches this type of treatment in the sequel "Zum Traum-Gesicht," where a human being from Jupiter voices his surprise at the vanities and meannesses of human life. But Brockes only touches briefly on the satiric possibilities. He returns to the philosophical problem of man's status in the inhabited universe, and in this poem too he cannot bear the idea that, in the light of plurality, this status is only relative. For, paradoxically, he makes the inhabitant of Jupiter admit that the very combination of opposites in man—good and evil, perfections and errors—gives one "almost a more marvelous piece of work to admire" than wholly perfect creatures do (6:295). Clearly, no sooner has the critical philosopher released man from the special status accorded to him by Christianity than he seeks new arguments to prove that man does not thereby sink into complete insignificance and disregard in the boundless, inhabited universe. Precisely by the creation of mankind on Earth, "God's power is recognized and exalted anew." Yet it is also true that

Es zeigen sich in jeder Welt,
Die er ohne Zahl erschuf, und noch erhält,
Von seiner Lieb und Weisheit neue Proben.
(6:295)

In each of the countless worlds that he created and still preserves,
new proofs of his love and wisdom appear.

In the poem "Vier Welte" (Four worlds), written at the same time,
contemplation of the "creatures of a thousand forms" in the universe
leads, in even more rapturous tones, back to the Creator-God, to
"reverence, praise, and gratitude, admiration and devotion" (6:282).
But Brockes seems to need to assure himself over and over again
of this consolation (which was suggested to him particularly by the
Newtonians' speculations on plurality);[140] he is overwhelmed by
doubts about the dignity of man in the face of the immense variety of
possible forms of rational life in the cosmos, even if the universe filled
with an infinity of life-forms is a proof of God's greatness. Here we
see the typical dialectic of the assumption of myriad cosmic humankinds: by this assumption man dispels the horror of being alone in
the "empty" universe (5:17) but simultaneously condemns himself to
anonymity in a collective, and consequently to unimportance or even
inferiority, and so awakens the doubt as to whether this all-powerful
God is still as mindful of him as when a single world was assumed.
The greatness of God and the dignity of man seem to be mutually
exclusive.

It is no coincidence, therefore, that the same volume of *Irdisches
Vergnügen* also contains "Vergleichung" (Comparison), a poem that
considers the question of man's identity from the Archimedean standpoint of the extraterrestrial: "What would such a race [on a more
fortunate planet] think of the doings of men?" The "hurly-burly on
Earth" might look to them like the swarming movements of a shoal of
fish, but this only gives rise to the question

Ob sie, wofern sie uns nicht dümmer,
Doch wilder, mördrischer und schlimmer,
Als wie die Fische, würden achten?
(6:322)

---

[140]See *Irdisches Vergnügen* 4:524–25 (speaking of Johann Albert Fabricius's translation of Derham's *Astro-Theology*).

Would they think us perhaps not more stupid than the fish, but more savage, murderous, and wicked?

The question is left unanswered. The reader waits in vain for man's honor to be restored, and equally in vain for praise of the God of love. Instead there is only uncertainty about man's status, and fear and shame in relation to superior beings in outer space: the question is formulated in a way that admits hardly a glimmer of hope.

Reading this poem, one cannot fail to be aware that the theme of the plurality of (different) worlds also implicitly raises the question of theodicy. In the eighteenth century, Leibniz was not the only one to see the connection between the two (see sec. 3 above). Leibniz avoided the question the assumption of happier creatures on other planets inevitably raised, and his disciplines among the lyric poets—for instance Johann Friedrich von Cronegk in his "Lob der Gottheit" (Praise of the godhead), Ewald von Kleist in "Unzufriedenheit des Menschen" (Man's discontent), or Johann Peter Uz in his poetic "Theodizee" (Theodicy)—tended rather to trivialize it by the readiness with which they resorted simply to praising, in formulaic, empty phrases, the goodness and justice of God as Lord of the "countless Earths" (Kleist). The one exception is Albrecht von Haller. Whereas Uz in his "Theodizee" was able to draw comfort from the idea that "an evil that we in the dust complain of may be of benefit to entire worlds,"[141] it is this very idea that awakens doubt in Haller, who was more well versed not only in theology but also in the science of the "thousand new suns"[142] than the other, anacreontic poet. From an early stage, the plurality of worlds with its "creatures of a higher kind"[143] plays a significant part in those of his poems that are concerned with the status of man. But when Haller links plurality with theodicy, his attempt to prove God's goodness is somewhat reminiscent of a feat of Münchhausen's. When he explains that we have to be unhappy *in order* that other beings on remote stars may be happy, he is attempting to pull himself, by his own hair, out of the metaphysical bog. The question remains unanswered; man's dignity, placed in doubt by the plurality of worlds, cannot be vindicated, and neither can God's benevolence:

Vielleicht ist unsre Welt, die wie ein Körnlein Sand
Im Meer der Himmel schwimmt, des Übels Vaterland!
Die Sterne sind vielleicht ein Sitz verklärter Geister,

[141]*Sämtliche poetische Werke*, ed. August Sauer (Stuttgart, 1890), p. 138.
[142]Albrecht Haller, *Gedichte*, ed. Ludwig Hirzel (Frauenfeld, 1882), p. 45.
[143]*Von und über Albrecht von Haller*, ed. Eduard Bodemann (Hanover, 1885), p. 132. See also *Gedichte*, pp. 151–52.

Wie hier das Laster herrscht, ist dort die Tugend Meister,
Und dieses Punkt der Welt von mindrer Trefflichkeit
Dient in dem großen All zu der Vollkommenheit[.]
(*Gedichte*, p. 141)

Perhaps our world, that floats like a grain of sand in the sea of the heavens, is the homeland of evil! Perhaps the stars are the seat of transfigured spirits; as vice rules here, so there virtue holds sway; and this dot, this world of lesser excellence, contributes to the perfection of the great universe as a whole.

Teleological thinking provides the motive force for such argument. In the passages quoted from Brockes's "Das Sonnenreich," it is simply an attribute of God's omnipotence that he is never inactive; here, instead, a *purpose*, an intention, is assumed as the cause of the creation of a plurality of worlds. But what intention? It is characteristic of the Enlightenment that Haller, in the lines quoted, assumes a purpose that does not, or not exclusively, relate to mankind on our planet (see sec. 2 above, at n. 63). Nor does he, like some others, attempt surreptitiously to restore the anthropocentrism rendered untenable by the plurality of worlds. But in the passage quoted, this anti-anthropocentric teleology has a profoundly disturbing effect on him, as it did on Brockes when, in "Nothwendigkeit der Betrachtung," he turned from the idea of plenitude to teleology. In this, Haller differs from many of his contemporaries, including for instance the young Lessing, whose didactic poetry is in many ways strongly influenced by Haller and who, inspired by Whiston and Huygens, had taken up the subject of plurality in one of his very first poems, entitled "Mehrheit der Welten" (Plurality of worlds). This work, planned on a grand scale, was never completed, but Lessing returns to the subject in a slightly later poem, "Die lehrende Astronomie" (The lessons of astronomy, 1748). Although the teleological thinking here is in principle pluralistic, Lessing nevertheless implicitly represents man as the purpose of Creation, on the basis of the theory of the transmigration of souls familiar from the philosophy of the day. Like many of his contemporaries, Lessing holds to the concept of a graduated series of beings, which had proved problematic in Newton's day, and speculates that man will ascend in future metamorphoses "from one star to another" toward ultimate perfection. Seen from this point of view, as opposed to Haller's, the supposition that the beings on other planets enjoy more happiness, virtue, or truth is not a cause for anxiety:

Dort seh ich, mit erstauntem Blick,
Ein glänzend Heer von neuen Welten;
Getrost, vielleicht wird dort das Glück
So viel nicht, als die Tugend, gelten.
Vielleicht dort in Orions Grenzen
Wird, frey vom Wahn, die Wahrheit glänzen!

«Das Übel, schreyt der Aberwitz,
«Hat unter uns sein Reich gewonnen.»
Wohl gut, doch ist des Guten Sitz
In ungezählten größern Sonnen.

. . . . .

O nahe dich, erwünschte Zeit,
Wo ich, frey von der Last der Erde,
In wachsender Glückseligkeit,
Einst beßre Welten sehen werde!
O Zeit, wo mich entbundne Schwingen
Von einem Stern zum andern bringen![144]

There, with astonished gaze, I see a shining host of new worlds and am comforted to think that there perhaps good fortune does not count for as much as virtue. Perhaps there, in the region of Orion, truth will shine forth, free from delusion.

"Evil," cries the voice of Folly, "has established its dominion among us." That may be so, but goodness is enthroned in unnumbered greater suns. . . .

O draw near, longed-for time when, free from Earth's burden, I shall, in ever-growing bliss, see better worlds. O time when unfettered wings shall carry me from one star to another!

Lessing's friend Christlob Mylius expounded this teleological consolation more fully in his treatise, "Gedanken von dem Zustande der abgeschiedenen Seelen" (Thoughts on the condition of departed souls). (On the subject of transmigration see also sec. 1 above.) Haller, as we have seen, does not, however, take this path. For him, in the passage quoted, teleology does *not* lead via the hypothesis of transmigration to that primacy of man which also salvages the belief

---

[144]*Sämtliche Schriften*, 3d ed., ed. K. Lachmann and F. Muncker, vol. 1 (Stuttgart, 1886), p. 125.

in God's goodness (cf. Carl Friedrich Drollinger: "Wo bliebe sonst des Schöpfers Liebe?"—Otherwise what becomes of God's love?)[145]

Although Haller is untypical in leaving open the question of theodicy, his teleological approach to the subject of plurality *is*, however, essentially characteristic of the eighteenth century. Such teleological thinking in poetry is particularly common in Gottsched's circle, with which Haller in his early period felt himself to be closely linked. Gottsched's ode on his fiftieth birthday (1750) is typical, not least in its banality. The observable existence of many solar systems (explained by Gottsched, as was by now usual, by means of Newton's laws of gravitation) leads at once to the question For what purpose? For the wise Creator "truly, does not do anything in vain."[146] "What wise man would light torches in barren fields?" he writes with reference to the moons of Jupiter discovered by Galileo's telescope. As our moon shines for us, so Jupiter's moons shine for the inhabitants of Jupiter. Their existence is thus deduced from the purposeful action of God toward a particular end, and not for instance from the idea of plenitude. An analogous approach is used in relation to planetary systems beyond our sun, and also (following Johann Heyn) in relation to comets.[147]

The approach of Gottsched's pupil Christlob Mylius in his "Lehrgedicht von den Bewohnern der Kometen" (Didactic poem on the inhabitants of comets) is similar ("A wise builder does not, after all, act without a specific intention"; "this is the judgment of reason").[148] Altogether, Mylius's poems and essays show an unusually strong preoccupation with the exciting and disturbing implications of the "thousand worlds" (p. 351) of the expanded Newtonian cosmology, always from the viewpoint of the anti-anthropocentric teleology characteristic of the time. He quite systematically included the dissemination of knowledge about these worlds and their inhabitants in his project of educating the public, publishing some of his essays on the subject in weekly periodicals of his that were aimed at a broader readership. The most significant writings from our point of view are the two chatty but serious pieces, never reprinted, that take up almost

[145]Carl Friedrich Drollinger, *Gedichte* (Basle, 1743), p. 22 ("Über die Unsterblichkeit der Seele").

[146]*Ausgewählte Werke*, ed. Joachim Birke, vol. 1 (Berlin, 1968), p. 227. See Christian Benjamin Schubert's poem "Vielheit der Welten" (Plurality of worlds): "Nichts, was ein weiser thut, vollzieht er ohne Grund" (A wise man does nothing without reason) (from Junker, *Weltraumbild*, p. 90; quoted by Junker from Schubert's *Lehrgedichte* [1751]).

[147]*Ausgewählte Werke* 1:229.

[148]*Vermischte Schriften*, ed. G. E. Lessing (Leipzig, 1754), pp. 354–55. Page references are to this edition. In relation to the following sentence see ibid., pp. 36–37, 40–41, 93–94, 100–101.

the whole of the seventy-fourth and seventy-fifth issues of his periodical *Der Naturforscher* (The natural philosopher). (Apart from them, these issues contain only Lessing's poems about plurality.) Boldly anticipating later conjectures, Mylius already imagines the amazing rational beings that inhabit the pages of today's science fiction at every level: "Perhaps some planet dwellers have fewer senses, and some have more, than we have; perhaps on some planets the rational inhabitants fly through the air like our birds; perhaps on some they swim in the water like our fishes; perhaps on some they crawl around in the interior of the planet like our worms" (p. 592).

One astonishing feature of these writings by the notorious "freethinker" Mylius is that the questions about the "purposes" and "intentions" that lie behind the creation of this magnificently enlarged range of animate beings invariably lead him back to the omnipotence and goodness of the Creator, to the yet greater "glory of God," even where he does not enlist the aid of palingenesis. The same is true of many of Mylius's contemporaries (if not of Haller, whom he cites at the beginning of his science fiction vision). For them, the question of God's justice in view of the diversity of the worlds, and that of the status and dignity of man on our planet, are not problematic. When dealing with this subject, lyric poetry in the eighteenth century almost always resounds with rapturous praise of the glory of God.[149] The poets add their voice to the acclamations that ascend to the Creator from myriad worlds. "Is it not consistent with God's wisdom so to arrange things that his majesty is glorified in infinitely many places?" Mylius, the suspected atheist, asks rhetorically; while J. F. W. Zachariae recommends that Christians above all should meditate on the idea that "so many heavens and worlds" were not created for us.[150] Remarkably, this hymn to God's greatness as the Lord of the innumerable worlds is offered up even in contexts where the scientific basis of the idea of the plurality of earths brings the writer close to a mechanistic conception of evolution such as that of Kant in the period before the *Critiques*. Thus in Gottsched's poem to Johann Christoph Benemann: even if it is the "way of the world, the nature and order of all things" that suns become burned out and worlds perish like comets hissing as they die, the poet is still able at the end to praise God and his "wisdom," in other words, to hold fast to the transcendent meaning

---

[149]See Jones, *Rhetoric of Science*. On German literature see Junker, *Weltraumbild*, pp. 81–82; specifically, see Gottsched, *Ausgewählte Werke* 1:231; Mylius, *Vermischte Schriften*, p. 362.

[150]Mylius, *Vermischte Schriften*, p. 100; J. F. W. Zachariae, "Die Nacht," in his *Tageszeiten* (1756), in *Poetische Schriften*, vol. 4 (Vienna, 1765), p. 175.

embodied in teleological thinking.[151] In the lyric too, then, the "heresy" has become a new gospel that continues to be proclaimed right up to the late odes of Klopstock and the cosmic visions of the young Schiller.[152] The "gospel" of the Enlightenment, now coming of age but still pious in the manner of deism, is proclaimed even in the lecture hall, in Christian Fürchtegott Gellert's *Moralische Vorlesungen* (Lectures on morality): "How these conceptions extend man's understanding, and how greatly they glorify the omnipotence, wisdom, and goodness of the Creator! If that region of the heavens called the Milky Way alone contains over forty thousand stars, and if these are all populated by living creatures, then, great God, what myriad nations praise your creating and preserving hand!"[153] That dangerous dialectic, last mentioned in connection with Brockes—of the fear of being unique in the universe and the shudder at the thought of cosmic anonymity—disappears from view as the soul, deeply moved, sings the praises of the great Architect and becomes enraptured by its own eloquence.

6. Recantation: The Universe "Within"

The occasion for such praise is, as Bodmer says about Brockes's "glorification of the Creator," the experience of the "greatness of the material world."[154] Kant, in the period of his *Critiques*, was to call this the *mundus sensibilis* as distinct from the *mundus intelligibilis*. The former is nothing more than a function of the subject: everything that occupies space is a projection of the self and is in that sense not real but "ideal." True greatness and sublimity are to be found not (as in Kant's *Allgemeine Naturgeschichte*) in the "starry heavens" with their "host of worlds" (Gottsched) so much as "in me," in the famous phrase of the "Conclusion" of the *Critique of Practical Reason*.[155] In "An die Astronomen" (To the astronomers), Schiller puts this idea directly to the astronomers:

[151]*Ausgewählte Werke* 1:120–25.
[152]See Wolfgang Düsing, "Kosmos und Natur in Schillers Lyrik," *Jahrbuch der Deutschen Schillergesellschaft* 13 (1969), 196–220.
[153]In *Sämmtliche Schriften* (Leipzig, 1784), 7:29.
[154]*Critische Betrachtungen*, p. 226.
[155]Kant, *Werke* 1:369; 5:174. See Rudolf Unger, " 'Der bestirnte Himmel über mir . . . ,' " *Gesammelte Studien*, vol. 2 (Berlin, 1929), pp. 40–66. The Gottsched quotation: *Ausgewählte Werke* 1:120.

Schwatzet nicht so viel von Nebelflecken und Sonnen,
Ist die Natur nur groß, weil sie zu zählen euch gibt?
Euer Gegenstand ist der erhabenste freilich im Raume,
Aber, Freunde, im Raum wohnt das Erhabene nicht.[156]

Do not babble so much of nebulae and suns; is Nature great only
because she gives you something to count? Your subject is admittedly
the most sublime among spatial things, but, my friends, it is not in
space that the sublime resides.

Whereas Friedrich von Hagedorn, doubting the value of "learning" (Wissenschaft), had prized the telescope less highly than the
plough (virtually echoing Socrates),[157] Schiller removes the telescope
from the poet's hand and refers him instead to that universe "within"
which is proclaimed in Goethe's "Was wär' ein Gott, der nur von
außen stieße" (What kind of a God would that be who only gave an
impulse from outside?). "Telescopes," Goethe notes in his Maximen
und Reflexionen (Maxims and reflections), "really only confuse man's
own intuition [den reinen Menschensinn]";[158] and in Wilhelm Meisters
Wanderjahre (Wilhelm Meister's travels), from which he took this senti-
ment, he develops the idea more fully:

How can a man face up to the infinite, except by gathering together
all his mental powers, which are pulled in many directions, in the
innermost and deepest part of himself; by asking himself, "Can you
even have a conception of yourself in the midst of this eternally living
order, unless it becomes apparent that within yourself, similarly,
there is something in constant motion, circling around a pure center?
And even if it were difficult for you to discover this center within
your bosom, you would recognize it by the fact that a benevolent,
beneficent effect emanates from it and gives testimony of it.[159]

This variation of the precept "Know thyself," which is akin to but
not identical with the "sublime" heightening of the self (see sec. 1
above, at n.44), is the conviction that is fundamental to German classi-

---

[156]Schiller, Sämtliche Werke, ed. Herbert G. Göpfert and Gerhard Fricke, 2d ed., vol.
1 (Munich, 1960), p. 253.
[157]Poetische Werke (Hamburg, 1825), 1:24 ("Die Glückseligkeit" [Happiness]). A
similar view is expressed by Drollinger, Gedichte, p. 22 ("Über die Unsterblichkeit der
Seele" [On the immortality of the soul]) and also, though he did not act upon it, by
Haller, Gedichte, pp. 24, 46–47 ("Die Alpen" [The Alps], "Gedanken über Vernunft,
Aberglauben und Unglauben" [Thoughts on reason, superstition, and unbelief]).
[158]Werke, ed. Erich Trunz, 2d ed., vol. 12 (Hamburg, 1956), p. 430.
[159]Ibid., vol. 8 (Hamburg, 1955), p. 119.

cism and German idealism around 1800.[160] As a result, the subject of
the plurality of worlds is not discussed at this period. "There is also
a universe within." "We dream of voyages through the universe. But
is not the universe in ourselves?"[161] The dignity of the human being
proudly conscious of his uniqueness exceeds that of the universe
with its "worlds." In Goethe's essay in celebration of Johann Joachim
Winckelmann (1805), this new anthropocentrism that is part of the
classical ideal of "Humanität" looks back almost in incomprehension
at the "Milky Way speculation" (in Thomas Mann's phrase) of earlier
generations:

> If a man's healthy nature functions as a whole, if, in the world, he
> feels himself to be within a great, beautiful, worthy and valued whole,
> if a harmonious sense of well-being affords him a pure, free delight—
> then if the universe were able to have feelings about itself, it would
> rejoice at having attained its goal and would marvel at this culmination
> of its own formation and being. For what is the point of all this
> elaborate array of suns and planets and moons, of stars and Milky
> Ways, of comets and nebulae, of worlds made and worlds in the
> making, unless in the end a happy man instinctively enjoys his exis-
> tence?[162]

Anyone familiar with the development of scientific themes, especially that of the plurality of worlds, in both science fiction and other
literature of the nineteenth and twentieth centuries, will in turn find
this creed of the classical age surprising and alien. Here once again it
is the Enlightenment that seems closer to the present. Goethe and
Schiller would have thought it an exaggeration or eccentricity on
Galileo's part to address his *Sidereus Nuncius* (the work through which
the plurality of worlds entered the history of thought as a serious,
topical issue, indeed as one crucially relevant to the question of man's
identity) to "everyone." Not so the Enlightenment, nor the present
day. As though echoing Galileo, the July 1978 issue of the *Smithsonian*

---

[160]Enlightenment figures such as Pope and Samuel Johnson, though they subscribe
to this view, still show a certain ambivalence: they cannot quite resist the fascination of
the infinity conjured up by the new science. See Pope's letter to John Caryll of 14 August
1713, in *Correspondence* 1:185–86. On Johnson's ambivalence see Richard B. Schwartz,
Samuel Johnson and the New Science (Madison, Wis., 1971), pp. 120–25. This attitude on
the part of classicism is also largely shared by (German) Romanticism; see Zöckler,
Geschichte 2:430–32, on the pride and the pathos of man's uniqueness in Schelling,
Hegel, Steffens, Baader, and others. See also n. 161.

[161]Goethe, "Was wär' ein Gott . . ."; Novalis, *Blütenstaub*, fragment no. 16.

[162]*Werke*, 12:98. Goethe comes closest to a belief in extraterrestrial life in his remarks
to J. D. Falk, but even these are inconclusive.

(p. 36) comments, with reference to the "very big cosmological questions" not the least of which is the one that concerns us here, "The
man in the street might well ask what this has to do with him. The
answer is—maybe everything."

That this is the answer given today is at least partly due to the
tradition of the idea of the plurality of worlds taken over from the
eighteenth century and carried on and transformed by the nineteenth
and twentieth centuries.

CHAPTER 5

Modern Times: Man and Superman— The Origin of Species in the Cosmos

1. From Metaphysics to the Science of Evolution: Chalmers, Whewell, Brewster, Flammarion, and Others

From the Copernican revolution onward, the leap from the scientifically established plurality of planets identical to or resembling Earth to the assumption of a plurality of inhabited worlds was justified by analogical reasoning such as is commonly used in the exact sciences. The baroque period was preoccupied with the question of whether and how this assumption could be reconciled with Christian dogma based on the revelation of the Bible. The Enlightenment tried to master the problem of plurality by means of teleology, a mode of thought belonging to natural philosophy but also often described as theological: God is said to have created the myriad planetary worlds for myriad humankinds, by whom his glory—magnified a myriadfold—is extolled. In the succeeding period, right up to the present day, the baroque and the Enlightenment approaches are still represented in the discussion of whether or not man is "alone" in the universe. What is new about this period, however, from the point of view of intellectual history, is that in those discussions of the subject that are historically relevant, the theological and metaphysical approaches referred to above are absent; at most they are very marginally present, as premises, usually not expressed explicitly, that are no longer felt to be legitimate and conclusive. The method of argument is, so far as is possible in philosophical discussion, of a purely scientific nature: To begin with, teleology is no longer accepted as a principle by which natural phenomena in astrophysics and biology may be interpreted.

More significantly, however, it is science (unconcerned with any such metaphysical purposes underlying the Creation) that supplies a radically new way of thinking that invests the idea of the plurality of worlds with a previously unimagined credibility that allows it to penetrate deep into the consciousness of the age.

This new mode is evolutionary thinking, which applies in two fields, astrophysics and biology. On the one hand, there is the evolution of the heavenly bodies in the universe: according to Laplace's nebular hypothesis (1796, 1798–1827), matter in the form of dust or gas becomes denser through a process of cooling and forms the solar planetary systems, which in turn are subject to a process of cooling and aging (which varies according to mass, position, and age); on the other hand, there occurs, at a particular stage in the development of such planets, the evolution of life—from microorganisms to complex animal and human creatures, and beyond *Homo sapiens* to even higher forms of life, in accordance with the model of the origin of species presented by Darwin (1859) and his precursors. This bipartite evolutionary schema, which envisaged the possibility of a "superman," was still able to exert its influence on Nietzsche and, at the other philosophical extreme, on Teilhard de Chardin.[1] The drawing of analogies between our Earth and the other planets of our solar system and of other solar systems still forms the basis for all these theories; and from about 1860 onward it acquires a wholly new conclusiveness through the momentous advance made by Robert Kirchhoff (and Wilhelm Bunsen) in perfecting spectrum analysis. For this supplied proof not only that the fixed stars are suns, but also that in them and the areas surrounding them—in other words, everywhere in the universe—the same chemical elements are present as on Earth. This confirmed in a scientifically unassailable manner Copernicus's abolition of the Aristotelian-Christian division of the universe into an earthly ("sublunar") region and a higher one composed of wholly different matter, and so it confirmed the basis on which, in postmedieval times, every theory of the plurality of worlds rested, and still rests. Among literary writers, it was above all Anatole France who, in his *Jardin d'Épicure* (Paris 1894, followed by an exceptional number of editions), linked the identical

[1]It was quite customary from the mid-nineteenth century onward to see the two evolutionary processes as analogous. For just one example, see Alfred Russel Wallace, *Man's Place in the Universe* (New York, 1903), p. 99. Yet Wallace is one of the few scientifically trained believers in a *single* inhabited world! On Nietzsche see Donald G. Sanderson, "Nietzsche and Evolution" (Ph.D. diss., University of Texas, 1974); see esp. *Morgenröte*, no. 45. Teilhard de Chardin, *Le phénomène humain* (Paris, 1955). The English translation (London and New York, 1959) was published with an introduction by Sir Julian Huxley, which draws attention to parallels in his own work.

nature of the Earth's chemical and physical composition and that of other planets, proved by spectrum analysis, with the idea that the innumerable planetary worlds of the universe harbored life (10th ed., 1895, p. 7). Not coincidentally, spectrum analysis is quite frequently one of the technological motifs found in science fiction of the nineteenth and twentieth centuries, for instance in John Munro's *A Trip to Venus* (London 1897, chap. 1). The impact of the three scientific advances mentioned caused the idea of the plurality of worlds to undergo a tremendous upsurge from the midnineteenth century onward, in literature, subliterature, and works of popular science. The second half of that century is the Golden Age of the idea of plurality.[2] Even in the first half of the nineteenth century the idea still retained its widespread credibility, which dated from the eighteenth century, especially in English-speaking countries. It speaks volumes that in 1835 the whole civilized world was taken in by a hoax perpetrated by the American journalist Richard Adams Locke, who in a New York daily newspaper, and then in a pamphlet, alleged that Sir John Herschel had made telescopic sightings of vegetation, animals, and (winged) humans on the moon.[3] We have to look far back into the eighteenth century, for example to the early work of Kant, to see the first instances of the combined theories of astrophysical and biologic evolution being brought to bear, if only in a rudimentary way, on the subject of the plurality of worlds; yet it took a considerable length of time for these theories, singly and combined, to displace theological and natural-philosophical ways of thinking and to establish themselves in the nineteenth-century discussion of the subject as a new approach that

---

[2]William C. Heffernan, "The Singularity of Our Inhabited World: William Whewell and A. R. Wallace in Dissent," *Journal of the History of Ideas* 39 (1978), 90, n. 25. For surveys including paraphrases see Camille Flammarion, *Les mondes imaginaires et les mondes réels*, 12th ed. (Paris, 1874), chap. 13, and Ralph V. Chamberlin, *Life in Other Worlds: A Study in the History of Opinion*, Bulletin of the University of Utah 22:3 (1932), chap. 4.

[3]On the "moon hoax" see for instance Roger Lancelyn Green, *Into Other Worlds: Space-Flight in Fiction, from Lucian to Lewis* (London and New York, 1958), chap. 7. On the idea of plurality in the early nineteenth century see Flammarion, *Mondes*, chap. 12, and Chamberlin, *Life in Other Worlds*, chap. 3. Chamberlin quotes (p. 28) a revealing review of Thomas Chalmers's *Discourses* in a theological journal: "The probability that the other orbs of our system are inhabited worlds, must appear so great, that a direct revelation from Heaven disclaiming the fact, would make but little difference in our assurance of it" (1818). In the nineteenth century, works of playful fantasy and of somnambulant clairvoyance also begin to make use of this theme. See in particular J. J. Grandville's grotesque drawings in *Un autre monde* (Paris, 1844), and Philippine Bäurle, *Reisen in den Mond, in mehrere Sterne und in die Sonne: Geschichte einer Somnambüle in Weilheim an der Teck* (Augsburg, 1834; 14th ed., 1872).

was relevant to, and indeed characteristic of, that particular era in the history of thought.[4] But it is only with this victory of the modern scientific way of thinking that the natural philosophy of the Enlightenment became completely secularized, embracing the mechanistic view already hinted at in Descartes or Kant. The stages of this development should now be briefly sketched.

The first major controversy in the nineteenth century about whether the universe is inhabited was backward looking rather than forward looking. In 1817 the Scottish clergyman Thomas Chalmers, who became widely known through his activities in church politics and his pronouncements on current social issues, published his *Discourses on the Christian Revelation, Viewed in Connection with the Modern Astronomy*. By 1818, nine editions had appeared in Britain alone, and in subsequent decades, writers of popular scientific works often refer to Chalmers's conception of the plurality of worlds, either agreeing with it or—more rarely—rejecting it.[5] Yet for Chalmers, the battle lines are still drawn up exactly as they were around 1800, when Thomas Paine deployed the theory of plurality as his heaviest gun against the Christian world view. Like so many others, Chalmers argues with a clergyman's eloquence but gives little attention to scientific detail. His position is that "unbelievers" are wrong to accuse Christianity of being geocentric and anthropocentric; God in fact created the many planetary worlds of the Milky Way and other galaxies neither for us nor "in vain," but for the many other humankinds that pay homage to him just as we do in our remote corner of the universe. God's greatness is thus infinitely heightened, and according to Chalmers, precisely this greatness, unimaginable to men, must allay the fear that the Creator cannot be thinking of us at every moment—an argument in which the unknown greatness of the God who is above all reason is the only datum![6] As for detailed speculation about God's relationship with the humankinds on other planets (Fall? redemption? re-

[4]On the predominantly metaphysical arguments, which still survive, see for instance Otto Zöckler, *Geschichte der Beziehungen zwischen Theologie und Naturwissenschaft*, vol. 2 (Gütersloh, 1879), pp. 428–39; also Flammarion, *Mondes*, chaps. 12, 13.

[5]One of the less well known publications is the pamphlet by the clergyman Samuel Noble, *The Astronomical Doctrine of a Plurality of Worlds Irreconcilable with the Popular Systems of Theology, but in Perfect Harmony with the True Christian Religion* (London, 1828). On the whole, he follows Chalmers's line of argument, but his explanation for Christ's appearance among us is the idea, later repeated by the Catholic Joseph Pohle, that the terrestrial human race is the most sinful (p. 44). See n. 4 above.

[6]*Discourses* (New York: Robert Carter and Brothers, 1851), esp. pp. 21–41. Page references in the text are to this edition.

demption by Christ?), Chalmers simply vetoes it (pp. 57–58). Of course such a rejection of dogmatic preoccupations is, like the embracing of teleology, nothing new.

But, this paternalistic certainty only imperfectly conceals the insecurity that may be characterized as the individual's sense of being lost and insignificant in the cosmic collective. And indeed the urge to accord some special distinction to man on Earth—a mere leaf in a vast forest—soon shows itself. Casting to the winds his own warnings against dogmatic speculation, Chalmers finally soothes the possible fears of his readers and hearers by assuring them that although the drama of salvation—the Fall and the redemption by Jesus Christ—was enacted on our relatively insignificant planet (p. 69), it is known and taken notice of in all the "worlds" of Creation, as can be inferred from passages in the Bible; and indeed the Crucifixion on Golgotha has meaning and validity for all other humankinds, even if these have not fallen prey to sin. For it acts as an object lesson to all worlds, an exemplary manifestation of God's love, which, in the immeasurable domain of the myriad worlds, still seeks out this grain of dust, our "fallen" planet. The drama on Earth thus acquires "an importance as wide as the universe itself" (p. 102). By sending his only begotten Son to Earth, God united everything with himself, both on Earth and in heaven (p. 104), gloriously demonstrated his power over the forces of evil, and mended the rift in Creation. The service that this Protestant theologian renders to Christians who are scientifically enlightened and also troubled cannot be overlooked: the centrality of man on this Earth is vindicated in a most impressive and indeed flattering way—in our sin lies our glory! But this in effect takes the debate back, if in an uncommon way, to the sort of questions paramount in the seventeenth century:

> If, by the sagacity of one infernal mind, a single planet has been seduced from its allegiance, and been brought under the ascendancy of him who is called in Scripture, "the god of this world;" and if the errand on which our Redeemer came, was to destroy the works of the devil—then let this planet have all the littleness which astronomy has assigned to it—call it what it is, one of the smaller islets which float on the ocean of vacancy; it has become the theatre of such a competition, as may have all the desires and all the energies of a divided universe embarked upon it. It involves in it other objects than the single recovery of our species. It decides higher questions. It stands linked with the supremacy of God, and will at length demon-

strate the way in which He inflicts chastisement and overthrow upon all His enemies. (Pp. 140–41)

Surely it is no more than being wise up to that which is written, to assert, that in achieving the redemption of our world, a warfare had to be accomplished; that upon this subject there was, among the higher provinces of creation, the keen and the animated conflict of opposing interests; that the result of it involved something grander and more affecting, than even the fate of this world's population; that it decided a question of rivalship between the righteous and everlasting Monarch of universal being, and the prince of a great and widely-extended rebellion, of which we neither know how vast is the magnitude, nor how important and diversified are the bearings: and thus do we gather, from this consideration, another distinct argument, helping us to explain why, on the salvation of our solitary species, so much attention appears to have been concentrated, and so much energy appears to have been expended. (Pp. 146–47)

Chalmers's *Discourses* provoked counterarguments that harked back to an even earlier phase of the plurality debate. Within the same year, a bookseller named Alexander Maxwell published *The Plurality of Worlds,* in which with pious simplicity he reminded his errant opponent in Christ that the revealed Word of God was also an infallible guide to cosmology and that it referred to and admitted of only the one world of the Old Testament (2d ed. [London 1820], p. 203). Bible-based orthodoxy of this kind (still represented even at the end of the century by Père François Xavier Burque in his *Pluralité des mondes considérée au point de vue négatif* [Montreal 1898], which was specifically directed against Darwin) underlines by its very anachronism how welcome the consolation of pluralistic teleology must have been right up to the beginning of the twentieth century (even though this teleology itself became obsolete in the light of intellectual developments very soon after the appearance of Chalmers's work) to Christians who were of relatively modest intellect but who no longer believed in the literal truth of the Bible.

As a curiosity we may note that this metaphysical consolation based on God's supposed intentions could even be dispensed with mathematical accuracy. In 1837, Thomas Dick presented a cosmic population census, based on the surface area of the planets and the population density of England, which arrived at a more exact figure than any terrestrial census: a total of 21,894,974,404,480 souls in our planetary system alone, not counting any that there might be on the sun.[7] Truly an unsurpassable proof of the omnipotence of the Creator: for here too the analogical and teleological demonstration culminates in the praise of God. (Nor does this minister of the Secession church fail to remind his readers that the belief in plurality accords with Scripture.)

Finally, pluralistic teleology is given a particular twist in works such as the *Physical Theory of Another Life* (1836) by Isaac Taylor, a British scholar with wide-ranging interests, and the *Consolations in Travel: or, The Last Days of a Philosopher* (London 1830) by Sir Humphry Davy, in his day an eminent chemist and president of the Royal Society. Both revive the idea of interplanetary migration of the soul, Taylor on a nominally Christian basis and tacitly following Thomas Wright, Davy indulging in pure fantasy that for concrete detail outdoes some of the creations of present-day science fiction. In the *Consolations*, a sleeper dreams that a guardian spirit shows him the higher modes of existence God has destined for the soul after death, when it will migrate from planet to planet and ultimately to the sun, each time in a new and more perfect incarnation (pp. 44–59). Thus for instance the intelligent beings that live in the atmosphere of Saturn are a sort of splendidly colored winged sea horse; they know no wars, and their capacity for intellectual enjoyment greatly exceeds ours.

A book by the eminent Danish physicist Hans Christian Oersted presents the plurality of worlds less imaginatively but still wholly in a religious context. A decade before *On the Origin of Species,* he infers from the ubiquity of matter and its laws in the universe that life too must be ubiquitous, present on all planets favorable to it, and at different stages of development, some of which may well be further advanced than the stage reached by man. Thus, unlike Taylor, Davy, Chalmers, and others, he works strictly on the basis of the empirically obtained chemical and physical data. He does, however, feel it necessary to counter the all-too-justified suspicion of materialism with an assurance that this evolution may also be seen as the continuous activity of the "eternally creating Spirit"; in another paraphrase he speaks of all happenings being directed by the "eternal, all-powerful Reason."[8] Thus a connection is made between scientific and religious truths, and in a vague way the two become merged. Oersted's remarks about the physical appearance of extraterrestrial life-forms are also vague: like most scientists of his day, even Flammarion, he refrains

---

[7]*Celestial Scenery,* "6th thousand" (London, n.d.), p. 405.

[8]*The Soul in Nature* (London, 1852), pp. 108–9, 133. Page references in the text are to this edition. In Danish and German, 1850.

from speculating on the external characteristics of these rational creatures, except that he too adheres to the principle of diversity arising
from the different environments (p. 129). Conjecturing "that an endless number of degrees of development may exist above the point we
have reached," he is, none the less, aware of the danger that this truth
may be a devastating one for mankind. The consolation he offers is
that our human race is still young and thus still has before it a long
future of "higher development" on our planet and "elsewhere" (p.
130). He is confirmed in this belief by Providence (also only referred
to in vague terms). In the more scientific parts of the book, Oersted
does not speak in teleological terms of a God pursuing his purposes;
but at the end it becomes clear that it is the concept of such a Being
governing everything in the universe that underlies and guides Oersted's thinking (p. 130). What is most interesting about this is that
while discussions of plurality still involve God's intentions, such arguments now have to be kept out of the strictly scientific sections of a
book such as Oersted's.

Only in 1853 is a new and higher level of discussion reached, in
the anonymously published treatise *Of the Plurality of Worlds* (5th ed.,
1859), which sparked off a lively and far-reaching controversy about
plurality. Its author was the influential English scientist and philosopher William Whewell, who from about 1850 onward also rose to
prominence as a moral theologian. This duality of interests explains
the odd combination in *Of the Plurality of Worlds* of progressive method
and reactionary result. For Whewell, unlike his friend and colleague
John Herschel, comes to the conclusion that the plurality theory is
untenable, and yet his mode of argument represents a decided advance on that of Chalmers and in fact of all earlier supporters of the
idea of plurality, in that he insists that the subject be discussed with
rigorous scientific exactitude. In Chalmers's book, all that really remained of science was a vaguely analogical way of thinking (the Earth
analogous to the planets, the sun to the stars); Chalmers's approach
was primarily that of a theologian, judging on the basis of God's
purposes, and concerned, when studying the universe, first with the
glory of God and second with man's significance in the history of salvation. Whewell, in contrast, undertakes, as he announces
at the outset, to discuss the subject in the light of modern science.[9]
And indeed his great, historic achievement was that he drew attention

[9] *Of the Plurality of Worlds,* "with an Introduction by Edward Hitchcock, D.D."
(Boston, 1854), p. iii. Page references are to this edition. Herschel was a teleological
pluralist; see his *Outlines of Astronomy* (Philadelphia, 1849), p. 520.

to the need to put the analogical argument for plurality more precisely and in greater detail than had been usual up to that time. The mere fact that the Earth, with its living creatures, is a planet among other planets was from now on no longer a sufficient basis for a scientifically legitimate conclusion that life and intelligence can be assumed on other planets; it now became necessary to show that the many, precisely defined conditions necessary for life here were reproduced elsewhere.

Still, even with Whewell it would not be true to say that theological-teleological and scientific modes of thinking, after going hand in hand for so long, have at last parted company.[10] This separation first occurs in Whewell's opponents, especially Thomas Collins Simon. Whewell himself, in practice, constantly mixes empirical scientific points of view with theological or at least metaphysical ones. As far back as 1841, Whewell, a Kantian idealist, had written to the empiricist Herschel that one must, and could, pursue science only in such a way as to confirm the a priori truths of reason.[11] In *Of the Plurality of Worlds*, Whewell's unshakable dogmatic premise is that man on our planet represents the supreme purpose of Creation and is therefore unique in the universe and incapable of being superseded. Even in the preface, which promises not a theological but a scientific and philosophical demonstration, Whewell admits that some of his results correspond "remarkably" to particular points of "religious doctrine"—and he adds, as though apologetically, that certainty in such matters cannot of course be provided by philosophy but only by religion, for which he is deeply grateful to his Creator (pp. v–vi). In the opening chapter he also refers back several times to Chalmers's specifically religious fear that the findings of science, which make it necessary to assume a plurality of worlds, may rob man of his unique status, vouched for by the Bible, as the creature of whom the Creator is most mindful. But whereas Chalmers's efforts were directed only at assuring us of a privileged position among countless cosmic humankinds, Whewell has the more radical aim of showing that his opponents argue for the plurality of worlds only on theological or at any rate metaphysical

[10]According to Heffernan, "Singularity of Our Inhabited World," p. 98. He too, however, stresses the methodological importance of Whewell's book, a point already made by Richard A. Proctor, *Our Place among Infinities* (New York, 1876), pp. 48–50; 1st ed. (London, 1875).

[11]Isaac Todhunter, *William Whewell* (London, 1876), 2:298. On Whewell's epistemological Kantianism see Curt J. Ducasse, "William Whewell's Philosophy of Scientific Discovery," in *Theories of Scientific Method*, ed. Ralph M. Blake et al. (Seattle, Wash., 1960), pp. 183–84, 217. See also David B. Wilson, "Herschel and Whewell's Version of Newtonianism," *Journal of the History of Ideas* 35 (1974), pp. 79–97.

grounds (p. 234), whereas *"modern* science" does not admit those many humankinds even as a hypothesis. On the contrary, it vouches for the uniqueness of man and his planet—a uniqueness that has no element of horror at being an anomaly, as it is also a uniqueness in the enjoyment of God's love and concern. The greatness of God, who according to the teleologists among the pluralists created the myriad stars for myriad intelligent "races" who all praise him, is exchanged for the unique, preeminent dignity and significance of man in Creation. When Chalmers tried to create a special position for man despite the plurality of worlds that threatened to condemn him to anonymity, it was on a dogmatic basis (the cosmic significance of *our* Fall). Whewell, however, believes that he is offering *scientific* proof of man's special position—even though he had accepted it as a dogmatic premise from the outset! (The reaction of his pluralist opponents was not to accept Chalmers's way of salvaging man's status but to return to the by-now-familiar downgrading of man.)

Thus from the start, Whewell's scientific demonstration lays itself open to the suspicion of being mere shadowboxing. His motivation is essentially theological. It is striking that while criticizing the pluralists' traditional analogical reasoning, he himself distorts logic in drawing an analogical inference of surpassing eccentricity, when he argues that just as the history of man amounts to only "an atom of time" in the history of the Earth, so only "an atom of space" is necessary for the location of intelligent life in the universe, and therefore our Earth is to be regarded as the only abode of rational beings in the cosmos (p. 121 and chaps. 5, 6). He concedes that lower animals may exist on other planets; but it is such a leap from animal to man that it is necessary to assume direct intervention by the Creator—just a single intervention, on our planet (p. 57).

From the geology of the "atom of space," Whewell turns to astronomy. Here he tries, as already stated, systematically to discredit all the analogies that suggested the existence of a plurality of worlds: The cosmic nebulas are for the most part merely clouds of gases and cannot therefore be separated out into clusters of stars even by the strongest telescopes; the nebular hypothesis, which is acceptable for the origin of our system, does not permit the conclusion that other planetary systems came into being in the same way; the fixed stars are not necessarily suns, since most of them are double stars that for reasons of gravitation could not have stable planetary systems; and even where there are single stars like our sun, the existence of planets cannot be proved (chap. 8). Finally, physical conditions on the other planets of our own system are so different from those on Earth that even Mars can at best have animal inhabitants (chap. 9). Only the Earth is in that moderate zone of the system where solid, liquid, and gaseous matter, heat and cold, humidity and dryness combine in such a way as to offer the only conditions favorable to life, or at least to higher forms of life (chap. 10).

Not surprisingly, therefore, the scientific demonstration culminates in a theological apotheosis of man as an intellectual, moral, and religious being. That God should have left the countless celestial bodies uninhabited and empty, that he did *not* create them "for" other living beings, need not concern us once we accept the idea that those vast quantities of matter are as nothing compared to the soul or the mind of man. The glory of the Creator rests on his creation of man; man, as the pinnacle of Creation, is worth, and worthy of, all the cosmic outlay of millions of stars: "Even if the Earth alone be the habitation of intelligent beings, still, the great work of Creation is not wasted." Man alone is "a worthy object of all the vast magnificence of Creative Power" (p. 280). In him Creation has "a worthy end" (p. 284). In terms of its *significance*, then, the Earth is once again the center of the universe (p. 273), and teleology has once more become geocentric and anthropocentric, as in the Middle Ages; indeed, anthropocentrism and theocentrism have become one. For the mind of man, glorified in what might appear to be a blasphemous way, is "in some measure, of the same nature as the Divine Mind of the Creator" (p. 276). Any step beyond man is unthinkable. It is not belief in a plurality of worlds, as Chalmers and many others had maintained, but belief in the uniqueness of the world and of man that accords with true Christian religious feeling (pp. 286, 290). That God redeemed mankind by sending his Son to Earth—the time-worn dogmatic argument makes a surprising reappearance—does more honor to his true greatness than the mistaken postulation, supposedly enhancing his glory, of a multiplicity of humankinds on planets in space.

So science here leads to a confession of faith. In a letter, Whewell admitted that the subject of *Of the Plurality of Worlds* was not really plurality but man's relationship with God, so that his book should be seen as a theodicy.[12] It would be more accurate to describe it as a Protestant anthropodicy. For the psalmist's question "What is man, that thou art mindful of him?" (Psalm 8) is the question at the heart of Whewell's book, and his answer accords with the creed of his church. In confessing his belief in the singularity of man, he repudi-

---

[12]To Sir James Stephen, 4 November 1853, in Todhunter, *William Whewell* 2:393.

ates the fear of anonymity in that collective of humankinds that, in his view, only a false theology could accept.

This solution to the old problem naturally unleashed a storm of protest not only among the representatives of science (which Whewell had so insistently claimed for his cause) but among theologians too. Seldom had the plurality of worlds been such a fashionable topic as in the years after Whewell's attempt to disprove it. His book won support from only a few writers, who themselves added no new perspectives.[13] But, equally, the attacks directed against Whewell on both flanks often reiterated the long-familiar arguments. If the controversy over Whewell's book nonetheless marks a new stage in the history of the topic, it is because, in the course of it, scientific argument gradually detached itself from the prejudgments of theology and "natural philosophy" and in particular from the teleological approach.

The response to Whewell that had the most lasting impact, judging by the number of editions, was *More Worlds Than One: The Creed of the Philosopher and the Hope of the Christian* (London 1854) by the Scot Sir David Brewster, at the time a very well known experimental physicist. His line of argument reflects the fact that he had come to Newtonian physics from theology. His first sentence—"There is no subject within the whole range of knowledge so universally interesting as that of a Plurality of Worlds"—no doubt expresses the feeling of the time. But his book, while delivering an imperious rebuke to Whewell, is still characterized by that traditional mixture of scientific analogical reasoning and anti-anthropocentric teleology, of empiricism and a priori metaphysical judgments, whose logical weaknesses Whewell, however vulnerable to attack his own logic was, had in fact pointed out quite effectively.

For Brewster, the planets in our solar system remain physically more or less analogous to Earth; at least one planet analogous to Earth is to be assumed in each of the other solar systems, even binary ones; and the only possible reason for the creation of so many "worlds" is that God intends them to be, either now or in the future, the home of living creatures, including beings endowed with reason and with a

[13]On contemporary reviews and other writings for and against Whewell, see ibid., 1:184–210. See also Zöckler, *Geschichte* 2:432–33. In later editions Whewell added the "Dialogue," in which he himself argues with his critics (e.g., Boston, 1855). Here once again theology and science are mingled. Whewell claims that his arguments (unlike those of the pluralists) are based not on revealed truth but on physical observation and calculation. Nevertheless he derives satisfaction from the fact that Scripture speaks against the plurality of worlds and that his view has the approbation of those whose attitudes are derived "from religious principles" (pp. 351–52).

soul. Moreover, not only are we not "alone," but we are also accorded no especially eminent status.[14] For since Brewster assumes a priori the inexhaustible diversity of the divine Creation, the probability is not only that the rational beings of other worlds are totally unlike man in physical form but also that some of them have been endowed with far higher intelligence and greater happiness than *Homo sapiens* (pp. 70–73, 158–59). They may even have reached a cultural level that puts Newton, Shakespeare, and Milton in the shade (p. 158). And there can be no doubt at all that for Brewster, the descendants of Adam and Eve are decidedly inferior morally to other cosmic "races" (p. 160).

Although all this may be the "creed of the philosopher" referred to in the title, Brewster's more detailed definition of our moral inferiority in the same passage reveals that it is also seen as sin in the Christian sense, as the inability to obey the commandments of Yahweh. In this book, scientific or philosophical arguments and Christian ones go hand in hand. There is no contradiction between reason and revelation (p. 131). Indeed, reason enables Brewster to restore man's position of preeminence—undermined, as we have seen, by reason itself—in a way that accords with Christian dogma. For even if the other human-kinds are "more glorious" than the descendants of Adam and Eve, analogical reasoning makes it unavoidable to assume that they too are sinful—and that they are all, without exception, redeemed by the crucifixion of Christ on *our* Earth (pp. 135–37, 145–47). This idea, which is not new, is put forward in opposition to Chalmers (who tended to see the other humankinds as being without sin), and yet it leads Brewster, the man of science, to the same conclusion, in essence, as that reached by the man of the church: terrestrial man is once again of central significance. Faced with the choice inherent in the plurality debate between the greatness of God (plurality) and the dignity of man (singularity), both desperately grasp at a solution that combines both (our sin as the source of our distinction). Significantly, both authors succeed in this only by returning to Christian dogma, which in terms of the history of thought is a regression to a stage even preceding pluralistic teleology.

Of course not all the responses to Whewell, reasserting the pluralistic consensus of the scientific establishment, revert to Christian dogmatism, especially not those whose argument is predominantly scientific. But even Whewell's more liberal opponents, such as William

[14]*More Worlds Than One* (New York, 1854), pp. 62–63, 80, 96–97, 130, 189–91, 252, and freq. Page references in the text are to this edition.

Williams or Sir John Herschel, do not as a rule progress beyond a method that unashamedly combines scientific argument by analogy with metaphysical teleology.[15]

As early as 1855, however, this combination was forcefully rejected in Thomas Collins Simon's anti-Whewellian *Scientific Certainties of Planetary Life* (London). Simon makes it very plain that, for him, only science is entitled to a voice on this subject; and science confirms, with complete certainty, the assumption that the planets of the sun and the stars are inhabited by intelligent beings that must be imagined to be humans like ourselves. Simon's proof rests entirely on the traditional twin analogies of sun and stars, Earth and planets; but it now becomes apparent that science has taken Whewell's lesson to heart—and can defeat him with his own weapons. For Simon demonstrates the sunlike character of the stars, and the Earth-like character of the planets, by means of mathematical and physical calculations relating to light, temperature, air, and gravity, which in precision and detail (covering some 250 pages) far exceed those of any of his predecessors. Simon does indeed establish "analogy in its strictest sense," as Whewell had demanded (pp. ix–x). This constitutes a radical rejection of any metaphysical-teleological mode of argument on the subject of the plurality of worlds, and Simon is fully aware of its historic significance: "Those who prefer to examine scientific questions without reference to their religious bearings, will not regret to learn that science alone solves wholly those before us . . . and does this in so peremptory a manner as to leave no room whatever for the moral, metaphysical and theological considerations which it has lately been attempted to introduce into these questions" (pp. x–xi). This is directed against both the anthropocentric teleology of Whewell and the pluralistic teleology of Brewster. Simon too, however, for all his bravado, still thinks it prudent to mention that although in his argument he excludes all theological considerations, he is convinced that his conclusion that other suns possess their own planets and that these have their own rational inhabitants is not contradicted by revelation. Yet it says much about the climate of thought at the time that he continues: "But it does not seem credible that any considerable portion of the reading public will ever need such explanations" (p. xv). (As early as 1818 an English

---

[15]William Williams, *The Universe no Desert, the Earth no Monopoly; Preceded by a Scientific Exposition of the Unity of Plan in Creation*, 2 vols. (Boston, 1855, anon.); 2:164–67 lists those who speak for the pluralist scientific establishment. John Herschel, in his obituary for Whewell, considers Whewell's anti-pluralism as a mere jeu d'esprit ("William Whewell," *Proceedings of the Royal Society* 16 [1866], 60–61); see n. 9 above.

theological journal had stressed that the belief in the plurality of worlds was so firmly established that not even a revelation from heaven could shake it now.)[16]

Simon's conception of the human being whom he envisages everywhere in the universe, from Mercury to Neptune and beyond, has, however, one surprising feature. One can occasionally detect, reading between the lines, that Simon takes it for granted that this human being recognizes God (e.g., pp. 6,238) and thus understands, in particular, the providence of the "Great Architect," whom Simon himself refers to at least twice, if only in passing (pp. 227, 230). Nowhere in this scientific treatise, though, is this mode of thinking used as a key to the recognition that we are "not alone" but part of a great cosmic community.

In the same year (1855) the scientist, theologian, and Oxford professor of geometry Baden Powell, in his treatise *The Unity of Worlds and of Nature*, similarly distances himself from both Whewell and Brewster. Here, however, detailed astrophysical investigation is replaced by a discussion of the principles of scientific theory. In his philosophical analysis of the controversy between Whewell and Brewster, Baden Powell shows that both are influenced in their discussion of the scientific problem by theological factors, namely prejudices about "final causes," and that on the basis of these they advance teleological arguments each of which invalidates the other: one starts out from the assumption that God intended the world and humankind to be unique, while the other takes it equally for granted that the divine plan encompasses the many planetary worlds with their intelligent, manlike inhabitants.[17] What is to be inferred from this? Powell does not accept one and reject the other (though he argues strongly against Whewell's singularism); nor does he declare the teleological approach to be bankrupt, but he does define its limits. Like any metaphysical approach, he says, it is appropriately employed in any discussion concerning man's "spiritual" nature and "spiritual truths." In all other matters, the appropriate philosophy is one that draws its conclusions from facts established by science. Mixing the two approaches can only lead to confusion (pp. 313–17). For Powell, the question of the plurality of worlds definitely belongs to the second category. It is not for teleological metaphysics (and still less for dogmatism) to establish anything in this area, nor has teleological metaphysics anything to fear, since it

---

[16]See n. 3 above.
[17]Second ed. (London, 1856), p. 278; see also pp. 250, 266. Page references in the text are to this edition.

is concerned with "another order of things" (pp. 324–26). If it does involve itself in scientific matters, it becomes embroiled in arbitrary contradictions, like Whewell's assumption that the divine intervention that brought about the origin of man can have taken place only on our Earth and not elsewhere in the universe too (p. 259).

Specifically, the scientific approach Powell sees as the only suitable one for dealing with the problem of the plurality of worlds is described as "inductive analogy" (p. 269) or "analogical probability" (p. 280). For him, it is virtually self-evident that such a strictly scientific method can lead only to the conviction that many, if not all, systems are inhabited and hence that *Homo sapiens* is only of minor importance in the universe. Although in questioning the validity of analogical reasoning, Whewell was methodologically pointing in the right direction, his arguments are not sufficiently convincing: "But the true spirit of the inductive philosophy tends to teach a modest and just estimate of man's place on one of the smaller planets in a subordinate solar system of a subordinate stellar cluster, his whole race occupying but a speck in space, and *as yet* a speck in time: while it points by analogy to other similar worlds, possibly in different phases of development, whether corresponding to *future* stages of the earth's progress, or to *past*" (p. 255).

It is also noteworthy that even at this pre-Darwinian time, Powell, like Brewster and others, reckons not only with the evolution of cosmic "worlds" but also with the "evolution of organized life" on the various globes as one stage of that cosmic evolution. In the period after the appearance of *On the Origin of Species* (1859), this was to become a *leitmotif* in the plurality debate, often with explicit reference to Darwin. But the most significant result of Powell's analysis of the philosophically and rhetorically confused dispute about plurality is the fact that his lucid formulations at last cleared the air: in the later nineteenth century, the Golden Age of the discussion about plurality, the supposed intentions and purposes of the Creator no longer figure in the debate, or at least they are not used, by the most influential participants, as evidence. The decisive role of the analogical method of inductive science—now used with greater precision in response to Whewell's criticism—is not seriously questioned from now on.

Though this was a major advance for science, it exacted a human price that, especially in the first decades after Baden Powell had spoken out for the autonomy of science, was all too evident. First the anti-anthropocentric teleology that took account of the multitude of humankinds had robbed the Christian of his belief that he was the only child of the Father, and the pinnacle of Creation; when this teleology too was rejected, that put an end to the certainty (which a number of thinkers had still preserved for Christianity) that the plurality of humankinds in the universe owed their origin and their significance to the wisdom and omnipotence of a purposeful metaphysical principle. This assumption of metaphysical wisdom made it possible, in a pinch, to bear even the sense of inferiority resulting from the probability that the inhabitants of other bodies were superior to us in intelligence and happiness. Compared with this, the new idea that man on our Earth has a definite place in the cosmic and biologic evolutionary process, which itself perhaps possesses a grandeur of its own, could at first offer only a meager consolation or substitute for religion–even when the universal process was seen as extending beyond man, so that there was the promise of a sort of secularized "destiny" in the prospect of the higher development of the human species, a destiny that had perhaps already been achieved on other, older planets. The horror of mechanism remained, and it was more potent (because more scientific) than when it had been foreshadowed from time to time in the previous two centuries.

It is therefore most instructive to see how, in the latter half of the nineteenth century, those who urge most strongly that the logical *basis* for the plurality of inhabited worlds must be scientific and not teleological nevertheless allow a religious (if not dogmatically Christian) *interpretation* of the resulting world view to creep in once again. A good illustration of this is provided by Richard A. Proctor, the enormously influential British popularizer of contemporary astronomy. In his collection of essays *Our Place among Infinities* (1875), he repeatedly demands the absolute separation of scientific and theological approaches. He considers metaphysical-teleological thinking to have had a particularly baneful influence when used as a guiding principle in the study of the natural world. "Great mischief may be done, . . . in fact great mischief has often been done, by the too frequent attempt to refer all things to some special design in the interests of such and such creatures. The reader of works in which such attempts are made is apt to regard these special indications of divine economy (so to speak) as forming a necessary part of the evidence on which he is to base his belief in the wisdom and benevolence of God" (New York 1876, p. 37). In cases of doubt, whichever explanation of a natural phenomenon best fitted the preconceived idea of the intentions and purposes of God would be assumed to be the correct one (p. 35). But it is not only for such pragmatic reasons that the study of nature must free itself from the preoccupation with the divine plan of Creation. A philosophical premise is the decisive factor here: on the one hand, Proctor considers that science cannot alter or detract from the belief that the universe is the realization of the wise and benevolent intentions of God; on the other hand, human reason is not in a position to know what the divine plan is and to confirm it by means of scientific findings (pp. 37–39). Hence Proctor's opposition to the search for "evidence of design" and his advocacy of the absolute separation of science from theology or metaphysics. And for Proctor, as for Baden Powell, the question of the plurality of worlds is indisputably a scientific one. Failure to recognize this was the error of Chalmers, Brewster, Dick, and generations of others who blandly assumed that the planets were "intended" by Providence to be the abode of living creatures and then set out to prove this scientifically (pp. 35–36). In the same way, Whewell's contrary thesis, that God's intention is limited to the Earth and life on it, and thus to a single human species, is in Proctor's view equally untenable (pp. 45–70).

In arguing for his own position, which unmistakably favors plurality, Proctor strictly follows his own guidelines, as is suggested by the very title of his book, *Other Worlds Than Ours: The Plurality of Worlds Studied under the Light of Recent Scientific Researches* (1870). He rejects arguments based on supposedly discernible intentions on the part of God, instead taking physical and chemical analogy as his sole guide.[18] His discussion is in fact one of the first to apply the findings of Kirchhoff's spectrum analysis to the question of the plurality of worlds, and thus to draw the analogy more precisely than Whewell could have thought possible: all the planets in our solar system are made up of the same molecular components, and the same is true of the planets, whose existence can be inferred, of other stars (pp. 150–51, 250–53). In the case of Mars, Proctor pursues the analogy so far as to speak confidently of its mountains and valleys, oceans and rivers, and finally also of its higher forms of life (p. 122). Characteristically, however, the same determination to confine himself to science that makes him certain that other worlds are inhabited also prevents him from even speculating on the particular characteristics of the creatures of other worlds. When imagination is allowed to override this prohibition, he says, as with Sir Humphry Davy (bizarre inhabitants on Jupiter), Whewell (jellylike fauna on Jupiter), or Brewster (life-forms on Jupiter that are a kind of Nereids, or else similar to eagles or doves), one enters the realm of the grotesque, the horrific, or the ridiculous

---

[18]This is also stated programmatically in *Our Place among Infinities*, p. 53. Against teleology, see *Other Worlds Than Ours* (New York, 1871), pp. 95, 132, 137, 165–67. Page references in the text are to this edition.

(pp. 129–30). We can certainly assume that the extraterrestrials will be different from us; but because Proctor declines to speculate on *how* they differ, he never runs the risk of frightening us with the possibility that other manlike creatures might be more highly developed than we—and once the premise of a Divine Providence (at least one with aims that we can have any inkling of) is abandoned, this possibility might indeed be frightening.

A particular feature of Proctor's argument is the idea, already touched upon by Baden Powell, that every planet in every planetary system is inhabited for a certain period. At the appropriate stage in the development of the planet (while it is cooling), life will come into being as a result of the sun's radiation, evolve by adaptation in accordance with Darwin's theory, and finally become extinct.[19] Proctor avoids, however, the sensitive question of whether further development beyond man is conceivable, or whether, on the contrary, we represent the highest point reachable by evolution.

This too is a measure of his scientific determination to abstain from speculation and teleology. Yet he is unable to excluded them totally. *Other Worlds Than Ours* ends with an expression of confident belief that the natural laws, and thus the innumerable worlds, are in the hands of an omnipotent, omniscient Being who has the whole under his supervision (pp. 333–34; see also p. 229). Proctor says that this belief *is derived from his knowledge of nature;* it is not a premise and does not influence the process of *acquiring* knowledge. Yet teleological formulations do slip in here and there even in this book that, taken as a whole, is a genuinely empirical study. This illustrates how slowly and gradually a priori teleological thinking disappeared. The other suns—the fixed stars—cannot have been "made in vain"; they must have planets, which in their turn must be "meant" to harbor life (pp. 256, 252); Jupiter must be "intended" to become, one day, "the abode of noble races" (p. 157), and being inhabited is the only possible raison d'être of Jupiter's moons (p. 154). There is mention of a "scheme of Creation" and even of "design" (p. 159); of the wisdom of the Almighty (p. 257) and even of his benevolence, as evidenced by the Earth's atmosphere, which, in a way wholly explainable in terms of natural laws, gives us a milder climate (p. 112). In this last instance the author more or less apologizes for the questionable nature of the "argument"; but basically these teleological phrases are no more than stylistic for-

---

[19]*Other Worlds*, pp. 252–53; *Our Place*, pp. 67–69. On Darwinism see, in particular, *Other Worlds*, p. 130.

mulas that do not in fact betray an a priori teleological approach on the scientist's part. What they do show is his personal belief in the guiding presence of the Creator in the workings of the laws of nature. Thus even an emphatically scientific book, whose guiding methodological principle is the separation of empirical and metaphysical ways of looking at things, betrays the power of that way of thinking which for over a century had dominated discussion of the plurality of worlds.[20]

In the next decade, a remnant of this mode of thinking creeps in, even less conspicuously, in Alexander Winchell's rigorously scientific discussion of plurality.[21] In his book *World-Life or Comparative Geology* (1883), this American professor of geology reaches the conclusion, purely on the evidence of the physical surface conditions of the other planets, that we have neighbors in the planetary systems of the universe (and that some celestial bodies such as, presumably, Mars and the outer planets, may, according to the theory of cosmic and biologic evolution, have been inhabited in the past). Winchell too believes that the physical appearance of the life-forms there must be the result of adaptation, on Darwinian lines, to physical environmental conditions, and since these would in some cases be very different from those on Earth, we can make virtually no assumptions about the forms of those creatures except that they can hardly be the same as man (pp. 497–508). Among the planets of the sun, Winchell considers that only the Earth is at present at the right stage of the cooling process to foster life. Unlike Whewell, however, he does assume analogous moderate zones in other systems: the universe is therefore "densely populated."

For Winchell, there was something so satisfying in this conclusion (which for Percival Lowell at the turn of the century was still to have the impact of a Copernican shock) that it had to be expressed in religious terms. The opposite idea—"oppressive silence of the planetary solitude"—"seems to sunder a bond of sympathy with the universe, and isolate mankind on an island rock where no message can ever arrive." His rejoicing over the fact that we do not live in such a cosmic desert celebrates a quasi-religious cosmic community of equals, and also—in this otherwise wholly scientific book!—leads directly back, if only fleetingly, to the Creator, who is close to "all of us" in all worlds and indeed allows us to discern his "purpose" in the visible cosmic "mechanism":

[20]The same may be said of John E. Gore, *The Worlds of Space* (London, 1894), p. 34.
[21]*World-Life or Comparative Geology* (Chicago, 1883), p. 500: "scientific consideration."

We have neighbors; they live beyond impassable barriers, but they gaze on the same galaxy, and we know they are endowed with certain faculties which establish a community between them and us. However conformed bodily, whatever their modes and means of organic activity, we know that they reason as we reason, and interpret the universe on the same principles of logic and mathematics as ourselves. The orbits which their planetary homes describe are ellipses; they have studied the same celestial geometry as ourselves; they have written their treatises on celestial mechanics; they have felt the impact of the luminous wave of ether; they have speculated on the nature of matter and energy; they have interpreted the order of the cosmical mechanism as the expression of thought and purpose; they have placed themselves in communion with the Supreme Thinker, who is so near to all of us that his voice is audible alike to the ear of reason in all the worlds. (Pp. 507–8)

This feeling of a universal bond is heightened to a passionate, mystical outpouring of emotion in the writings of the scientist who did more than any other in the nineteenth century to popularize the idea of the universality of humanity in outer space. This was Camille Flammarion, the prodigiously prolific astronomer and author, president of the Société astronomique de France, founded by himself, and director of the observatory at Juvisy-sur-Orbe, donated by an admirer. Flammarion's influence on the popular mentality of the nineteenth century is comparable to that of Erich von Däniken today. In 1862, when he was only twenty and an assistant at the Observatoire de Paris, he published his lengthy work *La pluralité des mondes habités: Étude où l'on expose les conditions d'habitabilité des terres célestes, discutées au point de vue de l'astronomie, de la physiologie et de la philosophie naturelle.* This was followed in 1865 by *Les mondes imaginaires et les mondes réels: Voyage pittoresque dans le ciel et revue critique des théories humaines scientifiques et romanesques, anciennes et modernes sur les habitants des astres;* and from then until his death in 1925, a plethora of works appeared— books of popular science, novels, and fictional accounts of excursions into the universe—through which the idea of the plurality of worlds must have won a cult following numbered in millions throughout the world. By 1892 there had been thirty-six French editions of his first book, and in the same period his second, consisting chiefly of an uncritical account of the history of the idea of plurality, passed through no fewer than twenty-one French editions (according to *La grande encyclopédie*). The translations are almost too numerous to count. Flammarion himself recorded in *Les mondes imaginaires* that *La pluralité des mondes* had been translated into almost all the languages "of the two continents," and the same is probably true of *Les mondes*.[22]

Flammarion prides himself on having presented, in his very first work, irrefutable proof that man's ancient dream of life on other stars is a reality,[23] and he can justly claim to have derived his "certainty" solely from the exact sciences.[24] Looking back over the history of dogmatic and metaphysical approaches to the subject of the plurality of worlds, Flammarion too reaches the conclusion that the introduction of such perspectives into this purely scientific problem is superfluous, inappropriate, and sterile. Astronomy and theology each have their own truth, and he advises the theologians, secure in the possession of their dogmatic certainties based on revelation, to regard the findings of our "poor human science" as beneath their notice. But when he cites the very similar demand made by Galileo, whose scientific truth had, over the intervening two hundred years, been triumphantly vindicated against the error of the "only saving Church," this indicates clearly the direction of his own thrust. Addressing the theologians, he writes in *La pluralité*: "If it happens that our human science, feeble as it is, from time to time makes a disastrous breach in your edifice, this fact must be an unequivocal sign to you that that edifice is not eternal" (p. 372). One such breach is of course made by the scientific conclusions about our neighbors in the universe. Flammarion does not make a polemical point of this, with the aim of provocation; he prefers not to attack the authority of the church directly. He makes his opposition known all the more effectively, however, by incidental remarks in inconspicuous places, as when in his critical bibliography of recent books on plurality, in *Les mondes imaginaires*, he remarks in passing that the God and heaven of his day are not the God and heaven of Moses, Jesus, Saint Paul, Saint Augustine, and Saint Thomas Aquinas. "That is a truth as clear as daylight" (p. 584). Similarly, the general metaphysical approach of teleology is left on one side rather than specifically refuted; only on the odd occasion does Flammarion set out in so many words the classic choice faced by modern science: the point is not, he says, to ask whether Mercury was created "in order to be inhabited by men" *(hommes)* but "the question

[22]*Les mondes imaginaires*, 12th ed. (Paris, 1874), p. 569. Page references in the text are to this edition.

[23]Ibid., p. 568; and see pp. 4, 590.

[24]*La pluralité des mondes habités*, 17th ed. (Paris, 1872), p. 328. Page references in the text are to this edition.

is to know if the physical condition of the world of Mercury would not hinder the development of the intellectual faculties of its inhabitants" (*Mondes*, p. 36).

For all his tendency to enthusiasm and verbosity, Flammarion adheres consistently to this principle of experimental, inductive science. Whatever planet or solar system is under discussion, he asks to what extent the chemical and physical conditions of a particular body are favorable to life. If it is habitable (according to terrestrial conceptions of the conditions necessary for life), then it is assumed to be inhabited: the two are in effect equated. Where the necessary conditions are present, life is brought into being by a "vital force" (*force vivificatrice* or *vitale*) inherent in matter—in other words, spontaneously. Flammarion speaks of "the life-giving force whose influence gave rise to the occurrences of spontaneous generation which brought living beings into existence."[25] He has more precise notions of the further evolution of life from the single-celled organism to the highest ones: following Darwin, he sees this development as taking place through progressive adaptation to the environment, with selection of the species best able to survive in the prevailing conditions.[26] It follows from this that "l'humanité dans l'univers" (this is the title of Book 5 of *La pluralité*) must have very different forms as a result of adaptation to the very different environmental conditions—and that we can therefore form no scientifically justifiable conception of the physical appearance of the human species elsewhere in the universe.[27] Even Christian Wolff's inference about the size of the inhabitants of Jupiter, based on the weakness of the light reaching the outer edge of our solar system, is derided by Flammarion, as it was by Voltaire and d'Alembert in the previous century (*Pluralité*, p. 228). As a scientist, Flammarion scorns cosmic anthropomorphism and relegates all detailed speculation on the external appearance of cosmic beings to the domain of the imagination and of novels about imaginary worlds (a genre of which he himself was an enthusiastic practitioner); nevertheless, he takes it for granted that the inhabitants of other celestial bodies include "men" (*hommes*).[28] In his analysis where he makes such a point of renouncing

[25]"La force vivificatrice dont l'influence fit germer les générations spontanées à l'origine des êtres": *Pluralité*, p. 242; see also p. 144; also *Mondes*, p. 98.

[26]*Pluralité*, pp. 144–45; *Mondes*, pp. 117–120.

[27]*Pluralité*, pp. 222, 242; *Mondes*, pp. 60–61, 123–24.

[28]E.g., *Pluralité*, pp. 251–55. On the distinction between science and literature see for instance *Mondes*, p. 5. Flammarion's own "novels," especially *Lumen* (1887) and *Uranie* (1889), need not be discussed in detail here. For one thing, they give such free rein to visionary fantasy that they more or less lose contact with science and instead adopt spiritualism and telepathy as the means of learning about the inhabited worlds.

all "imagination" in the name of scientific "reasoning" on the basis of empirically obtained facts (*Mondes*, p. 590), Flammarion clearly assumes that the natural laws operate in an automatic manner that will in every case necessarily lead to "man."

To man—but not always, and in every place, to humans just like us. For although Flammarion refuses to speculate even on the physical form of the "men" of other worlds, he is, strangely enough, prepared to assert that compared with the men of other "worlds," the inhabitants of Earth do not constitute the highest form of life. As in almost all theories about the plurality of worlds, however diverse they may be in other respects, man is demoted, this time in the context of evolutionary thinking. The Earth is "inferior," and so is that miserable pygmy, *Homo sapiens*.[29] "Any man is pessimistic when faced with the state of the world" (*Pluralité*, p. 258). But precisely the misery of man is a guarantee of the higher perfection of the extraterrestrials: "From whatever point of view one considers the question of man, one discovers irrefutable evidence of the inferiority of our world and the proof of extraterrestrial superiority; all the teachings of philosophy and of ethics are united in bearing witness to this" (*Pluralité*, p. 281). Clearly a philosophical belief is here superimposed on the conclusions reached by naturalistic argument. Flammarion is not referring to higher development in the evolutionary sense, as du Prel does (see sec. 2 below), although evolution is an element in his thinking. Rather, he is arguing that there *must* be a hierarchy rising above the defective level of perfection of *Homo sapiens* (*Pluralité*, p. 325). Curiously, it is the principle of analogy, which he usually uses with such caution, that confirms him in this postulation; our suspicions are aroused at once by the high-flown style of the passage:

> Just as here below, in our modest abode, the inmost constitution of all living things is marked by a *natural tendency toward the light,* from the plants that grow at the bottom of rocky cavities to the infant in the cradle who turns toward the light, so, in all Creation, beings ascend toward a higher destiny. In all the worlds of the universe, humankinds do not stop at the same level; they climb, they establish an infinite diversity in the heavens, and each has its own appointed

---

In addition, the author uses the stages of their rudimentary plots as starting points for the textbooklike exposition of the views he has already set forth in *La pluralité* and *Les mondes*. The one possible exception is *La fin du monde* (1894). Although here too the motif of psychic communication with the inhabitants of Mars plays a part, the main theme is taken from cosmic, not biologic, evolution—the dying of the Earth as it grows cold, and thus the end of the human race, in ten million years' time.

[29]*Pluralité*, pp. 56, 258.

place in the unity of the divine plan formed by the Eternal One at the beginning of the world. (*Pluralité*, p. 284)

It is truly astonishing to see Flammarion, with his naturalistic views that include the idea of spontaneous generation, suddenly reverting to an expression of belief in God's plan of Creation. But this does not mean that he relapses into pre-Darwinian theological teleology. He is not saying that God, acting with a certain purpose in mind, made the other heavenly bodies "for" other beings, or for our future incarnations. In another passage he pointedly applies the adjective "mysterious" to the "divine plan," and unlike his predecessors, who would have indulged in insistently self-assured speculation, he refrains from speculating about it (*Pluralité*, p. 254). Nevertheless, such (very rare) signs do indicate that for Flammarion, the empiricist, the scientifically discovered truth of the plurality of worlds leads to an overall world view that can only be described as religious.

He himself describes it in this way when he accuses Whewell's Christian singularism of being contrary to the "true religious spirit" (*Mondes*, p. 541). It is pluralism, and not the "egoistic" and "shabby" assumption of only one world (*Pluralité*, p. 313), that leads to such true religious feeling. "The doctrine of the plurality of worlds has brought us to the threshold of a religious belief built upon the true system of the world" (*Pluralité*, p. 319). Flammarion's religion is a vague, emotional mysticism. Nature—that is, the infinity of worlds that is "our paradise"—is "the visible expression of infinite power"; "everything has become divine; God himself has seemed to us even greater, more powerful, and more majestic; and we have felt all the beauty, all the truth of that spectacle" (*Pluralité*, pp. 313, 326). Thus in this transcendent interpretation of the universe, God is seen to be active in the coherent functioning of the natural laws; but there is more than this. An equally strong element in Flammarion's religiosity is his ecstatic feeling of brotherhood toward all the "humankinds in the universe." "The philosophy of science" enjoins us to believe that the human races in the great heavenly archipelago are our "family" (*Mondes*, p. 113). Indissoluble ties unite the stellar "humanities": "It is the divine law of the family. We are all brothers" (*Pluralité*, p. 316). There is no sense of being lost in the collective of human species, but instead a passionate eagerness to "embrace the millions" in the universe. And accordingly *La pluralité des mondes* ends with a shout of rejoicing at the existence of life throughout the cosmos:

Oh, let us cherish this doctrine as a treasure of the soul, let us consecrate it to the God of the Stars;—and when sublime nights, surrounding us with their magnificence, light up their diamondlike constellations in the east, and spread out their mysterious lights across the boundless sky . . . across the immensity of the worlds, among the starry heavens, beneath the silvery veil of the distant nebulas, in the immeasurable depths of infinity and beyond that to the unknown regions where the eternal splendor is revealed . . . shall we greet them? My brothers, let us all greet them: those are our sister humanities [*les Humanités nos sœurs*] passing by! (P. 328)

When the plurality of worlds became a gospel in the age of Enlightenment, starting in the late seventeenth century, what were accepted a priori as the intentions of the Creator of the many worlds formed part of the basis for that gospel. This is not the case with Flammarion's pluralistic "religious spirit": he rejects theological teleology as a method of approach and, as a believer in experimental science, condemns the outdated assumption of "God . . . a priori" and "intention" (*destination*) or "aim [*but*] . . . in the works of nature."[30] It is wholly by the inductive method that empirical science discovers that we have a "family" in the universe; and it is this insight that then suggests, as a "result a posteriori," a transcendent power, a "God-spirit" as the apotheosis of nature and its laws.[31] Flammarion, who traces back the history of the idea of plurality to the beginning of human intelligence, claims to be the first to have given it, by means of "the examination of the astronomical facts," that scientific incontrovertibility which demonstrates its truth once and for all (*Mondes*, pp. 567–68). The triumph of this truth in his own day, after man has taken thousands of years to come close to it, marks, for Flammarion, the onset of a new era in man's understanding of the world and of himself; and this new understanding, despite its scientific nature, has a (nondogmatic) religious corollary. And so, on the last pages of his (extremely superficial) history of the idea of plurality in *Les mondes imaginaires et les mondes réels*, Flammarion can hardly avoid representing himself, the scientist,

---

[30]*Dieu dans la nature* (Paris, 1867), p. 13.

[31]Ibid., pp. 14, 16. See also p. 493: "Sous quelque aspect que l'esprit méditatif observe la nature, il trouve une voie aboutissant à Dieu, force vivante, dont on croit sentir les palpitations sous toutes les formes de l'oeuvre universelle" (Whatever aspect of nature the thoughtful spirit observes, it discovers a path leading to God, the living force that one believes one can feel pulsating within every one of the forms that make up the universal creation).

as being at the same time the founder of the contemporary religion that is the culmination of the history of mankind.

2. The "Struggle for Existence in the Heavens" and the New Anthropocentrism: Du Prel, Lowell, Wallace, and Others

The works of Proctor, Winchell, and Flammarion illustrate the fact that explicit rejection of the wise "intentions" of the Creator as an argument for the plurality of worlds does not necessarily lead to a wholly naturalistic view of the universe based on the ideas of cosmic and biologic evolution. Religious attitudes still creep in: Proctor reverts (at least in his terminology, or in his personal beliefs) to theological teleology; Winchell and Flammarion look beyond natural phenomena to a transcendent Divinity. As the century progresses, however, it becomes more and more the norm for metaphysical considerations to be totally excluded, not only in arguing for, but also in interpreting, that "truth" (established by scientific methods) with which, according to Flammarion, human consciousness has entered a new phase. The enormous upsurge in the philosophy and literature of plurality around the turn of the century takes place wholly on the basis of secular science.[32] Much of the discussion at this time centered on the supposed high intelligence of the inhabitants of Mars. From his observatory in Milan the Italian astronomer Giovanni Virginio Schiaparelli had in 1877 observed the "canali" on that planet, and this irresistibly suggested the idea of a global irrigation system by means of which a civilization older than ours, and far more advanced technologically and also culturally, had gained total mastery of its planet. Flammarion, Jakob Heinrich Schmick, and others had already put forward this idea before it was popularized by the American astronomer Percival Lowell in the 1890s and became an international sensation that lasted for a considerable time.[33]

An interesting earlier representative of purely secular speculation on plurality (preceding the commotion surrounding Lowell's books) is the popular philosophical writer Carl du Prel. His accessible but scientifically grounded book *Die Planetenbewohner und die Nebularhypo-*

---

[32]See for example Garrett P. Serviss (one of the next writers to be discussed), *Other Worlds* (New York, 1901), p. 1: "Other worlds and their inhabitants are remarkably popular subjects of speculation at the present time."

[33]Flammarion, *Uranie* (1889), pt. 3, chap. 3; idem, *Fin du monde* (1894); Schmick, *Der Planet Mars eine zweite Erde, nach Schiaparelli* (Leipzig, 1879). See n. 54 below.

these (The inhabitants of the planets and the nebular hypothesis, 1880) was much discussed in the plurality debate of the late nineteenth century. Du Prel's position in relation to the ideas of his day is indicated by the title of the series in which his book appeared: Darwinistische Schriften (Darwinian writings). Seven years earlier du Prel had already published a book significantly entitled *Der Kampf ums Dasein am Himmel* (The struggle for existence in the heavens). In it he attempted to draw a more precise analogy than was usual between cosmic evolution and biologic evolution, which were constantly linked in the plurality theory of the day. According to du Prel, the Darwinian principle of the evolution of species through natural selection of the life-forms best adapted to environmental conditions was exactly paralleled, in the origin of stable solar systems, by the selection, through the action of the law of gravity, of planets that are advantageously placed and therefore capable of survival. Of greater philosophical significance was the step that logically followed, namely the transference of the Darwinian principle of the biologic evolution of species to the development of life on such planets. For du Prel, this step was an obvious and not a very risky one, given that the "view that all planets that have cooled down are habitable and are therefore inhabited [!]" has been provided "by spectrum analysis [with] a scientific foundation" whose soundness can be doubted only by anyone still subscribing to the "Ptolemaic-Christian world view."[34] How far du Prel's position has advanced beyond that of the middle of the century is shown by his deliberate forthrightness when an opportunity arises to draw attention to the ideological dynamite contained in the modern theory of plurality. Accordingly, he develops his ideas in a wholly scientific manner. Any teleology that "postulates the hyperphysical origin of efficient functioning," in other words attributes it to the intentions of an entity transcending nature, is mistaken (*Planetenbewohner*, pp. 59–62). Du Prel will not even accept an immanent teleology that assumes purposefulness in natural processes, since "life in the cosmos is far too limited, in space and in time, for us to regard the preceding development as a preparation for this purpose of producing life" (p. 62).

Given these premises, what is the likelihood of manlike extraterrestrial life? Analogical reasoning based on conditions on Earth is bound to lead to the conclusion that life in our solar system is confined

---

[34]*Entwicklungsgeschichte des Weltalls* (the retitled 3d ed. of *Der Kampf ums Dasein am Himmel*) (Leipzig, 1882), pp. 367–68; see *Die Planetenbewohner und die Nebularhypothese* (Leipzig, 1880), p. 73. Page references in the text are to *Planetenbewohner*.

to the inner planets (pp. 66–68). On the "major bodies in our system" the natural evolution from the unconscious to consciousness cannot take place (p. 67). What is most interesting is du Prel's extrapolation from this about the nature of the inhabitants of other cosmic worlds. Applying the Darwinian way of thinking to the life-forms existing on other planets, he is able to conclude that probably there "the further development of consciousness" has been "continued beyond the level that we have reached in an *organic* way" (p. 80). That is to say that in the further course of evolution, new organs have developed as part of the continuing adaptation to environmental conditions (pp. 129–30). What sort of organs? Here du Prel shows his greatest originality and sets out the actual thesis of his book. Simplified, it is roughly this: what our intellect produces by means of technology must at a higher stage of evolution be brought about by newly formed organs. These new organs and their senses, which relate to such things as electricity, magnetism, and other natural forces, replace our technical apparatus. What our technology can achieve by means of flying machines could be done on more highly evolved planets by winged human beings, while their technology in its turn would be able to overcome far more difficult problems, and so on. Here too, then, analogy operates. Du Prel conjectures that there may be "a certain parallelism between the stages of organic development there and technical development here" and suspects "that the products of the human mind offer veiled clues to the characteristics of those planetary organisms which we would recognize as higher beings" (p. 80). It is logical to assume that alongside the development of these additional senses and organs, the "intellectual nature of the planet dwellers" is also more highly developed (chap. 6). Neither biologically nor intellectually, then, does *Homo sapiens* represent the highest rung of the cosmic evolutionary ladder. Thus the demotion of man, difficult to avoid in the philosophy of plurality, and the old speculative notion of an increased number of senses appear here within a specifically Darwinian framework, as "conclusions drawn from scientifically obtained premises" (pp. 124–25). The fact that du Prel does not go one step further and describe the extraterrestrial organic developments concretely and in detail indicates precisely his position between Enlightenment speculation about a greater number of senses and turn-of-the-century science fiction with its use of bizarre, superior monsters.

Nevertheless, du Prel's answer to the question of man's position within the cosmic whole is clear, even if it is never spelled out: we are no more than an intermediate stage in the evolution of life; it is precisely the triumphs of our technology (and not the misery of the world, as is claimed by Flammarion, who is philosophically rather backward in this respect) that give us reason to think that organic evolution has progressed further in other, older worlds than on Earth. How human beings whose self-esteem has been enhanced in the nineteenth century precisely by the advances of science are to come to terms with or resign themselves to this scientifically established fact—when that same science has made it impossible to take refuge in the (perhaps not impenetrable) wisdom and justice of a "personal" Creator—is a question that falls outside the strictly scientific limits within which du Prel and others work. But it is there between the lines, inevitably accompanying the renunciation of teleology and of metaphysics of whatever kind. Unlike Flammarion and some others, du Prel refrains from surrounding the higher development on other planets with a mystical aura. There is not the faintest suggestion of cosmic religiosity. But du Prel seems to show awareness of the fact that his speculation poses a threat to man's self-image when he stresses, as though appeasingly, that there is "something reassuring" (p. 59) in the idea that the heavenly bodies are inhabited, though he does not indicate in what way it is comforting. (Lowell was to see the idea more as a shock to be overcome.) We learn that we do not represent the climax of the great drama of the evolution of life in the universe but that we do have a specific position within it. Whether this was really felt at the time to be a reassuring confirmation of man's status seems doubtful when one thinks of the number of novels from H. G. Wells's onward in which the Earth is invaded by the notorious "bug-eyed monsters." A different kind of blow is dealt to man's self-esteem by turn-of-the-century fiction where Martians are portrayed as more advanced culturally and morally as well as intellectually, and we as the colonized primitives of the universe; thus in Flammarion's *La fin du monde* (1894) and George Griffith's *Olga Romanoff* (1894) it is the Martian scientists who warn the earthlings of the danger of an approaching comet and give them the information necessary for survival. Du Prel, then, does not answer, and indeed hardly poses, the question of the nature and significance of man in the framework of cosmic and biologic naturalism—the question of his identity, given that in the context of the plurality of worlds (as opposed to Darwin's own theory, which applied to the Earth alone) he can no longer be seen as the highest product of evolution.

This is the dilemma posed by the philosophical anthropology, stripped of all metaphysics, of the second half of the nineteenth century, and (though it had occasionally been foreshadowed in the seventeenth and eighteenth centuries) it is this that gives the plurality debate of that time a wholly new aspect. This dilemma is again studiously avoided by the highly regarded American astronomer and successful writer Garrett P. Serviss, in his book *Other Worlds, Their Nature, Possibilities and Habitability in the Light of the Latest Discoveries* (New York 1901). Here all is physics: Serviss examines, with a wholly consistent approach and in accordance with the latest findings of astronomy, the temperature and atmosphere, gravity and surface conditions, orbital speed and seasons, age and direction of development of all the planets of the solar system, generally following the evolutionary approach of Laplace's nebular hypothesis and Darwinian biology. What is the outcome of Serviss's extrapolation from these scientific data? On the premise that conditions favorable to life do in fact produce life (pp. 64–65), his comparison of the physical conditions of other planets with those of Earth leads him to the conclusion adopted by a large proportion of the science fiction of various countries at the turn of the century. According to Serviss, the outer planets of our system are sunlike formations that are burned out, are in the process of becoming worlds, and are *not yet* inhabited (a view very different from Winchell's); the innermost planet, Mercury, is uninhabitable by terrestrial standards; Venus is a younger world than the Earth: "We are at liberty to imagine our beautiful sister planet as now passing through some such period in its history as that at which the earth had arrived in the age of the carboniferous forests, or the age of the gigantic reptiles who ruled both land and sea" (p. 59). As for Mars, it was formed, by separation from the condensing solar nebula, earlier than the Earth, cooled down more rapidly because of its smaller mass, and must therefore be a biologically older world than ours. Its manlike, intelligent beings must be several evolutionary stages ahead of us; by their "superhuman powers" of intellect they have brought their planet under technological control, as is shown by the canals we can see (chap. 4). Finally, the moon, which because of its small mass is now cold, dehydrated, and without an atmosphere, may at some earlier cosmic period have been the home of living creatures (p. 237).

Serviss was able to invoke the support of Lowell's theories, already mentioned, about the development of the planets and especially about the advanced state of the intelligent beings on Mars, theories that from roughly 1895 to 1910 set the whole world agog—not only the scientific world, which was plunged into violent controversy, but above all, that of literature. Lowell, a widely traveled polymath and gentleman-businessman of a leading Boston family, an adherent of the nebular hypothesis and of Darwinism, was led by Schiaparelli's telescopic discovery of the lines on Mars to study with obsessive intensity the question of whether Mars, our closest neighbor, is inhabited. In 1894 he had an observatory built at Flagstaff, Arizona, where he and his assistants worked for decades (Lowell died in 1916), observing and mapping the surface of Mars in an attempt to find answers to the most diverse questions. Lowell presented his conclusions in innumerable lectures in both America and Europe and also in a number of books, which aspired to scientific seriousness though they were written in a popular style: *Mars* (1895), *Mars and Its Canals* (1906), *Drawings of Mars* (1906), *Mars as the Abode of Life* (1908), and as the general basis for his theory, *The Evolution of Worlds* (1909).

His thesis can be briefly summarized as follows. The straight lines on Mars, visible through the telescope, cannot be of natural origin. The regularity of their "system," the geometric structure of the network of lines, can only be artificial. The telescope reveals not geologic formations but strips of vegetation alongside the irrigation canals by which, in the universal "struggle for existence," intelligent beings are trying by better economy to improve their chances of survival on a planet that is growing colder and drying out.[35] (Lowell emphasizes that the physical form of these intelligent beings is not that of "men": in accordance with the different conditions in their world, to which they would have had to adapt, they might for instance be frogs or lizards.)[36] "Oases" can be seen at the intersections of canals. The variations in the appearance of the areas of vegetation are seasonal, caused by the freezing and thawing of water in the polar zones of Mars. Lowell believes he can make a good scientific case not only for water but for a moderate temperature and an atmosphere. Thus the conditions necessary for life, judged by the terrestrial standard, are present, and in Lowell's view, as in that of du Prel and many biologists of the day, such favorable conditions occurring at a particular stage of a planet's development will always necessarily lead to the development of organic life from inorganic matter: this is "an inevitable phase of planetary evolution." Thus life is "an inevitable outcome of the cooling of a globe," and for Lowell too it arises by "spontaneous generation."[37] According to Laplace's nebular hypothesis, Mars, a planet more distant from the sun, must have become separated from the sun's mass of dust and gas earlier than the Earth; it therefore follows that on Mars we may reckon not only with life but with comparatively older life

[35]*Mars as the Abode of Life* (New York, 1908), p. 204. See Otto Dross, *Mars, Eine Welt im Kampf ums Dasein: Eine gemeinverständliche Studie für Freunde der Himmelskunde* (Vienna, 1901). Unlike Lowell, Dross supposes the Martians to be anthropomorphous.
[36]*Mars* (Boston, 1895), p. 211.
[37]*Mars as the Abode of Life*, pp. 37, 66.

that has progressed further in evolutionary terms but is still decidedly "akin" to ours.[38] (Venus, however, is still in the carboniferous age.)

Not only, then, have the intelligent beings on Mars achieved the technological and intellectual feat of effectively harnessing natural forces and resources by means of their global irrigation system, and advanced so far in their technology that, for them, inventions like the telephone and television have long since become museum pieces;[39] in order to achieve that worldwide organization of irrigation reaching from pole to pole, they must also have made greater social and political progress toward global cooperation than the quarrelsome nation-states of Earth. (Thus science contradicts mythology!) All the same, this advanced civilization, which in the interests of the "survival of the fittest" has taken the practical and rational course of organizing itself into a single pacifist society, must, by the inexorable law of cosmic evolution, of progressive cooling down and dehydration, be approaching its end. This, moreover, provides a grim object lesson for humankind on Earth: it prefigures the future and the final decline of terrestrial civilization, which is at present at a stage of cosmic evolution that on Mars already belongs to the distant past.[40] (Back in 1894, Flammarion had described the Earth's becoming cold and dying in his novel La fin du monde.)

By 1909 at the latest, Lowell's theories about extraterrestrial civilization, extrapolated from his observations of Mars, were called into question, at least scientifically, by the advent of more powerful telescopes.[41] In that year, the sixty-inch telescope at the Mount Wilson Observatory in Southern California revealed the ordered network of canals to be an irregular hodgepodge of geologically explainable canyons caused by erosion. Lowell's claim that his theories were not speculations but logically sound inferences from empirically obtained facts,[42] though made in all honesty, eventually collapsed. But this does not deprive his life's work of all value. At the heart of it was Lowell's fascination with the idea of the possible existence of highly evolved extraterrestrial life-forms,[43] and this was not dependent on the reality of the canals. Today the American Mars probe has for the time being made life on Mars appear unlikely, but this does not mean that life may not have existed there in the past, and still less that planets in

[38]Ibid., p. 196.
[39]Mars, p. 209.
[40]Mars as the Abode of Life, pp. 216, 135.
[41]William Graves Hoyt, Lowell and Mars (Tucson, Ariz., 1976), p. 170.
[42]Mars as the Abode of Life, p. 214.
[43]Hoyt, Lowell and Mars, p. 55.

other systems may not harbor life. Moreover, there is another element in Lowell's books that has importance for the history of thought and is still relevant: unlike other evolutionary "naturalists" among the pluralists (for instance du Prel, Serviss, and even Flammarion), Lowell does give intensive thought to the dilemma of man when suddenly confronted by science with the fact, or at least a high degree of probability, that he is "not alone" and, furthermore, does not represent the highest form of life in the cosmic development of species.

In his books on Mars, though his arguments are wholly scientific, Lowell regularly returns to this subject toward the end. Here theological elements, still present in the work of a scientist such as Oersted, are totally lacking; Lowell, addressing himself to a large readership that includes nonscientists, is representative of his time in staying firmly within the realm of anthropology, from which all metaphysics is excluded. The question he asks is What is man in the light of the presence, on even one other planet, of manlike intelligent beings whose development has evolved further than ours? Like Fontenelle, he deals a blow to man's pride: "On earth, for all our pride of intellect, we have not yet progressed very far from the lowly animal state, that leaves no records of itself."[44] Unlike Fontenelle, however, Lowell does nothing to restore man's self-esteem. Nor does it occur to him to compensate for, and neutralize, the acknowledgment of Mars's higher technological civilization by maintaining that the Earth has a higher *culture*, perhaps because for him technical and cultural progress go hand in hand: further intellectual and social development is a precondition of the global technology that achieves mastery over nature. He also lays little stress on the idea emphasized by his disciple Mark Wicks, that man, not having progressed very far as yet, has still a splendid evolutionary future before him *(To Mars via the Moon,* 1911). In human terms, Lowell says, the effect of his discovery of the Martians, whose intellect is not only superior but perhaps quite different, is a shock: it is, as he himself expresses it, a second Copernican shock at the dethronement of man. Man's "unique or self-centered position" is no more, and this is a blow that must, first of all, be absorbed by man's consciousness and then integrated into a new awareness.[45]

According to Lowell's anthropology, the shock is even more psychological than philosophical. The encounter with something that is unknown and thought to be superior, he says, is always uncanny and liable to induce fear and uncertainty. The instinct of self-preservation

[44]*Mars and Its Canals* (New York, 1906), p. 364.
[45]*Mars as the Abode of Life,* p. 216.

would then go to any lengths to prove the possibility to be an impossibility. The highly civilized Westerner who has raised savage colonial peoples to his own level is now, paradoxically, in the position of the savage in the cosmic context. Of the psychology of the encounter with extraterrestrial life-forms, Lowell writes: "Like the savage who fears nothing so much as a strange man, . . . the civilized thinker instinctively turns from the thought of mind other than the one he himself knows."[46] When writers in the age of conquest used the comparison with the European colonization of primitive regions of the world in the context of the plurality of worlds, they took it for granted that "we" would be the conquerors. When Fontenelle, the classic exponent of the idea of plurality in the Age of Enlightenment, made the same comparison, it was all the same to him whether we were the colonial masters or the savages. With Lowell, the tables have been well and truly turned. We are not only technologically inferior but also psychologically underdeveloped. It is puerile, Lowell thinks, to resist the idea of higher manlike "races" in the universe; it is time to grow up (*Mars*, p. 210). This also means rising above terrestrial provincialism and thereby—at last—renouncing man's claim to be the measure of all things. The devastating impact of this cosmic-anthropological realism is only a little softened by the consolation of growing up and of becoming a member of the great family in the cosmos—an inadequate compensation indeed; for Lowell has evidently no use for the idea, embraced by science fiction writers from Flammarion (*La fin du monde*) to Arthur C. Clarke (*A Space Odyssey*), that the advanced intelligent beings might act as guardians guiding our development. Lowell's view is less reassuring: *Mars* ends with the following passage, which summarizes this amateur astronomer's reflections on the cosmic dimension of nonmetaphysical anthropology:

> That we are the sum and substance of the capabilities of the cosmos is something so preposterous as to be exquisitely comic. We pride ourselves upon being men of the world, forgetting that this is but objectionable singularity, unless we are, in some wise, men of more worlds than one. For, after all, we are but a link in a chain. Man is merely this earth's highest production up to date. That he in any sense gauges the possibilities of the universe is humorous. He does not, as we can easily foresee, even gauge those of this planet. He has been steadily bettering from an immemorial past, and will apparently

---

[46]*Mars*, p. 210. For Lem's idea that there may be a "quite different" kind of intelligence see chap. 1 above, at n. 50; see also the discussion of Kepler (at the end of chap. 2 above).

continue to improve through an incalculable future. Still less does he gauge the universe about him. He merely typifies in an imperfect way what is going on elsewhere, and what, to a mathematical certainty, is in some corners of the cosmos indefinitely excelled.

If astronomy teaches anything, it teaches that man is but a detail in the evolution of the universe, and that resemblant though diverse details are inevitably to be expected in the host of orbs around him. He learns that, though he will probably never find his double anywhere, he is destined to discover any number of cousins scattered through space. (P. 212)

Understandably, in the light of the attitudes of the day, this cosmic "humanism," which demanded such a high degree of maturity on man's part without the support of a total world view that was even in the broadest sense religious, did not meet with universal acclaim (though it was enthusiastically received by some). The adverse reaction to Lowell (and also to du Prel, Serviss, and their followers, who did not spell out the anthropological inference implicit in their position) could lead in two directions—back to teleology as the basis for belief in the plurality of inhabited worlds, or back to a single world—though in either case it had to be based on contemporary, indeed the very latest, scientific research. At this stage, purely theological-teleological argument is no longer a serious option.

The first of these paths was chosen by the Roman Catholic priest Joseph Pohle in his apparently widely read two-volume work, *Die Sternwelten und ihre Bewohner: Eine wissenschaftliche Studie über die Bewohnbarkeit und die Belebtheit der Himmelskörper nach dem neuesten Standpunkte der Wissenschaften* (The star-worlds and their inhabitants: A scientific study of the habitability and inhabitedness of the celestial bodies according to the latest state of the sciences, Cologne, 1884–85).[47] Though Pohle adopts an openly theological position, "the empirical basis" (1:56) is provided by the modern sciences, chiefly spectrum analysis. This basis is that many, if not all, heavenly bodies are habitable. From this, Pohle states, "philosophy" infers that they are actually inhabited (2:64), that Mars is indeed "a second Earth." What Father Pohle calls philosophy is, however, in reality the "atheistic materialism" of du Prel, Flammarion (!), and others who, by their assumption that life comes into being automatically when the necessary chemical conditions are present, project Darwinism on to the whole universe (1:vii). Pohle considers that assumption to be unprovable (as indeed

---

[47]Third ed. (1902); 7th ed. (1922). Page references in the text are to the 1st ed.

it has yet to be proved) and replaces it with his own premise, which depends for its credibility on faith—at least "philosophical" if not religious faith—and which could in fact do just as well without the support of spectrum analysis. It is the old proof, discarded in the nineteenth century but still alive to this very day in some Catholic circles,[48] that rests on the "purposes" of God: it would not accord with his wisdom to have left possible worlds barren and empty. Once again we are told of the Creator's glorification of himself by the praise issuing forth from all the humankinds in the universe. An intriguing addition is the further priestly argument (possessing, he claims, the power of incontrovertible proof) that, if only to "compensate" himself for the "sinfulness" of Adam and Eve's descendants, God had to create other, better species of manlike beings on other stars (2:201–3). It is of course easy for Pohle, with the wisdom of hindsight, to refute the dogmatic objections to the plurality of worlds[49]—despite such archconservative phenomena as Father Burque's wholly dogmatic antipluralist book of 1898, already mentioned.

It is worth noting, however, that the more significant attempts at this time to preserve the Earth's unique status as the only abode of life come not, or not directly, from the theological but from the scientific camp. How close the similarity between physical and chemical conditions here and on other planets needs to be to permit the conclusion that other heavenly bodies are inhabited is at bottom a scientific question; so is the one closely linked with it, namely, how far one can conceive of life-forms that are not, or not quite, the same as those with which we are familiar. It is therefore not surprising that, in the age of science, the combined forces of astronomy and biology should sometimes come down on the side of singularity. In these cases, however, an underlying theology soon shows itself, which was not the case with the scientific mode of argument of the pluralists du Prel, Serviss, and Lowell. Even where a tendency to think metaphysically in terms of a "purpose" does not direct the course of the argument (as it did with Chalmers and Pohle, where it helped to support the case for pluralism, and to some extent with Whewell, supporting the opposite view), the manner in which the ideas are developed does at least

[48]Kenneth J. Delano, *Many Worlds, One God* (Hicksville, N.Y., 1977).

[49]Vol. 2, 204–212. He uses the old "liberal" arguments in relation to dogma: the Bible speaks only of our world and thus does not necessarily rule out other worlds; the beings on other planets (Pohle envisages only beings similar to man) either have not suffered a Fall and therefore do not need salvation or they are saved by some means unknown to us. A similar position is adopted by de Montignez, who counters Flammarion with his "Théorie chrétienne sur la pluralité des mondes," *Archives théologiques* 9 and 10 (1865).

open up the prospect of an anthropocentric teleology, which makes it legitimate at any rate to suspect that beneath the surface, the whole discussion was theologically motivated all along.

This is the case with William Miller, an Edinburgh lawyer and the author of a book titled *Wintering in the Riviera*. In his study *The Heavenly Bodies: Their Nature and Habitability* (London 1883), he concludes, on the basis of "analogical deduction" (p. 343) and extensive discussion of the most recent research findings, that none of the planets of our system offers conditions comparable to those of Earth and that therefore none of them can possess life (life like ours, that is; he cannot conceive of any other kind). Mars is too dry and lacks air, Jupiter is too cold, Saturn too hot, and so on. As for the planets of other systems, they are merely a figment of inexact analogy, born of the *wish* that they might exist. The conclusion, predictably, is Christian man's tribute to himself in the age of empirical science: we, to whom God sent his son in human form, are the only objects of his care and love:

> And if we can imagine Earth so linked to Heaven, and think of its inhabitants as so highly favoured and so highly prized, we are enabled the better to comprehend our lofty position, and we can proudly rejoice to think that though our world, amidst the millions of celestial bodies vastly larger, is—looked upon with a material eye—but an insignificant and indiscernible speck, yet, as not merely the sole dwelling-place of rational life, but the seat of the most amazing of all wonders, a being destined to a glorious immortality clothed for the present in a mortal frame, but conscious of his high prerogative; and therefore, and by reason of all that, because of its inhabitant Man, has supervened, it is a greater and grander globe than the largest and brightest orb which gleams upon the midnight sky. (Pp. 346–47)

Another attempt to refute atheistic pluralism, which carried more weight than this statement by an educated amateur, was that made by the Nestor of English science and cofounder of the theory of evolution, Alfred Russel Wallace. In 1907, when he was eighty-four, a controversial figure at the height of his fame, he challenged Percival Lowell with a polemical scholarly publication, *Is Mars Habitable?* His negative response to this question rests on his anthropocentrism, which has ultimately a religious or at any rate supranaturalistic basis. This had caused him, for some decades past, to modify the Darwinian version of evolutionary theory: Wallace accepted the organic development of the species through natural selection in the struggle for existence (he too uses these expressions), but only for those species that are below *Homo sapiens*; the origin of man, or at least of his consciousness, and therefore also of his brain, requires (as Whewell had already asserted) a special intervention by a transcendent intellectual principle, which amounts to an act of creation.[50] This claim to man's unique dignity was of course endangered by the prevalent idea of plurality. Logically, the next step for Wallace was therefore to reinforce his biologic claim by means of an astronomical proof of man's uniqueness in the universe. He had already done this before writing the book on Mars, in a work written for the layman, *Man's Place in the Universe: A Study of the Results of Scientific Research in Relation to the Unity or Plurality of Worlds* (1903).

Wallace too emphasizes that his demonstration is conducted wholly scientifically, by the inductive method, and that it accords with the latest findings of astrophysical and biologic research.[51] He claims that it was precisely the nonscientific, a priori arguments—theological or at any rate metaphysical—that for centuries led to the assumption of plurality (pp. 9–10). Whewell had made the same criticism. Whewell's thesis of the uniqueness of the Earth as the abode of higher life-forms, overtaken in the meantime by the scientific research of two generations, is underpinned by Wallace in two ways.

As an *astronomer*, he provides a cosmologic framework for the view of the Earth as a unique phenomenon. In so doing he shows his greatest originality—and lays himself most open to attack. Unlike Whewell, he is primarily concerned with the *structure* of the universe. Adhering closely to contemporary astrophysics (Simon Newcomb, Norman Lockyer, and others), he argues that the visible universe is not infinite but limited and that it has the form of a (single) cohesive system of concentric rings; the Milky Way forms the outermost. Its midpoint is the absolute center of the universe; close to and arranged around this center is a spherical grouping of stars, and within this grouping, separated by an inner space with far fewer stars in it, is

[50]*Darwinism* (London, 1889). See Heffernan, "Singularity of our Inhabited World"; Malcolm Jay Kottler, "Alfred Russel Wallace, the Origin of Man, and Spiritualism," *Isis* 65 (1974), 145–92. Kottler gives an account of certain developments of Wallace's views which took place over a number of decades. These include in particular the variations of Wallace's assumption that the transcendent principle which guides the process of evolution had already made a supernatural intervention before the origin of *Homo sapiens*, namely at the inception of life and of the consciousness of animals. Kottler links this departure from Darwin with Wallace's spiritualist convictions. See Kepler's similar ideas, chap. 2 above, early in sec. 5.

[51]*Man's Place in the Universe* (New York, 1903), pp. v, 23, 276–77. Page references in the text are to this edition. The book went into its 4th ed. in 1904.

another conglomeration of stars, at the periphery of which our solar system is located (see the diagrams in *Man's Place in the Universe*, pp. 296–97). This places us not at the precise center of the cosmic structure, at that center of gravity about which ultimately every heavenly body rotates, but at a distance of some thirty light years from it (p. 300). Still, this position—"our central position (not necessarily at the precise centre) in the stellar universe"—has for Wallace a "meaning" that can only be described as symbolic: only here is life possible (p. 169).

This brings him to his second line of argument. As a *biologist* who has taken to heart Whewell's insistence on the precision required for a scientifically legitimate inference by analogy, Wallace rejects the idea that the identical nature of matter (proved, subsequent to Whewell, by spectrum analysis) and the identical nature of the chemical and physical laws everywhere in the universe can, or must have, led more than once to the development of life, or at any rate of its higher forms (p. 258). Wallace plays by rules that reduce the opposition's chances even more than Whewell's did, but the principle remains the same: life (in its higher forms), its origin, and the evolution of its species is conceivable only under conditions absolutely identical to those on Earth. These conditions include the relatively limited variation in temperature; the precise tilt of the Earth's axis toward the ecliptic, which regulates our seasons; the exact chemical composition of the atmosphere; the size and physical state of the planet; and the distribution of its landmasses and surface water. In addition the system must have remained stable for a long period, in order to allow evolution to proceed undisturbed. According to Wallace, this last factor rules out all the systems of the Milky Way—the great majority of heavenly bodies—as environments hostile to life (p. 280). But what of the star systems within the ring formed by the Milky Way? Here Wallace is dealing with probabilities that for him amount to virtual certainties: the most recent telescopes give reason to believe that sunlike stars, unlike double stars, are rather exceptional; whether the exceptions possess planetary systems cannot be empirically determined, and if they should have them, it is unlikely that they would include Earth-like planets, since Wallace has infinitely multiplied and complicated the preconditions absolutely essential for life, and life of any other kind than that which we know is not, for him, a realistic possibility (p. 283). It is more than unlikely, he concludes, that all the necessary conditions could ever be, or have ever been, reproduced elsewhere in exactly the same combination, the only one conducive to life (pp. 306–13). Man, confirmed in his "central position," is a unique phenomenon in the universe.

While in principle Wallace's biologic argument—with refinements reflecting the present state of microbiologic and genetic research—still (or once again) merits consideration today, his cosmologic argument against the plurality of inhabited worlds has too many weaknesses to have had much impact even in his own day. His speculations on the structure of the universe were both too bold and too precise in relation to their basis in empirically determinable fact.[52]

Nevertheless, one must grant the justice of Wallace's repeated and emphatic claim that, unlike his predecessor Whewell, he keeps strictly to scientific facts and modes of thought in developing his thesis. He applies these modes of thought, however, to an inadequate body of observed data, and only this lack of rigor makes it possible for him to conclude that "we are alone" and effectively at the center. This does arouse the suspicion that his relapse into a sort of modified pre-Copernicanism (at a time when Lowell is demanding that people at last adjust themselves psychologically to Copernicanism!) is theologically motivated after all. There is something disarming about Wallace's assurance, *after* the résumé of his results obtained strictly by the inductive method, that these results contain nothing that could cause disquiet to the "religious mind" (p. 314). In this passage, freed from his self-imposed restraint now that the purely scientific discussion is concluded, he repeats his supranaturalistic view, already touched upon, that consciousness, or the human mind, cannot be a product of matter. What is more, he now also makes mention of the view "that the universe was actually brought into existence for this very purpose," namely to make it possible for man to come into being as the highest form of life (p. 315). True, he presents this view— anthropocentric teleology of the first water—as that of religious thinking. But there is no mistaking his own sympathy with this interpretation. "There is no incongruity in this conception," he writes in the very next sentence. In other words, in the idea that the whole history of the entire universe was aimed at producing man on our small planet, Wallace sees no disproportion between the outcome and all that has gone into preparing for it. The overwhelming temporal and spatial dimensions of the universe only serve to heighten the dignity and significance of man as its highest form of life. Indeed, like Whe-

---

[52]On the adverse reactions at the time, see Heffernan, "Singularity of Our Inhabited World," pp. 92–93. But Heffernan for his part acknowledges the higher level of scientific argument that raises Whewell and Wallace above the pluralists who argued from metaphysics (pp. 95–96). In the late nineteenth century even those who deny that our solar system is inhabited commonly assume life in other systems; see Zöckler, *Geschichte* 2:438; John E. Gore, *The Worlds of Space* (London, 1894), p. 39.

well, and, much earlier, Klopstock, Wallace reverses the thrust of the comparison: "And for the development of such a being what is a universe such as ours?" (p. 318). Man is not "of no importance," as the theory of the plurality of worlds so often asserts; on the contrary, "What a piece of work is man! How noble in reason! How infinite in faculty! . . . In apprehension how like a god!" (pp. 317–18).

This Renaissance view (in a cosmic framework) of *Homo sapiens* certainly leaves no room for the fear of being alone in the universe.[53] The confident self-image is manifestly a secular one: Wallace does not, for instance, use man's central role in God's plan of salvation, the idea that his incarnate Son was sent to us, as secondary evidence. (Unlike Whewell, Wallace never even alludes to this.) And yet, as has already been suggested, this self-confidence opens up a religious perspective. Is it not legitimate for man as such a "piece of work," as a product of nature, to see himself as the realization of a *divine* purpose, of which nature is perhaps the tool? Is Wallace not in fact declaring his support for the "argument from design"? At the key point of his epilogue he expressly leaves this possibility open: "And is it not in perfect harmony with this grandeur of design (if it be design), this vastness of scale, this marvellous process of development through all the ages, that the material universe needed to produce this cradle of organic life, and of a being destined to a higher and a permanent existence, should be on a corresponding scale of vastness, of complexity, of beauty?" (p. 317). In fact the equivocal "if it be design" is not Wallace's final word on the matter. He speaks of the creative principle as the "infinite power" (p. 318), and he closes with a quotation from Proctor in which, in the same context, "infinite power" is equated with "infinite purpose," the infinite creative purpose of the Almighty: "Infinite power, . . . Infinite Purpose . . . , Infinite Being, Almighty" (pp. 319–20). Nevertheless one is struck by the *veiled* way in which Wallace introduces his metaphysics, his anthropocentric teleology. This shows once again the extent to which scientific thinking has freed itself from the grip of metaphysics—and the extent to which it has not.

3. Novels at the Turn of the Century: The End of the World—the Future of Mankind

The space literature of the late nineteenth and early twentieth centuries, consisting mainly of novels and short stories, was of a quite

[53]Noble, *Astronomical Doctrine*, considers that the pluralists deserve credit for mak-

different kind from that of earlier centuries. Its epoch-making newness and its vitality sprang from the meeting of literature with the evolutionary ideas in biology that were making such an impact through Darwin's work. In the new era in astronomy, inaugurated at the same time by Kirchhoff's spectrum analysis, space literature applied the new biologic ideas to the life-forms of other "worlds" in the universe, which were themselves seen as undergoing physical evolution in accordance with the nebular hypothesis (see sec. 1 above). The combination of these scientific developments, in biology and in astrophysics, offered to writers a new field of undreamed-of possibilities in which to work. "Science," writes John Munro in his science fiction novel *A Trip to Venus* (London 1897), "far from destroying, will foster and develop poetry . . . , serve the poetical spirit by providing it with fresh matter. . . . Consider the vast horizons opened to the vision of the poet by the investigations of science [he specifically mentions the nebular hypothesis and spectroscopy, pp. 78–81, 20] and the doctrine of evolution" (p. 105). The theoreticians' dilemma is that man, with his confidence and pride in his recent technological achievements, may possess the highest status on his own planet but can, in the context of the many inhabited worlds, be accorded at best a precarious intermediate rank in the development of species. There is also the puzzling duality of the prospect before man: a glorious evolutionary future or collective degeneration, an ideal global state or the end of our Earth as it dries out and cools down like Mars. This provocative undermining of man's certainties, in a cosmic natural world no longer subject to the wise guidance of a Creator, has from the final decades of the nineteenth century onward (up to the present day, really) provided a continuing inspiration for those works of the literary imagination that participate in the adventure of planetary or "Copernican consciousness" (Fritz Usinger). This is the start of a new chapter in the history of imaginative literature, which is only inadequately described by the usual label "science fiction." For it also opens an exciting new era in the history of Western man's view of himself, when new answers to the question of his identity begin to emerge.

The immense growth in space literature in the wake of the doctrine of evolution and spectrum analysis (and also, in conjunction with these, the nebular hypothesis) is evident on a purely statistical level. From the closing decades of the nineteenth century up to the First World War, European and American space fiction, ranging in quality

---

ing it impossible to feel "that sense of loneliness and destitution" induced by the assumption of a single world (p. 14).

from cheap thrillers to philosophical novels, attains a volume that is barely measurable, especially in the English-speaking countries.[54] By comparison with this, the space literature of the first two-thirds of the nineteenth century is of little significance, either in quantity or in content. It does not seriously explore the theme of the encounter with life-forms unknown to us, and in other ways, too, the few novels of this period that might be classed as space literature are pitched at an extremely low intellectual level, as well as being scientifically anachronistic. They tend on the whole to be utopias, satires, allegories, dream visions, mythological romances, or exotically fantastic tales of "other worlds," which might just as well have been set on some remote island, for instance, as in space. If they are sometimes categorized as science fiction, it is only in an excessively loose sense of the term; any science fiction motifs that they use are merely vehicles for, say, social criticism or political concerns.[55]

Of course these varieties of space literature (in the broadest sense), which are of only marginal interest in relation to the plurality of worlds, did not die out even after Darwin and Kirchhoff.[56] (Some of them still flourish today.) In the second half of the nineteenth century

[54]This emerges clearly for instance, despite the alphabetical arrangement, from George Locke's *Voyages in Space: A Bibliography of Interplanetary Fiction 1801–1914* (London, 1975). Locke includes only novels and stories written in English (and a few translated into English). The chronological diagram on p. 68 shows the enormous increase in this kind of literature around the turn of the century. The amount produced between 1800 and the 1870s is minute by comparison. The statistical survey, on p. 67, of the planets dealt with in the novels shows the overwhelming fascination exerted by Mars. Locke outlines the contents of the works, many of which are now exceedingly rare. These summaries of contents show clearly what were the recurrent motifs in these works of light fiction (which in general were evidently of low quality) for juveniles and adults.

[55]George Fowler, *A Flight to the Moon: or, The Vision of Randalthus* (Baltimore, 1813); Thomas Erskine, *Armata*, 2 vols. (London, 1816–17); "Joseph Atterley" [George Tucker], *A Voyage to the Moon* (New York, 1827); "F. Nork" [Friedrich Korn], *Die Seleniten* (The Selenites, Pirna, 1834); A. E. Papinga, *Das Leben und Weben im Planeten Venus* (The living world on the planet Venus, Jüterbog, 1835); Joseph Emil Nürnberger, *Astronomische Reiseberichte* (Astronomical accounts of voyages, Kempten, 1837); Anon., *A Fantastical Excursion into the Planets* (London, 1839); Tirso Aguimana de Veca, *Una temporada en el más bello de los planetas*, which appeared in 1870 in the periodical *Revista de España*, but was written in the late 1840s (see Brian J. Dendle, "A Romantic Voyage to Saturn," *Studies in Romanticism* 7 [1968], 243–47); C. I. Defontenay, *Star* (Paris, 1854); Sidney Whiting, *Heliondé: or, Adventures in the Sun* (London, 1855).

[56]See the bibliography by Locke (n. 54); the brief summaries usually give a good idea of the contents. German and French works of this kind that might be added are: J. G. Pfaff, *Die Reise in den Mond* (The moon voyage, Kassel, 1864); Achille Eyraud, *Voyage à Vénus* (Paris, 1865); "H. Graffigny" [Raoul Marquis], *Les voyages merveilleux* (Paris, n.d. [c. 1883]); *Passyrion über Deutschland: Beobachtungen und Kritiken eines Marsbewohners* (Passyrion on Germany: The observations and criticisms of a Martian). "Aus dem Marsischen übersetzt von Intrus" (Rostock, 1905).

these were joined by several further variations: the purely technological fantasies of Jules Verne, in which no encounter with extraterrestrials takes place; the futuristic fantasy stories of the Russian space
pioneer Konstantin Ziolkovski, which aim to present the physics of
space in a concrete and vivid manner and are thus popular astronomy
in the guise of novels rather than exoanthropological explorations of
the universe; the light fiction that, like *Star Wars* in our own day,
transposes the cops-and-robbers schema to outer space; children's
books; and even, in the case of George du Maurier's successful book
The Martian (1897), the society novel (about a phenomenal writer who
has the soul of a Martian).[57]

Around the turn of the century certain novels stand out above this
deluge of popular fiction (which may of course be of interest in its
own right as the reading matter of the masses in industrialized society). These are novels on a higher literary and intellectual level, in
which philosophy and imagination combine to take up the challenge
with which Darwin's theory of evolution, against the background
of the nebular hypothesis and the revolutionizing of astronomy by
spectrum analysis, confronted those among the contemporary public
who were educated and philosophically open minded. Among works
of this kind, the most significant from the perspective of the intellectual
historian are the novels about Mars by Kurd Lasswitz and H. G. Wells,
Auf zwei Planeten (On two planets, 1897) and *The War of the Worlds*
(1897). They enjoyed considerable public success and are still among
the most widely read works of science fiction, and among the most
substantial in content. By treating the theme of man's encounter with
higher cosmic forms of life in opposite ways, they provided, as it were,
a dual paradigm that still influences present-day science fiction of the
more sophisticated kind (and indeed pulp fiction too). It is important
to realize, however, that these two novels are not isolated phenomena
but only two out of a multitude of thematically related space novels
that at the time were successful as well; but of all those novels from
around 1900, these two were the "fittest," the most able to survive.

[57]On Verne see for instance Peter Costello, *Jules Verne, Inventor of Science Fiction*
(London, 1978). Ziolkowski, *The Call of the Cosmos* (Moscow, n.d.); idem, *Beyond the
Planet Earth* (New York, 1960). Cosmic adventure stories include such works as George
Griffith, *Olga Romanoff, or The Siren of the Skies* (London, 1894); Frederick Jane, *To Venus
in Five Seconds* (London, 1897); "André Laurie" [Paschal Grousset], *Les exiles de la terre*
(Paris, 1888). Books for juveniles: Fenton Ash, *A Trip to Venus* (London, 1909); Oscar
Hoffmann, *Unter Marsmenschen* (Among Martians, Breslau, 1905); August Niemann,
Aetherio: Eine Planetenfahrt (Aetherio: A planetary voyage, Regensburg, 1909); Albert
Daiber, *Die Weltensegler: Drei Jahre auf dem Mars* (Navigators to other worlds: Three years
on Mars, Stuttgart, 1910).

We will look first at those other novels, which are less well known, though several of them were republished in the 1960s and 1970s. It will be seen that motifs and forms of thought that we find in the works of Lasswitz and Wells were in the air at the turn of the century but that only these two authors portray the extreme possibilities inherent in the "encounter" in a way that is powerful both as literature and as philosophy. The other authors treat the theme more tangentially or do not follow it through intellectually to such disturbing or stimulating effect. In this sense *Auf zwei Planeten* and *The War of the Worlds* do represent a new departure.

All these novels written before and after 1900 reflect the three great scientific and philosophical advances, Laplace's nebular hypothesis, Darwin's theory of evolution, and Kirchhoff's spectrum analysis, and quite frequently mention them directly and in technical terms. Together they suggest that different planets are chemically and physically alike and that when a given planet is at the stage in its "geologic" evolution that is most favorable to life, life comes into being and develops there. This is the scheme of things that the novels take as their basis. Often particular attention is paid to Mars as the world of a civilization superior to ours. This occurs surprisingly early—not merely earlier than the Martian novels of Lasswitz and Wells, but even before the idea of an advanced civilization on Mars was made popular by Lowell. (This view of Martian civilization is almost always based on the misinterpretation of Schiaparelli's *canali*—"channels"—as *canals*.)

The other planets of our solar system are usually presented in novels of this period as being younger in evolutionary terms than the Earth or, at best, of the same age. Their forms of life thus offer no threat to the status of *Homo sapiens*. The possibility of a challenge to the self-image of terrestrial man is not even mentioned; his rank in the evolutionary scale is not questioned. We will turn to just a few examples of this kind of novel before focusing on the novels about Mars, which are more problematic and more interesting from this point of view. (One feature shared by those novels of either kind that aim at scientific realism is the technological requirement of comfortable spaceships driven by a mysterious force that counteracts gravity—whereas for instance in Edwin L. Arnold's *Lieut. Gullivar Jones: His Vacation* [1905], a light novel with no serious pretensions of any kind, Mars is reached by means of a magic carpet.)

In 1894 the American multimillionaire and inventor John Jacob Astor published *A Journey in Other Worlds*, a novel set in the year 2000. The intellectual framework is indicated by the use of Darwinian terminology ("survival of the fittest," "struggle for existence," and so on) as well as by references to the nebular hypothesis and to spectroscopy.[58] But Jupiter and Saturn, the only planets visited by the team of scientists on a highly official mission in their spaceship, offer no competition to man. Here nature has so far produced only a monstrous fauna, including such things as giant tortoises, flying jellyfish, dinosaurs, giant ants, and dragons: similar life-forms, if they ever existed on Earth at all, died out at periods of evolution that are long past. Depicting an endless succession of encounters with these dreadful creatures in an alien and uncongenial environment, Astor fashions his theme into a planetary novel for game hunters: rifle in hand, the scientists roam through the unfamiliar terrain like explorers of the African continent at the time. Their philosophy is simple: "Man is really lord of creation" (p. 221), not only in the biblical sense paramount in Kepler's very similar formulation, but also in the special sense characteristic of the Victorian age. "Is there any limit to human progress on the earth?" asks one of the travelers on Saturn. "Practically none," comes the reply, given by the spirit of an American bishop that resides there (p. 314). For Astor feels the need for man's unique status to be further underpinned by Christian theology. Like Klopstock, Astor has a star for the blessed, one for the damned, and one for purgatory. Moreover, the fact that man is created in God's image now means that his striving for progress, toward ever greater mastery of nature, is a process, desired by God, of coming closer to the Creator (p. 317). Surely, a strange and no doubt unique return to dogmatic concepts in the context of enlightened modern scientific and social thinking!

No such theological maneuver is felt to be necessary by the American physician and author of many kinds of books, Gustavus W. Pope, who in his evolutionist novel *A Journey to Venus* (Boston 1895) depicts similar primeval conditions in the "young world" of Venus. Amid primeval jungle vegetation, in a climate suggestive of the end of the world, the travelers have to survive one adventure after another with all kinds of grotesque monsters, some drawn from mythology and the Bible, others from prehistoric life on Earth. The novel reaches its climax with the appearance of "Venusian man," an anthropoid ape eighteen feet tall—the "missing link" in person (pp. 336, 413). Naturally, the status of the observers from Earth is not threatened by this encounter with their primitive ancestor; on the contrary, it is confirmed by the emphasis laid on how far removed they are from him.

[58](New York), e.g., pp. 13, 169, 234, 290, 384, 406.

Edwin Pallander takes over this schema in *Across the Zodiac* (London 1896), a planetary adventure story with a philosophical element. In a spaceship equipped with every refinement from a piano to a library of belletristic literature, a Flying Dutchman of a captain, who has designs on the lives of his passengers but also gives them scientific explanations of all astrophysical and biologic phenomena, lands first of all on the moon. Here there was once a civilization as advanced as that of contemporary Europe, but it has long since died out. By contrast, in the Volcanic world of Saturn, billowing with poisonous clouds, evolution has progressed only as far as reptiles, flying wolves, mammoths, and sea monsters; it can be calculated that biologic development will produce *Homo sapiens* only in the far distant future. Venus presents a similar picture: it too is a world still in the making, with giant flies, carnivorous trees, and other surprises characteristic of young worlds. From these dangers the passengers only just manage to escape back to their luxurious craft, where the sinister captain weaves his deadly plots. On the whole journey there is no sign of humans or of even higher life-forms. *A Columbus in Space* (New York 1909) by Garrett P. Serviss (the notable theorist of "other worlds" already mentioned) is another novel about Venus that does nothing to call into question the colonialist pride of the nineteenth-century European. Serviss divides the population of Venus with precision into a subhuman and a human tribe and thus avoids enunciating seriously the question of man's status in the universe. That is the fundamental difference between these novels and those about Mars, since on Mars a "superhuman" civilization is assumed to exist. How, then, does *Homo sapiens* view himself when confronted with that "world"?

Mars had already been the focus of attention in Henri de Parville's half-serious space novel *Un habitant de la planète Mars* (Paris 1865), which treats rather playfully the idea that all planets of the solar system are inhabited. In a meteorite taken from an American mine, the fossilized corpse of an anthropomorphous Martian is found. This is not the starting point of a speculative cosmic adventure novel, however, but merely the beginning of a guide, in the form of a novel, to the astrophysics and biology of our solar system, in which the author assigns to each planet its present, past, or future inhabitants. Parville already has the gradation that was soon to become de rigueur: the moon is an extinct world, Venus a younger world similar to ours, and Mars an older world with an older human race, while all the other planets have progressed only to more or less primitive organisms. The exception, oddly, is Mercury, which has humans, though they are less advanced than we are (pp. 139–40). According to the schema of cosmic and biologic evolution, the Martians ought to be superior to us. But Parville makes an illogical attempt, which is interesting and perhaps not surprising at this early stage (only half a dozen years after *On the Origin of Species*), to preserve the uniqueness, and with it the unassailable status, of *Homo sapiens* on our planet. Parville's Martians—who are later to become the typical representatives of a higher rung on the evolutionary ladder, futuristically perfected or even over-perfected—are "inferior" to us, even though they belong to an older human race. Why? Parville is vague: "Life is swifter, and a being is less capable of being perfected" (p. 129). This evasive tactic is revealing.

No such tactic is adopted in the next three space novels of the pre-Lowell years that are of philosophical interest, and this is the more noteworthy as they are all set wholly on Mars; Mars is not merely one of several planets introduced, as in Parville. (Their relative importance in the history of the genre can be gauged from the fact that all three novels have in recent years been reprinted in the series Classics of Science Fiction.[59]) They are *Across the Zodiac* (1880) by the English scientist and historian Percy Greg; *A Plunge into Space* (1890) by Robert Cromie, which passed through many editions and laid the foundation for its author's fame as one of the pioneers of the interplanetary novel; and *Journey to Mars* (1894) by Gustavus W. Pope (just mentioned), who is regarded as the precursor of Edgar Rice Burroughs and his cosmic adventure novels. In his preface to *Journey to Mars* (New York 1894), Pope explicitly defends the new genre of extraterrestrial "scientific romance" against the criticism that its plots and themes, like those of stories about fairies or gods, fall outside the scope of human sympathy and interest. In his counterargument he pinpoints the crucial assumption embodied in these three novels about Mars (and many other space novels of the day): that since science, philosophy, and religion force one to assume that the nature of "Humanity" on every habitable planet must necessarily always be essentially the same, it is superfluous to invent fantastic, grotesque, or monstrous creatures for the "other worlds," and especially for Mars, which differs from the Earth no more than Texas differs from Arizona.[60] Of course this is what Pope *wants* to believe as a writer, since it means that his portrayal

---

[59](Westport, Conn.: Hyperion Press, 1974, 1976), with introductions that briefly place each work in context. On these and other novels mentioned in this section see Green, *Into Other Worlds*; also J. O. Bailey, *Pilgrims through Space and Time* (Westport, Conn., 1972); Mark R. Hillegas, "Victorian 'Extraterrestrials,' " in *The Worlds of Victorian Fiction*, ed. Jerome Buckley (Cambridge, Mass., 1975), pp. 391–414. All three give little more than summaries of contents.

[60]This comparison is made by Pope in *Journey to Venus*, p. 115.

of Mars can present everything as being more or less comme chez nous, and hence, he thinks, "interesting." It is the same with Greg and Cromie. Nevertheless, despite the familiar "human" characteristics of the Martians, the three novels are also alike in depicting them as inhabitants of a cosmologically older planet and therefore as superior in every way to the inhabitants of Earth—that is to say, they represent a more advanced stage of evolution than that which we have reached. The Martians, in other words, afford us a preview of our future, and in every case this future of ours, the Martian present, is regarded critically as a development that, while it is inevitable, leads in a false direction, toward a perfecting of the species that is undesirable because it eliminates what "we" see as the essentially human values. Whereas the pessimism of Lowell's books about Mars and Flammarion's *La fin du monde* relates to the inexorable destruction of our world by the physical cooling down and death of our planet, here the fear is of that evolutionary phase of emotional, spiritual, and cultural decline or degeneration which begins after or even at the optimum stage, robbing man of his humanity—and which will perhaps force him ultimately to cede his dominant role in the world to a quite different species, a terrifying prospect that the young Wells, in particular, conjures up again and again.[61]

Of the three Martian novels, Pope's *Journey to Mars* is the least ambitious in its treatment of this theme: the reader's attention is gripped above all by the lively plot, in which episodes follow one another with breathtaking speed. The Martians have a space station at our South Pole, and from here a naval lieutenant travels with them to their world, where in a subtropical paradise he has one adventure after another. His love for a princess takes him to banquets and court balls but also leads to a duel with a princely rival; the astronaut's yarn ends with a dramatic rescue from imprisonment and escape back to Earth. This action could just as well be taking place somewhere on Earth, but for the fact that Pope repeatedly refers to the schema of cosmic and biologic evolution that forms the background to the story: while Venus is still only a world in the making, and Pluto one that is already extinct, the Earth and Mars are at different stages in between these two extremes. The yellow, blue, and red races on Mars are thousands of years ahead of us in all branches of knowledge and technical inventions—in other words, in everything that the nineteenth century meant by "progress." But they have passed their peak

---

[61]See H. G. Wells, *Early Writings in Science and Science Fiction*, ed. Robert M. Philmus and David Y. Hughes (Berkeley and Los Angeles, 1975), chap. 5.

and have entered the phase of decline and degeneration that the laws of nature have ordained for every planetary "world." The Earth and terrestrial mankind, however, are still at a youthful stage of evolution; Mars shows what is to come. We can expect "a great and glorious future," until the future humans on Venus finally witness the gradual decline of man on Earth as it grows cold and its landscape turns to steppe, just as we see the decline of the Martians and they, thousands of years ago, saw that of the "great and glorious planet Pluto" (pp. 107, 278, 298, 335–36). For the Earth and its human inhabitants, then, the author projects "progress" forward into a future measured on an astronomical scale. This initially has the effect of boosting the self-esteem of *Homo technologicus*, because it is primarily advances in the applied sciences that have made the world of the Martians, with its network of canals, so superior to that of Earth. The Martians' attainment of total control over nature, and the permeation of all areas of life, even religion, by scientific ways of thinking (p. 195), have made life on Mars everything that could be desired. "Happiness" would be complete if only one could protect oneself from the showers of meteorites (pp. 196–97). This in turn throws an unfavorable light on Earth, with its living conditions unfit for human beings, its barbaric tribes, warring religions, and so on (p. 194). Yet the utopian, urbane world of Mars, with its higher or at least more advanced civilization, is also a decadent world. The exotic luxury and the effortless and aimless way of life also have the effect of encouraging the selfish pursuit of pleasure: the reader is to see it as symbolic that, at the end, the most dissolute and corrupt of all the princes on Mars gains the ascendancy.

The juxtaposition of *Homo sapiens* with the form to which he can be expected finally to evolve thus acquires, in this novel and also in the other two set on Mars, an odd duality of perspective: there is the jubilant expectation of progress stretching forward into a future measured on an astronomical time scale, but also the anticipation of spiritual and moral degeneration as the final stage, according to the laws of nature, of the cosmologic and biologic development of our species. Either way, the uniqueness of man in the universe, or even in our solar system, becomes an untenable hypothesis.

Percy Greg's *Across the Zodiac* (London 1880) is more strongly critical of the stage of evolution reached on Mars, which for us is still to come. But the underlying conception is the same. Here too the world on Mars, where the narrator lands in his luxuriously equipped spaceship, is inhabited by human beings like ourselves. History has led to "over-cultivation" (1:136), to a way of life that is effortless, elegant, and gracious, and technically, socially, and politically rationalized to the last degree, in a "material paradise" complete with cinema and telephone, where all work is done by machines (1:138). The price paid for this is a loss of vitality, zest for life, and humanity. Human relationships have become totally impersonal, the keeping of harems is the norm, a wife is a chattel, and individuals live selfishly and hedonistically with no thought for the morrow (2:121). Religion has died out; bringing up children is a nuisance and is therefore left to institutions; young people are intellectually highly developed but have no moral principles or goals in life (2:182). The few who have preserved that humanity which the Martian population in general "overcame" thousands of years ago have formed a secret organization and gone underground; most of the action in this eventful novel of love and adventure has to do with this secret society. *Across the Zodiac* has therefore, with some justification, been called a dystopia,[62] an antiutopia—but one firmly anchored in the scientific thinking (astronomical, physical, and technological rather than Darwinian biologic thinking) of the day. But the motif of the secret society represents an attempt to boost man's self-esteem such as we do not find in Pope's *Journey to Mars*, although the latter in some ways already shows the influence of *Across the Zodiac*. Despite the presumably automatic nature of cosmic and biologic evolution, a small group of Martians are able to keep alive the feelings and attitudes we think of as human, and this offers at least some hope that, in the name of what for centuries has been called "humanity," man will be able to resist the evolutionary compulsion to degenerate spiritually to a state that is no longer human.

In Cromie's novel *A Plunge into Space* (London 1890), which also unmistakably shows the influence of Greg's *Across the Zodiac*, the futuristic world of Martian civilization is still more forcefully rejected. Here too Mars is a "dying" planet prefiguring the future development of the Earth. The impoverishment of its vegetation proceeds inexorably, only temporarily held in check by the oases, the artificial paradises in which the Martians, with their superior technology for controlling nature, have been able to create settlements (2d ed., London 1891, pp. 90, 99). "The inhabitants that we shall find upon it will be creatures far surpassing ourselves in every attribute of mind and body. They will have developed social, moral, and physical conditions such as we cannot imagine. They are at the pinnacle of their perfection. Before them there is no further progress. Their only change must be towards

[62]Sam Moskowitz, Preface to the Hyperion Press reprint (1974), pages not numbered.

decay" (pp. 103–4). So we are told even before the landing of the steel spaceship that represents the peak of terrestrial scientific achievement; and when the travelers meet the Martians, this melancholy prediction is at once confirmed or even exceeded. At the very first meeting, a wise instinct prompts the travelers to approach the aliens with the submissive demeanor of Indians toward whites (see above, at n. 46, on the reversal of the old pattern). They find palatial houses in beautiful clean cities, attractive people who pursue their interests at leisure and hardly need to work, girls like angels, luxurious "automatic" banquets, exquisitely polite behavior, cultivation of literature and the arts, unimaginable technical conveniences from videotelephones to the abolition of weather by climate control, a single language, and a government that attaches supreme importance to the individual. No wonder the astronauts, who set off for Mars from the hostile conditions of Alaska and are here received with perfect hospitality, are deeply impressed and recall the Earth as a world "half barbaric, wholly restless and sorrowful" (p. 138). The story seems to have all the makings of a utopia, but it soon turns in the opposite direction.

For one major aspect of the evolution beyond the stage reached by terrestrial man is that the intellect has suppressed, indeed eliminated, everything "animal" (p. 112). In their first enthusiasm the travelers note this as something positive, but they soon come to realize that the "calming" of all passions (p. 180) is not wholly desirable. In this world, in which everything that could be wished for has long since been achieved, the sciences have no more problems to solve, there is nothing to fight for and nothing to defend from the brutality of life, and no way forward into the future can be imagined; not only have the grand emotions died out, but even curiosity has been extinguished. Thus for instance the Martians have long since become bored with space travel and given it up (all the civilizations they encountered proved backward compared to themselves); and they soon lose interest in the new arrivals, just as the visitor to a zoo gradually loses interest in its new acquisitions (pp. 182, 202–3). Where there is no sense of life's challenge, nor of the misery of imperfection, there can also be no happiness, remarks one of the travelers (p. 192). The condition of the Martians is "placid happiness" (p. 203), a completely empty existence. The travelers, bored in their turn and no longer attracted by the "sweet monotony of absolute perfection" in this ideal Martian world (p. 182), flee from the planet.

This stage of unsurpassable perfection is in fact already a phase of decay—of the decay of all that is human. A mirror is held up to *Homo sapiens* showing him the culminating point of his development—

the point the nineteenth century, with its faith in technological progress, took as its goal. That it is viewed critically and judged unfavorably does not alter the fact that development in that direction is part of the inexorable process of evolution, governed by natural laws. The astronauts' hasty return to Earth is thus an act of intellectual escapism, a flight from the future, a refusal to face up to the collective fate of mankind on the planet that will be the next to go the same way as Mars. In depicting this confrontation between man and the extraterrestrials who represent his own future, Cromie, unlike Greg, does not call on man to resist by preserving human values. As is clear from the quoted lines that anticipate, before the landing on Mars, what the Martians will be like, the basic outlook of the book is one of melancholy at the thought that the Earth's cosmic destiny should be decadence. And despite occasional formulaic references to "creation" (e.g., p. 138), this is a wholly secular destiny, that is, a future determined by the astrophysical and biologic laws of nature. We see here the potentially tragic dimension of this aspect of modern science.

The strongest impression created by all three of these novels, then, is of a fear of the future, which is expressed in the unfavorable presentation of the Martian world; and yet the overall impression they leave behind is ambivalent. The portrayal of the perfect world corresponds too closely to the wishes and dreams of the Victorian age, with its faith in progress, to be thought of *only* as a nightmare. In the novels about Mars by Wells and Lasswitz the two aspects are wholly separated: one author reacts with horror to the Martians' hyperintelligence that has degenerated into inhumanity, while the other looks up toward the higher level, not only of technological achievement, but also of shared human culture, attitudes, and behavior that has been attained. And once the polarity of degeneration and perfection had been presented in a way that could stand as a model, the same pattern was reproduced many times, right up to the present day. It appears in two of the most important science fiction novels of that same (pre-Gernsback) period: George Griffith's *A Honeymoon in Space* (London 1901) and Mark Wicks's *To Mars via the Moon* (1911). It will be best to look at these, too, from the thematic point of view, before discussing Lasswitz's and Wells's novels, since the latter's creation of a new variation on the plurality theme (the invasion from space) will thereby be thrown into sharper relief.

In *A Honeymoon in Space*, the degeneration of the humans on Mars is set more explicitly in the context of Darwin's theory of evolution than in the novels dealt with so far, in which the theory is presupposed as a matter of course and not actually discussed. In Griffith's novel, the eccentric earl of Redgrave takes his young American wife on a honeymoon tour of the worlds of our planetary system in a spaceship complete with club furniture and mahogany paneling. He is a declared supporter of Darwinism (p. 275) and explains all phenomena in the light of that theory. Accordingly, the technical terms of Darwinism recur again and again, from "survival of the fittest" to "degeneration," and the various planetary worlds illustrate the effect of biologic evolution on the nonterrestrial bodies according to the stage each one has reached in its development. Thus Jupiter is a world still being formed, a world of fiery volcanoes that is still hostile to life (p. 239); Saturn, in contrast, has already reached the stage of having giant reptiles and, simultaneously, manlike vertebrates living in caves and trees (pp. 272–74). On the moon, evolution has led to a civilized species of men "similar to us," but this has, in the struggle for existence, long since become degenerate and reverted to barbarism, its members appearing now as grotesque, brutish parodies of mankind. "I wonder," comments the aristocratic Darwinist, "how many hundred of thousands of years it will take for *our* descendants to come to that" (p. 112).

But *will* they come to that? Is evolution a process that continually repeats itself in an identical manner? Obviously much depends on physical conditions on the planet. Although on the moon the development was reversed as soon as the stage of *Homo sapiens* had been reached, Griffith's book shows altogether three kinds of being that are at an evolutionary stage beyond that of man; or rather two, if one excludes the higher beings on Venus. For these, winged humans with songlike speech and graceful movements, have, through adaptation to the environment (p. 169), developed not from a higher species of ape but from birds. Also, they represent not a higher level of evolution and intellectual dominance over the environment understood in Darwinian terms but, incongruously—like the inhabitants of Perelandra (Venus) in C. S. Lewis —a state defined in *theological* terms, the state of sinlessness in the Garden of Eden before the temptation by the serpent (p. 199). Griffith gives prominent treatment to the older civilization on Mars, however, and this is unequivocally represented as having undergone further evolution in the Darwinian sense, as in the earlier novels about Mars. Here civilization has passed its zenith. The human beings—anthropomorphous, but somewhat bigger than *Homo sapiens*—have relapsed spiritually and culturally into barbarism, though they are still in possession of their superhumanly developed intellect and their advanced technology—a technology of weapons of annihilation. As in Cromie, the course of "civilization," determined by evolution, has seen feelings and passions dwindle to nothing, overshadowed by the superior power of the intellect. On this dying planet, the "fittest," who have survived in the struggle for the diminishing resources necessary for life, are "over-civilized savages," "purely intellectual beings," "cold, calculating, scientific animals" (pp. 141, 142, 152). The Martian who comes on board the spaceship does not know what anger or love is, but he can name an exact purchase price for Redgrave's young wife. The suspicion that the inhabitants of Mars "may not be men at all, but just a sort of monster with perhaps a superhuman intellect with all sorts of extra-human ideas in it" (p. 134) is confirmed in the crassest manner: "That thing a man!" "What brutes!" (pp. 162–63).

But Griffith evidently does not see this decline in human qualities as the only possible direction for evolution beyond our present stage: we need not necessarily assume, by analogy, that this is the form that our own future on this planet will take. For the novel depicts a different group of anthropomorphous rational beings living in another society that is technologically far in advance of terrestrial civilization. This is located on one of the moons of Jupiter, Ganymede. Its neighbor, the ice-covered moon Europa, has already succumbed to the cold, and the same cosmic death threatens Ganymede; but this ominous prospect is accompanied not by moral and cultural decline but by a period of supreme cultivation and morality. Here too the development of "superhuman intelligence" represents the "survival of the fittest" (pp. 227, 232), but in the opposite sense. This society, in which the people live under huge glass domes, like greenhouses, harnessing all natural forces in order to exploit to the full the dwindling reserves of warmth on their globe, is not torn by a degrading, competitive Darwinian struggle for survival; instead the Ganymedeans live a life of peace, rationality, and beauty in their cultivated world. Whereas the Venusians have never come into contact with sin, the inhabitants of Ganymede (like the Venusians in Munro's *A Trip to Venus*)[63] have learned that sin is folly "and that no really sensible or properly educated man or woman thinks crime worth committing" (p. 231). In their struggle with nature they have, in the midst of the uninhabitable deserts that cover ever more of their globe, produced an "exquisitely refined civilization" that, it is suggested, can perhaps also be achieved in the far distant future on our Earth (p. 233).

Thus man's confrontation with the extraterrestrials offers two different visions of the possible future of *Homo sapiens:* on the one hand, an intellectualized society that cultivates human attitudes and feelings

[63](London, 1897), p. 163.

to an exquisite level of refinement; on the other hand, beings who combine high intelligence with bestiality. There is no hint as to *how* one or the other development comes about, whether any human influence can be brought to bear, or whether the process is subject to an evolutionary mechanism whose course is determined by the environment and cannot be influenced. This picture of the cosmic development of species leaves us with a half-fearful, half-hopeful prediction of the future form of man.

In *To Mars via the Moon* (London 1911), published ten years later, Mark Wicks sees these matters in a far more simple, even trivial light. To be sure, in this novel, while the biologic aspect is somewhat neglected, the author's account of the astrophysical basis shows more detailed knowledge of the subject than is found in most novels of this kind. The exposition of Percival Lowell's views is intended to set the meager action of the novel on the firm basis of the known facts and the conclusions that can be drawn from them, and thereby to exclude from the outset any merely arbitrary fantasy (in the preface Wicks speaks of "randomly imagining"). But his portrayal of life on Mars more closely approaches a utopia than an exobiologic world that is plausible in Darwinian terms; in this too he takes his cue from Lowell. The Martians are shown, as their planet's death from cold draws closer, living on the banks of the famous canals with fabulous comforts provided by their technology. They are rational and well balanced, with unfailing good taste and tact. The author constantly, to the detriment of the "action," makes pointed comparisons between them and the three Englishmen, culturally so far behind them, who have arrived in an airship that is the pride of imperialist technology. Thus for instance political and social conditions in England cannot compare with the radical democratic welfare state on Mars, where neither poverty nor divorce, illness nor religious disputes, ever occur or ever could occur. Thus in this novel, unlike *A Honeymoon in Space, we* are the uncivilized savages in comparison with the Martians (pp. 217, 300). The object of this confrontation with the extraterrestrials who prefigure our own future is not, however, to satirize the state of affairs on Earth but to project forward the certainty of progress over astronomical periods of time. Precisely because terrestrial man is physically, morally, and intellectually far behind the inhabitants of Mars, he has, cosmologically speaking, almost his whole life before him, and that is an uplifting, inspiring thought, with none of the ambivalence that it had in Pope's *Journey to Mars:*

Our world has seen the rise and fall of many civilisations, but fresh ones have risen, phœnix-like, from the ashes of those which have departed and been forgotten. "The individual withers," but "the world is more and more." As it was in the past, so will it be in the future—ever-changing, ever-passing, but ever-renewing, until the final stage is reached.

Since the earliest dawn of our creation the watchword of humanity has been "Onward!" and it is still "Onward!" but also "Upward!!" The possibilities of the development of the human race in the ages yet to come are so vast as to be beyond our conception; for, as Sir Oliver Lodge has remarked, "Eye hath not seen, nor ear heard, nor has it entered into the mind of man to conceive what the future has in store for humanity!" Then:

"Forward, forward, let us range,
Let the great world spin for ever down the ringing grooves of change!"

This, then, is the great lesson which Martian civilisation teaches us. Surely it affords no reason for the depression and pessimism in which some upon the earth are so prone to indulge; but rather should it stir them to a more earnest endeavour, by gradually removing the obstacles which now bar their progress, to improve the social conditions of the people; so that they in their turn may improve their intellectual conditions, and lend their aid to the general advancement of the world they live in.

Gloom, depression, and pessimism, of which we have had more than enough of late years, never yet helped any one. They have, however, proved disastrous to many.

Remember our world is young yet! (Pp. 300–301)

Thus while Flammarion, for instance, had focused in *La fin du monde* on the distant end of man's development, his extinction through the cooling of the solar system, Wicks instead concentrates on the glorious intermediate phase of development that for him promises the fulfillment of all of the nineteenth century's dreams of civilization and progress. Wicks knows that some contemporaries have reacted to projections of mankind's future with depression and pessimism, but he thinks such reactions may be postponed until far into the future ("ages hence," p. 299). And unlike Griffith and his predecessors, he links that pessimism not with the fear of the decline of culture and human values that threatens when those dreams are fulfilled, but only with the cosmophysical event of the Earth's cooling. That he chooses to ignore the more human problems that have emerged in our historical survey is perhaps a sign that here, as in some other cases, not all that is known, hoped for, or feared is actually expressed. Such hopes and fears are, however, brought openly to the fore in Lasswitz's and Wells's novels about the cosmic invasions of the Earth, which at the end of the century attracted considerable, and justified, attention.

4. Invasion Inspiring Hope or Fear: Lasswitz and Wells

Of the many novels (not only from the English-speaking countries) that reflect the obsession with Mars brought about by Lowell,[64] *Auf zwei Planeten* (1897) and *The War of the Worlds* (1897; in book form 1898) are the most significant in terms of ideas, the most powerful, and the best known today; they were also the most successful in their own day. Wernher von Braun remembers Lasswitz's novel as having "in a very short time" reached "several hundred thousand copies";[65] his memory is no doubt at fault, but the mistake is a telling one. Still, Franz Rottensteiner's calculation of seventy thousand copies up to 1930, together with translations into Danish, Norwegian, Swedish, Dutch, Spanish, Italian, Polish, Hungarian, and Czech, indicates a notable success; around the turn of the century *Auf zwei Planeten* was "probably the best known European novel of space fiction."[66] Wells's *The War of the Worlds* was published in several editions and translations around 1900; there were parodies, "sequels," and imitations as well,[67] so that it, too, clearly played a prominent role in the literary life of the day.

One reason for this success was without doubt the new way in which the theme is treated in these two books compared with the other space novels, particularly those about Mars, of the last years of the century. In all of these, the travelers conformed to the type of the Victorian conqueror, journeying through the solar system and

---

[64]See Mark R. Hillegas, "Martians and Mythmakers: 1877–1938," *Challenges in American Culture*, ed. Ray B. Browne et al. (Bowling Green, Ohio, 1970), pp. 150–77; idem, "Victorian 'Extraterrestrials,' " pp. 391–414; idem, "The First Invasions from Mars," *Michigan Alumnus* 66 (1960), 107–12.

[65]In the introduction to the new, abridged edition by Scheffler Verlag (Frankfurt, 1969).

[66]"Kurd Lasswitz: A German Pioneer of Science Fiction," in *Science Fiction: The Other Side of Realism*, ed. Thomas D. Clareson (Bowling Green, Ohio, 1971), p. 289. On the translations see Hillegas, "Martians and Mythmakers," p. 157.

[67]C. L. Graves and E. V. Lucas, *The War of the Wenuses* (Bristol, 1898); Garrett P. Serviss, "Edison's Conquest of Mars," (New York) *Evening Journal*, 1898. See David Y. Hughes, "*The War of the Worlds* in the Yellow Press," *Journalism Quarterly* 43 (1966), 639–46, p. 642 on Serviss.

emerging from the encounter with extraterrestrials with their self-esteem and sense of superiority triumphantly confirmed, since the aliens were either not yet human or else humans who were already degenerate. In Wells and Lasswitz, in contrast, it is the Martians who conquer space; they land on Earth and pose a most serious threat. (In Pope's *Journey to Mars* the motif of the Martian space station at the South Pole was merely a starting point for the action and was in no way threatening.) The fear aroused in the course of the other cosmic expeditions was not an immediate fear but the somewhat abstract fear of the distant future of mankind, of the degeneration of the species or of the Earth's cooling down and dying; moreover, it was held in check by the expectation that that distant catastrophe would be preceded by a period in which man would evolve to a high level of civilization. As late as 1911, Wicks responds to this expectation—chooses, for his own comfort, perhaps, to respond to it—with something approaching euphoria. In Wells and Lasswitz the fear is more real, more gripping: *Homo sapiens* is literally overpowered by the incomparably more advanced beings from his neighboring planet, who burst upon his complacent world and dominate with the force of an upheaval of nature. In *Auf zwei Planeten*, unlike *The War of the Worlds*, this fear, which is of the most primitive kind, does not last. But the initial effect of the extraterrestrials' superiority is to arouse terror, even though it later inspires hope, the hope that terrestrial man can develop to a higher level under the tutelage of the Martians. Only in this way can Lasswitz's novel end with the prospect of global "enlightenment," the moral education of man to a high level of civilization. In Wells's vision, though, the destruction of the world is averted only at the last minute, and we are left only with a sense of horror, especially inasmuch as the closing chapter stresses, with the reinforcement of scientific arguments, that this destructive invasion by the terrifying monsters could be repeated at any time.

This difference is in large measure due to the fact that the two authors, although both Darwinians, define the Martians' superiority, the result of evolution, in different ways. In intellect and technology the Martians in both novels have progressed far beyond the stage reached by humans. In Wells this evolution has led to the total elimination of everything human, so that it is quite impossible for humans to understand Martians and impossible, consequently, to communicate with them in any way. In Lasswitz, in contrast, the development of the human species to a higher degree of intellectuality has reached, on Mars, a level of moral and cultural perfection that can best be understood in terms of the concepts with which "we" are familiar.

Lasswitz, who taught philosophy and physics at the Gymnasium Ernestinum at Gotha, was a Kantian; his Martians, who in contrast to Wells's octopuslike monsters are entirely anthropomorphous, have wholly absorbed the categorical imperative into their nature. It has made them, the "Nume" as they call themselves, happy and contented, that is to say free from the tyranny of instincts and desires, in accordance with the teaching of their "immortal philosopher," Imm. It has also given them the sense of cultural mission that makes them undertake the colonization of man on Earth as a matter of course. The "worldwide best seller"[68] *Auf zwei Planeten* tells the story of the conquest of the Earth by these masters and civilizers of the solar system.

The Martians establish themselves on the Earth, in the first place, simply for reasons of survival: on their aging, densely populated planet they need the raw materials Earth has to offer, above all the solar energy, more plentiful here, which they know how to use. They have settled at the North Pole and set up a space station above the pole as a station for traffic between Earth and Mars ("Marsbahnhof der Erde"!), and their invasion could have been perfectly peaceful but for the facts that a misunderstanding brings about an armed conflict with a ship of the British merchant navy and that the Martians, seeing the barbarity of the terrestrial states' warfare among themselves, regard it as their duty to set up a protectorate over the Earth. By ethical instruction, terrestrial humanity is to be compulsorily raised to the level attained on Mars several hundred thousand years earlier. Ironically, however, this elevation is not achieved without some instances of brutality on the part of the Martians. Corrupted in this physically and morally alien environment, but also horrified by the uncivilized behavior of men, they finally operate a regime of terror and exploitation. A rebellion by the states of the Earth is, surprisingly, successful; but a truce opens up the prospect of friendly cooperation between the two planets, by means of which the aim of global education will be achieved. Cured of the "terrestrial rage" (*Erdkoller*) that had possessed them, the Martians, no longer misunderstood and now acting rightly, embark on a program of development aid for humankind whose evolution is seen as a moral ascent, as a development toward the Martians' "Kantian" outlook: there is no suggestion of any other possibility. The scientific achievements of the Martians, sensational though they are compared to the stage reached by terrestrial man, are only of second-

---

[68]Klaus Günther Just, "Über Kurd Lasswitz," *Aspekte der Zukunft* (Bern, 1972), p. 39. Volume and page numbers refer to the "Volksausgabe" (Leipzig: B. Elischer, n.d.).

ary importance. The book ends not with a technological utopia but with—as a symbolic pledge for the future—the marriage between a German polar explorer and a Martian woman.

Lasswitz's Kantian version of evolutionary thinking, in which the human species develops in the direction of "enlightenment," of the shaping of the personality by the autonomy of reason, naturally gives him the opportunity to cast satiric sidelights on European civilization at the end of the nineteenth century—to criticize its culture, morals, and social and political conditions by juxtaposing it with a contrasting society. But Lasswitz is concerned less with the state of affairs that needs to be overcome than with that which is to be attained and the means of attaining it. His book expresses the hope of more rapid evolution through the educative intervention of extraterrestrial guardian angels. Yet this vision lacks any theological overtones. With everything viewed in Darwinian terms, that would be anachronistic. That Lasswitz's conception is a secular gospel of salvation and in that sense religious (like Clarke's *Childhood's End* or the Däniken cult in more recent times) is another matter. But in this context too it is significant that according to Lasswitz, the causes of the qualitative difference between "Nume" and "Baten" (Earth dwellers) are *"biologic"* (much more so, incidentally, than in Clarke or Däniken).

This biologically determined difference has three aspects, which correspond to the three metaphors that recur throughout the novel: in the encounter with the aliens from Mars, human beings are children, Indians or "savages," and animals. Of course these images are always a cue for the humbling of man's anthropocentric pride, especially his image of himself as the colonial master. For man to look at himself in the context of the plurality of inhabited worlds, once he is familiar with Darwinism, is to find himself toppled from his pedestal; though he may become reconciled to this displacement by the prospect of his own higher development and, still more, by that of the species as a whole (the idea of individual transmigration of the soul has been discarded). Human beings are *children:* the Martians "feed" them, "treat them like children," "send them to school" (1:92, 157; 2:273). They are a *"race"* whose ability to become civilized, in the sense of exercising free individual self-determination, does not go unquestioned: "We are, after all, poor redskins" (2:114). "What savages you are" (1:251). There is doubt as to whether "any bond of relationship can be formed" between these "Indians" and the Nume. "Anyone who has seen that stupid face with those blinking dots that are supposed to be eyes, . . . and those coarse movements, will say to himself that we can only tolerate this race as a domestic animal that may perhaps be useful" (2:103). This is the view of those Martians who are anti-Baten, who also speak of the "wild animals that we must tame" (2:102). But even Martians who are well-disposed toward humans see them as *animals*: "For goodness' sake, he is only a man, after all! It is so terribly funny when he makes an effort to be really nice." "But you can't tell . . . whether it feels so terribly funny to him. An animal that we tease often seems completely ridiculous to us, but then I can't help thinking that perhaps it is suffering bitterly. And a man isn't *only* funny—" (1:162–63).

To be one of the Nume, by contrast, signifies (despite the occasional element of kitsch in their speech) possession of "reason, freedom" in the Kantian sense (2:111). The children, or savages, or animals, acknowledge this freely (2:274). But it instills in them the confident belief that they too can reach this stage of development, and the Martians are only too willing to help them. The end of the novel looks forward along the path "that the world can now tread toward freedom and peace."

At the same time, Wells too depicts the conquest of the Earth by Martians, and he uses the same images as Lasswitz to clarify the relative positions of humans and aliens. Humans are compared not only to uncivilized natives, in particular to the Tasmanians wiped out by the English, but also to animals—ants, frogs, rabbits, apes, and so on—while the invaders here too represent an overwhelmingly superior power. In addition, the difference between humans and Martians is once again expressed in evolutionary terms. Mars is older than the Earth, according to the nebular hypothesis, but has been shown by spectroscopy to consist of the same elements, and its creatures represent a stage in the further development of *Homo sapiens*—the pinnacle of terrestrial evolution—a million years ahead of him.[69] Lasswitz, the Kantian, could only imagine this development as an ethical ascent, but Wells, the disciple of the Darwinist T. H. Huxley, sees in it only (or at least primarily) the degeneration of what, in Lasswitz too, are seen as the essential human values. In *The Time Machine*, this is a relapse into physical and mental infantilism; in *The War of the Worlds*, the degeneration of man's nature into pure, immeasurably heightened intelligence, which pays as much attention to man as man would to "harmless vermin" (2:7). The life-forms most able to survive are not the best, the most desirable from an ethical and social point of view. Strictly speaking, Lasswitz's Martians may be so highly intellectual

[69]*The War of the Worlds*, bk. 2, chap. 2. In the following discussion references are only to volume and chapter, as the chapters are generally very short.

that they are perhaps deficient in feeling; but they do show under-standing and sympathy toward man, whereas Wells's monsters, intel-ligent and totally without feeling, act more like machines (2:2). This difference is paralleled by a difference in appearance between Lass-witz's and Wells's extraterrestrials: on the one hand, lordly human beings with shining eyes and light-colored hair; on the other, gro-tesque giant heads without bodies but with sixteen tentacles around the triangular mouth. For Wells, this detailed description of the alien life-form was not, as with many of his imitators, simply an arbitrary imaginative creation, but a plausible extrapolation from the Darwinian theory that Wells studied closely and that he also applied, like so many others at the time, to the planetary worlds whose existence he assumed as a matter of course. In the 1890s he explicitly justified this extrapolation in several scientific articles written for lay readers, in which he argued strongly against exobiologic anthropomorphism and the complacent assumption that man is the most highly evolved crea-ture in the universe.[70]

In the "scientific romance" The War of the Worlds, this illusion is destroyed much more forcefully but on the same astronomical and evolutionary basis, which is repeatedly stressed. One day, at an idyllic country town close to London, cylindrical spaceships fall from the sky, and the octopuslike Martian monsters crawl awkwardly out of them. Revulsion and astonishment soon turn to terror when the curi-ous humans who approach the visitors waving a white flag are effort-lessly and, as it seems, automatically killed by their "heat-rays"; and the terror increases when the Martians rapidly assemble three-footed machines, taller than a house, in which they can travel around at the speed of an express train. Not only do these giant spiders reduce one place after another to ashes and rubble by means of the sinister heat-rays, they also hunt humans: with mechanical precision their metal arms catch anyone they find out of doors. The explanation proves to be that the Martians feed on human blood, injecting it directly into their veins. The horror reaches its height when the machine-borne army of extraterrestrials, spreading destruction all around, advances on London, and the population flees in panic. But, as unexpectedly

[70]On this whole paragraph see Early Writings, chap. 5 (introduction and texts); Bernard Bergonzi, The Early H. G. Wells: A Study of the Scientific Romances (Manchester, 1961), esp. chap. 5; J. P. Vernier, "Evolution as a Literary Theme in H. G. Wells's Science Fiction," in H. G. Wells and Modern Science Fiction, ed. Darko Suvin and Robert M. Philmus (Lewisburg, Penn., 1977), pp. 70–89; Helmut Jansing, Die Darstellung und Konzeption von Naturwissenschaft und Technik in H. G. Wells' "scientific romances" (Bern, 1977), esp. chap. 6.

as they landed, the monsters themselves are brought to a standstill after a few days, struck down in their turn by bacteria that are new to them but to which humans are immune.

While the invasion lasts, the country that at the time ruled the seas and continents is subjugated in a more brutal and above all more unnerving way than any colony ever has been. More apt than the comparison with a colonial people (frequently made ever since the belief in a plurality of worlds acquired a scientific basis, though not always as a metaphor for *us*) is that with animals (already found in Cyrano, if not earlier), which Wells uses ever more prominently, so that it becomes the dominant leitmotif. "I was no longer a master," says the narrator, speaking of the devastation wrought by the super-brains, "but an animal among the animals, under the Martian heel" (2:6). Dogs eat human corpses, humans gnaw at the skeletons of sheep and cats.

How are we to interpret the use of this metaphor? More than once the narrator points out, in an admonitory tone, that like begets like: the civilized European has exterminated not only the Tasmanians but also the bison, and now a similar fate overtakes him, through these monsters who in the course of evolution have become "practically mere brains" and who are as far superior to him as he is to the animals (2:2). It is tempting (as it is in Cyrano, not to mention the allegories of imperialism in modern science fiction) to interpret this deliberate attack on the "complacency" or the "vanity" of man (1:1) as a purely terrestrial allegory. Justice is done, as it were, by a reversal of roles: the European is taught a lesson by being made to see how his brutal behavior toward animals and "lesser breeds" of men appears from the victims' point of view. This is a widely accepted interpretation of *The War of the Worlds*.[71] But one must not overlook another idea, which Wells himself expressly underlines in his final chapter. This is the lasting feeling of horror at being exposed and unprotected in a universe that must, in the light of the insights of contemporary science, be assumed to be inhabited—inhabited by more highly evolved beings, beings superior to us. This is not merely a fear of the possible destruction of the bourgeois order that considers itself unassailable—an attempt on Wells's part to *épater le bourgeois*.[72] Wells would be all in favor of it;[73] but if that had been his only object, he need not have troubled to bring in extraterrestrials. Some invented race of humans or species

[71]See n. 70 above.
[72]Bergonzi, *Early H. G. Wells*, p. 133.
[73]See the quotations from Wells in ibid., pp. 124, 132.

of animals, an unforeseeable epidemic, or a rebellion by a colonial people might have served his purpose even better because it would have been more "realistic." The real shock inflicted by *The War of the Worlds* is one of recognition, the recognition that contemporary science makes the position of *Homo sapiens*, even within his own planetary system, questionable and insecure. "Are we or they Lords of the World?" asks the motto taken from Kepler. And " 'What are these Martians?' 'What are we?' " (1:13). Lasswitz reinterpreted this recognition so that it ceased to be a shock and instead enabled man to feel confident that, although at the present stage he was indubitably of inferior status, he possessed the ability to develop, or to be educated, toward a more desirable state. Wells too suggests that this drastic blow to the self-confidence of civilization may have brought with it some benefits: it has made man less complacent, and therefore less likely to become decadent, and it may have brought a gain in scientific knowledge, an awareness of the possibility that in the far distant future man will be able to colonize Venus as the Martians colonized the Earth (2:10). But this is only a fleeting conjecture, and it is worth noting that Wells totally ignores the main idea of his teacher, Huxley, on this subject: that the essential thing is to combat cosmic and biologic evolution with all the means in our power, in the name of human values, so that not "the fittest" but "the best" survive.[74]

Instead, Wells's final chapter stresses the possibility that there could be more such invasions. This motif is (or was at the time) scientifically unimpeachable: according to the nebular hypothesis the cooling of Mars has advanced further, and the green Earth offers the Martians a chance of survival into the future. Thus the terror and the fear are not a product of mere fancy:

> We have learned now that we cannot regard this planet as being fenced in and a secure abiding place for Man; we can never anticipate the unseen good or evil that may come upon us suddenly out of space. . . . For many years yet there will certainly be no relaxation of the eager scrutiny of the Martian disk, and those fiery darts of the sky, the shooting stars, will bring with them as they fall an unavoidable apprehension to all the sons of men. . . .
>
> It may be . . . that the destruction of the Martians is only a reprieve. To them, and not to us, perhaps, is the future ordained.
>
> I must confess the stress and danger of the time have left an abiding sense of doubt and insecurity in my mind. I sit in my study writing

[74]T. H. Huxley, *Evolution and Ethics, Collected Essays* (New York, 1968), 9:81.

by lamplight, and suddenly I see again the healing valley below set with writhing flames. (2:10)

Like the comparable shock, however different, in Lasswitz's novel, the cosmic shock of recognition enunciated here in the motif of the extraterrestrial invasion is of course ultimately that of the question of man's identity, presenting itself in a new form at this stage in the history of consciousness.

As Arthur C. Clarke has pointed out,[75] it was Lasswitz and Wells who introduced the idea of the invasion of the Earth by extraterrestrials into world literature. If we look back over the history of the idea of the plurality of worlds, however, it is clear that their novels also include motifs that were widespread around 1900, some of which, in one form of another, go back to the Renaissance.[76] (Even the invasion motif had at least been suggested long before, in Swift and Voltaire.) Looking ahead to the development of science fiction in the twentieth century, we see that *Auf zwei Planeten* and *The War of the Worlds* have also acted like prisms, each generating a whole spectrum of themes. Hugo Gernsback, who founded modern science fiction in 1926, owed a crucial debt, as indicated above, to both Lasswitz and Wells;[77] and the great flowering of science fiction around the middle of the century is almost unthinkable without the thematic influence of Wells, whose *The War of the Worlds*, as Clarke has observed, seemed in some ways more "relevant" then than around 1900.[78] Lasswitz's influence around the middle of the century is less widespread but perhaps the more intensive for that: the novel of planetary invasion, *Childhood's End* (1953), which has already been mentioned several times as an exemplum of the type, and which for many people epitomizes the kind of science fiction that has both literary and philosophical aspirations, stands in a direct line of descent, as has been convincingly shown, from *Auf zwei Planeten*.[79] Several new editions of both Wells's and Lasswitz's novels were published in the fifties and sixties; *Auf zwei Planeten* even came out in an English translation, and *The War of the Worlds* with an introduction by Stanislaw Lem. The most plausible

---

[75]*Voices from the Sky* (New York, 1965), p. 215: "The Menace from Space (capitals seem unavoidable) was unknown before the time of Wells."

[76]Darko Suvin makes a similar point about Wells; see Suvin and Philmus, *H. G. Wells and Modern Science Fiction*, p. 28.

[77]On this and the following sentence see chap. 1, sec. 3 above.

[78]Clarke, *Voices from the Sky*, p. 214, and see p. 215; on Wells's influence see Mark R. Hillegas, *The Future as Nightmare: H. G. Wells and the Anti-Utopians* (Carbondale, Ill., 1967).

[79]Hillegas, "Martians and Mythmakers," p. 165.

historical explanation for the continuing interest in these novels is that they provided the paradigm that still serves as a frame of reference for most works of science fiction that take as their subject that philosophical and imaginative adventure of modern times, the encounter with the alien as either an enemy or a guardian.[80]

In treating the subject of the plurality of worlds specifically in the form of stories about Martian invasions of Earth, Wells and Lasswitz are, as it were, viewing the Earth from the outside. There is a suggestion of this imaginative perspective in the passage quoted from the final chapter of *The War of the Worlds,* and similarly in a passage in Wells's essay "From an Observatory" (1894):

> There is a fear of the night that is begotten of ignorance and superstition, a nightmare fear, the fear of the impossible; and there is another fear of the night—of the starlit night—that comes with knowledge, when we see in its true proportion this little life of ours with all its phantasmal environment of cities and stores and arsenals, and the habits, prejudices, and promises of men. Down there in the gaslit street such things are real and solid enough, the only real things, perhaps; but not up here, not under the midnight sky. Here for a space, standing silently upon the dim, grey tower of the old observatory, we may clear our minds of instincts and illusions, and look out upon the real.[81]

In a passage in *Auf zwei Planeten* the vantage point is, however, literally extraterrestrial. In a Martian spacecraft the polar explorers have reached the Martians' space station 6,536 kilometers above the North Pole. Looking back at the Earth, they become lost in thought, and their thoughts are no doubt similar to those that passed through the minds of the Apollo 8 astronauts on 24 December 1968—thoughts that for four hundred years have underlain the theme of the plurality of worlds and that have not lost their fascination even today:

> In the center, at their feet, floated the Earth, a shining disc. It had the appearance of the waxing moon just after the first quarter, but one could also see the part not lit by the sun, as the moon's light bathed it in a faint shimmer. The whole disc of the Earth appeared at a visual angle of sixty degrees and so just filled one-third of the sky below the horizon. The shadow line cut across the Arctic Ocean close to the

[80]See chap. 1, sec. 3 above.
[81]*Certain Personal Matters* (London, 1898), p. 266.

mouth of the Yenisey, so that most of Siberia and the western coast of America lay in darkness. The glaciers on the east coast of Greenland shone brightly in the midday sunshine, and Iceland stood out as a shining white patch against the dark waves of the Atlantic Ocean. . . .

The Germans stood deeply silent, wholly absorbed in this sight never before vouchsafed to a human eye. Never until then had they realized with such clarity what it means to be whirled around in space on the tiny grain that is called Earth; never before had they seen the sky beneath them. The Martians respected their mood. (1:230–32)

Between the lines we sense the question of man's own nature, of how his own identity is defined by this confrontation—imaginary but philosophically plausible—with extraterrestrial life-forms of comparable or higher intelligence, or of a comparable or higher species. This is the question from which, since the beginning, the theme of the plurality of worlds in all its many variations has derived its vitality and also its explosive charge. Today not only has it become fashionable but, as was shown by our introductory survey of the articulation of the theme in present-day astrophysics and microbiology, it is also a question that scientists, and others besides them, cannot avoid confronting. It may take generations for science to arrive at a unanimous answer; or, as Roland Puccetti's philosophical reflections have suggested (see chap. 1 above, at n. 25), it may prove logically impossible to find an answer at all. But the fact that the question has been asked and is still being asked shows a creditable readiness, which one cannot help admiring, to embark on that adventure of modern reason which sets out to explore the last frontier that we can conceive of. Whether this quest is merely quixotic, or has about it "the grace of tragedy,"[82] the image of man since the Copernican revolution would be decidedly the poorer without it.

[82]See Steven Weinberg, *The First Three Minutes* (New York, 1977), pp. 154–55: "The more the universe seems comprehensible, the more it also seems pointless. . . . The effort to understand the universe is one of the very few things that lifts human life a little above the level of farce, and gives it some of the grace of tragedy."

Index